Lessons in Environmental Justice

Lessons in Environmental Justice

From Civil Rights to Black Lives Matter and Idle No More

Michael Mascarenhas

University of California, Berkeley

Los Angeles | London | New Delhi
Singapore | Washington DC | Melbourne

FOR INFORMATION:

SAGE Publications, Inc.
2455 Teller Road
Thousand Oaks, California 91320
E-mail: order@sagepub.com

SAGE Publications Ltd.
1 Oliver's Yard
55 City Road
London, EC1Y 1SP
United Kingdom

SAGE Publications India Pvt. Ltd.
B 1/I 1 Mohan Cooperative Industrial Area
Mathura Road, New Delhi 110 044
India

SAGE Publications Asia-Pacific Pte. Ltd.
18 Cross Street #10-10/11/12
China Square Central
Singapore 048423

Printed in the United States of America

ISBN 978-1-5443-2195-0

This book is printed on acid-free paper.

Acquisitions Editor: Jeff Lasser

Editorial Assistant: Tiara Bennett

Production Editor: Olivia Weber-Stenis

Copy Editor: Amy Marks

Typesetter: Diacritech

Proofreader: Alison Syring

Indexer: Diacritech

Cover Designer: Candice Harman

Marketing Manager: Rob Bloom

MIX
Paper from
responsible sources
FSC® C008955

20 21 22 23 25 10 9 8 7 6 5 4 3 2 1

CONTENTS

PREFACE

As a first-generation college graduate—the only person in my entire extended family to graduate with a doctoral degree—and a person of color, born in the United Kingdom of refugees from South Asia, and then an immigrant and eventual citizen of Canada, and now the United States, I am intimately familiar with the experiences, challenges, and biases faced by the poor, immigrants, and people of color in Canada and the United States, where I currently reside. I have lived with displacement, poverty, and racism my entire life. As a student and later a faculty member, I have witnessed place, race, ethnic, class, gender, and able bias; lack of adequate representation of diverse backgrounds and perspectives; and other institutional variables (for example, lack of support networks and lack of diversity in leadership) that have led many students and faculty to question their existence in the academy, drop out, and leave. This intransigence of the academy has been hard to watch, but it has inspired me, through various research, teaching, and service activities and approaches, to make the academy a more diverse, fair, equitable, and inclusive place.

For me this book project marks a turning point for thinking more deeply about environmental justice and environmental racism. Its consequence is, in part, the outcome of my frustration with academia to theorize environmental injustice and environmental racism in a more meaningful way. The book is directed at undergraduate students. Many of them, like me, have experienced multiple forms of environmental injustice and environmental racism but struggle to make sense of what they are experiencing. This book is an effort to help them understand that they are not alone. This book is also directed at those who come from privileged environs, to help them understand that racism is so much more than intentional and hostile acts. As this book articulates, it is also rooted in white supremacy, white privilege, and institutional racism. I am deeply indebted to the authors who contributed to this pedagogical intervention. Their response to my poorly written outline was encouraging. And I am thrilled to write that the book is so much more than I imagined. I am sure you will agree.

ACKNOWLEDGMENTS

This book project would not have been possible without the extensive editorial oversight of Dr. Kelly Grindstaff, at the Lawrence Hall of Science, UC Berkeley. In addition to working with each author, she conceptualized many of the chapter teaching activities, drawing on her experience teaching Environment and Society, and Environment and Politics, as well as researching STEM and multicultural education. Her pedagogical contribution and creative insight is all over this book.

SAGE would also like to acknowledge the following reviewers:
Benjamin C. Brown, University of New Hampshire
Kishi Animashaun Ducre, Syracuse University
Penn Loh, Tufts University
Jonathan London, University of California, Davis
Kari Marie Norgaard, University of Oregon
Dianne Quigley, Brown University

ABOUT THE EDITOR

Michael Mascarenhas is a first-generation college graduate, anti-colonialist, anti-racism comrade. Professor Mascarenhas's scholarship examines the interconnections between contemporary neoliberal reforms, environmental change, and environmental justice and racism. This interdisciplinary body of research brings together concepts from critical race theory and environmental studies to help cultivate knowledge that contributes to political activism and coalition politics. Mascarenhas is an associate professor in the Department of Environmental Science, Policy, and Management at the University of California, Berkeley. He is the author of *Where the Waters Divide: Neoliberalism, White Privilege, and Environmental Racism in Canada* (2012) and *New Humanitarianism and the Crisis of Charity: Good Intentions on the Road to Help* (2107). He was an expert witness at the Michigan Civil Rights Commission on the Flint water crisis and an invited speaker to the National Academies of Sciences, Engineering, and Medicine's Committee on Designing Citizen Science to Support Science Learning.

ABOUT THE CONTRIBUTORS

Julian Agyeman is a professor of urban and environmental policy and planning at Tufts University. He is the originator of the concept of "just sustainabilities," the intentional integration of social justice and sustainability. He is the author or editor of 12 books, including *Just Sustainabilities: Development in an Unequal World* (2003) and *Sharing Cities: A Case for Truly Smart and Sustainable Cities* (2015). He is editor-in-chief of *Local Environment: The International Journal of Justice and Sustainability;* series editor of *Just Sustainabilities: Policy, Planning, and Practice;* and co-editor of the Routledge Equity, Justice and the Sustainable City series. In 2018, he was awarded the Athena City Accolade by KTH Royal Institute of Technology, Stockholm, for his "outstanding contribution to the field of social justice and ecological sustainability, environmental policy and planning."

J. M. Bacon is a visiting assistant professor in the Sociology Department at Grinnell College. Bacon's research explores the relationship between culture, identity, and eco-social practice with special attention to activism, emotions, and decolonization. Current projects include an ongoing consideration of settler solidarity with Indigenous-led environmental activism, an analysis of the experiences of LBGTQ+ environmental activists, and place attachment among Celtic cultural practitioners.

Carolina L. Balazs is a research scientist for the California EPA's Office of Environmental Health Hazard Assessment (OEHHA), where she leads the Human Right to Water project. Balazs is also the co-founder and co-lead of the Water Equity Science Shop at the University of California, Berkeley, a cross-institutional collaboration promoting community-based water equity research. Her studies on social disparities in drinking water contamination in California were among the first such studies in the state. Prior to joining OEHHA, she was a postdoctoral scholar at the University of California, Davis, and worked as a research scientist with the Community Water Center. She is the recipient of the Switzer Environmental Leadership Fellowship, the UC Chancellor's Award for Diversity and Community, a National Science Foundation Graduate Research Fellowship, and a Fulbright Fellowship. She holds doctoral and master's degrees from UC Berkeley in energy and resources, and a bachelor's degree in environmental science from Brown University.

Phil Brown is a University Distinguished Professor of Sociology and Health Science at Northeastern University, where he directs the Social Science Environmental Health Research Institute and its PFAS Project lab, which has grants from the National Science Foundation to study social policy and activism concerning PFAS, and from the National Institute of Environmental Health Sciences (NIEHS) to study children's immune responses to PFAS and community response to contamination. He directs an NIEHS T-32 training program, Transdisciplinary Training at the Intersection of Environmental Health and Social Science, and heads the Community Outreach and Translation Core of Northeastern's Children's Environmental Health Center (Center for Research on Early Childhood Exposure and Development in Puerto Rico/CRECE) and both the Research Translation Core and Community Engagement Core of Northeastern's Superfund Research Program (Puerto Rico Testsite to Explore Contamination Threats, PROTECT). His books include *No Safe Place: Toxic Waste, Leukemia, and Community Action*; *Toxic Exposures: Contested Illnesses and the Environmental Health Movement*; and *Contested Illnesses: Citizens, Science, and Health Social Movements*.

Robert D. Bullard is Distinguished Professor of Urban Planning and Environmental Policy at Texas Southern University in Houston. He is often called "the father of environmental justice." Bullard is the author of 18 books that address environmental racism, urban land use, housing, transportation, sustainability, smart growth, climate justice, and community resilience. His latest books include *Race, Place, and Environmental Justice after Hurricane Katrina* (2009), *Environmental Health and Racial Equity in the United States* (2011), and *The Wrong Complexion for Protection* (2012). In 2008, *Newsweek* named him one of the "13 Environmental Leaders of the Century." In 2013, Sierra Club honored him with its John Muir Award; and in 2014, the organization named its new Environmental Justice Award after him. In 2018, the Global Climate Change Summit named Bullard one of 22 Climate Trailblazers. And in 2019, *Apolitical* named him one of the world's 100 Most Influential People in Climate Policy.

Stella M. Čapek is Elbert L. Fausett Emerita Distinguished Professor of Sociology at Hendrix College. She has taught Environmental Sociology, Social Change/Social Movements, The Urban Community, Images of the City, Medical Sociology, Food/Culture/Nature, Sociology of Travel and Tourism, Sociological Theory, and Exploring Nature Writing, as well as sustainability-focused travel seminars in Costa Rica and in the U.S. Southwest. She co-authored *Community Versus Commodity: Tenants and the American City* and *Come Lovely and Soothing Death: The Right to Die Movement in the United States*. Her articles focus on environmental justice, tenants' rights, urban design, environmental health, and social constructions of nature and the self. She also publishes environmentally themed creative nonfiction and enjoys interdisciplinary collaborations focused on social justice.

Alissa Cordner is associate professor and Paul Garrett Fellow at Whitman College, where she teaches sociology and environmental studies courses. Her research focuses on environmental sociology, the sociology of risk and disasters, environmental health and justice, and public engagement in science and policy making. She is the author of *Toxic Safety: Flame Retardants, Chemical Controversies, and Environmental Health* (2016), which won the 2018 Schnaiberg Outstanding Publication Award from the American Sociological Association's Section on Environmental Sociology, and the co-author of *The Civic*

Imagination: Making a Difference in American Political Life (2014). She has conducted extensive research on the regulation, research, and activism related to industrial chemicals. She also studies the sociological aspects of wildfire risk management in the Northwest with a focus on firefighter safety, public safety, and resource management.

Cristina Faiver-Serna is a Chicana scholar and a doctoral degree candidate in geography at the University of Oregon. She has a master's degree in public health, and worked for several years as the director of health education and outreach at a nonprofit community health center in Southern California. Her current research project builds on her time spent working alongside *promotoras de salud* against the everyday oppressions of environmental racism, and critically examines the multiscalar geographical and historical infrastructure that requires their ongoing community care work. Faiver-Serna is an interdisciplinary critical race scholar, whose expertise and research interests in critical environmental racism and injustice, Latina feminist theory and praxis, Latinx geographies, and critical public health studies have developed both within, and outside of, academia.

Jill Lindsey Harrison is associate professor of sociology at the University of Colorado Boulder. Her research focuses on environmental justice, environmental politics, workplace inequalities, and immigration politics. In her research, she identifies the narratives, other interactive dynamics, and broader political economic structures through which people come to define highly inequitable circumstances as reasonable and unproblematic. She also identifies the practices through which other groups push the state to remedy those inequalities. She has done so through research on political conflict over agricultural pesticide poisonings in California, immigration policing in rural Wisconsin, and government agencies' environmental justice efforts. Her first book, *Pesticide Drift and the Pursuit of Environmental Justice* (2011), won book awards from the Rural Sociological Society and the Association of Humanist Sociology. Her second book, *From the Inside Out: The Fight for Environmental Justice within Government Agencies,* was published in 2019.

Elizabeth Hoover is associate professor of American studies and faculty chair of the Native American and Indigenous Studies Initiative steering committee at Brown University. Her first book, *The River Is In Us: Fighting Toxics in a Mohawk Community* (2017), is an exploration of Akwesasne Mohawks' response to Superfund contamination and environmental health research. She is currently working on her second book, *From Garden Warriors to Good Seeds: Indigenizing the Local Food Movement,* exploring Native American community–based farming, food and seed sovereignty, Native chefs in the food movement, and the fight against the fossil fuel industry to protect heritage foods. Hoover's other publications include co-editing *Indigenous Food Sovereignty in the United States* with Devon Mihesuah (2019) and several articles. Outside of academia, Hoover serves on the executive committee of the Native American Food

Sovereignty Alliance and the board of North American Traditional Indigenous Food Systems.

George Lipsitz is professor of black studies and sociology at the University of California, Santa Barbara. He is the co-author with Barbara Tomlinson of *Insubordinate Spaces: Improvisation and Accompaniment for Social Justice.* His single-authored books include *The Possessive Investment in Whiteness, How Racism Takes Place, A Life in the Struggle, Time Passages,* and *A Rainbow at Midnight.* Lipsitz serves as chair of the board of directors of the African American Policy Forum. He was awarded the American Studies Association's Angela Y. Davis Prize for Public Scholarship in 2013 and its Bode-Pearson Prize for Career Distinction in 2016.

Deniss Josefina Martinez is a doctoral student in the Graduate Group in Ecology at the University of California, Davis. She is also a health policy research scholar with the Robert Wood Johnson Foundation. Her work seeks to understand how California Native Nations navigate power differentials in varying natural resource stewardship collaborations with western institutions. She is passionate about increasing Indigenous representation in environmental stewardship in order to support environmental justice, health equity, sovereignty, and cultural vitality.

Beth Rose Middleton Manning is an associate professor and chair in the Department of Native American Studies at the University of California, Davis. Middleton's research centers on Native environmental policy and Native activism for site protection using conservation tools. Her books *Trust in the Land: New Directions in Tribal Conservation* (2011) and *Upstream* (2018) focus on Native applications of conservation easements and on the history of Indian allotment lands at the headwaters of the California State Water Project, respectively. Middleton is currently developing projects focused on tribal participation in the carbon market, on California Indigenous legal history, and on intersecting Afro- and Indigenous Caribbean histories. She is passionate about increasing under-represented perspectives, especially Indigenous perspectives, in academia and in environmental policy and planning.

Paul Mohai is a professor in the School for Environment and Sustainability at the University of Michigan and faculty associate at the Institute for Social Research. He is a founder of the Environmental Justice Program at the University of Michigan and a major contributor to the growing body of quantitative research examining disproportionate environmental burdens and their impacts on low-income and people-of-color communities. He has served on the U.S. Environmental Protection Agency's National Environmental Justice Advisory Council and has provided testimony on environmental justice to the U.S. House of Representatives, the U.S. Senate, and the Michigan Civil Rights Commission.

He is author of numerous articles, books, and reports focused on race and the environment.

Rachel Morello-Frosch is a professor in the Department of Environmental Science, Policy and Management and the School of Public Health at the University of California, Berkeley. As an environmental health scientist and epidemiologist, Morello-Frosch conducts research that examines race and class determinants of environmental health disparities among diverse communities in the United States with a focus on climate change and environmental chemicals. She also writes about the influence of community-based participatory research projects on environmental health science, regulation, and policymaking. She is the co-author of numerous books and articles. Her awards include the Chancellor's Award for Research in Public Service, University of California, Berkeley (2012), and the Damu Smith Environmental Health Achievement Award, Environment Section, American Public Health Association (2010).

Kari Marie Norgaard is professor of sociology and environmental studies at the University of Oregon. Over the past 15 years, Norgaard has published and taught in the areas of environmental sociology, Indigenous environmental justice, gender and environment, race and environment, climate change, sociology of culture, and sociology of emotions. She is the author of *Salmon and Acorns Feed Our People: Colonialism, Nature, and Social Action* (2019) and *Living in Denial: Climate Change, Emotions, and Everyday Life* (2011) and a recipient of the Fred Buttel Distinguished Contribution Award, a Sociology of Emotions Recent Contribution Award, and the Pacific Sociological Association's Distinguished Practice Award. Her work on climate denial and Indigenous environmental justice have been covered by *The Washington Post*, *National Geographic*, the British Broadcasting Corporation, the Canadian Broadcasting Corporation, National Public Radio, *High Country News*, and *Yes Magazine*, among others.

David N. Pellow is the Dehlsen Chair and professor of environmental studies and director of the Global Environmental Justice Project at the University of California, Santa Barbara. His teaching and research focus on ecological justice issues in the United States and globally. His books include *What Is Critical Environmental Justice?*; *Total Liberation: The Power and Promise of Animal Rights and the Radical Earth Movement*; *The Slums of Aspen: Immigrants vs. the Environment in America's Eden* (with Lisa Sun-Hee Park); *Resisting Global Toxics: Transnational Movements for Environmental Justice*; *The Silicon Valley of Dreams: Environmental Injustice, Immigrant Workers, and the High-Tech Global Economy* (with Lisa Sun-Hee Park); and *Garbage Wars: The Struggle for Environmental Justice in Chicago*. He has served on the boards of directors for Global Response, The Global Action Research Center, the Center for Urban Transformation, the Santa Clara Center for Occupational Safety and Health, Greenpeace USA, and International Rivers.

Kaitlin Reed (Yurok/Hupa/Oneida) is an assistant professor of Native American studies at Humboldt State University. Her research is focused on tribal land

and water rights, extractive capitalism, and settler colonial political economies. She is currently working on a book titled *From Gold Rush to Green Rush: Cannabis and California Indians*. This book connects the historical and ecological dots between the Gold Rush and the Green Rush, focusing on capitalistic resource extraction and violence against Indigenous lands and bodies. In 2018, she was awarded the Charles Eastman Fellowship of Native American Studies at Dartmouth College. Reed is an enrolled member of the Yurok Tribe in Northwestern California.

Sarah M. Rios is an assistant professor in the Department of Community and Environmental Sociology at the University of Wisconsin–Madison. Rios received a doctoral degree from the University of California, Santa Barbara, in 2018. Rios's ongoing research examines how farmworkers and former prisoners elucidated new ways of knowing about an environmental disease known as Valley Fever, its connection to cumulative vulnerabilities, and new ways of healing from its devastating effects. Rios is especially interested in how poverty, pollution, and prisons lead to health problems, and how community-based knowledge can address and redress cumulative health impacts. Over the next few years, her studies will develop a broad framework that sheds light on community-based knowledge and variations of community-based research methods to further explore the links among race, place, and health.

Oday Salim is a clinical assistant professor of law and director of the Environmental Law & Sustainability Clinic, as well as an attorney at the National Wildlife Federation in its Great Lakes Regional Center. Prior to joining the clinical program, Salim practiced environmental law in Pennsylvania and Michigan, focusing on stormwater management, water quality permitting, water rights, environmental justice, land use and zoning, utility regulation, mineral rights, and renewable energy. He has litigated in administrative and civil courts at the local, state, and federal levels, and has done transactional work for individuals and nonprofits. In 2018, he was named one of *Grist* magazine's 50 Fixers for his work on environmental and public health protection in minority communities.

João Costa Vargas is a professor in the Department of Anthropology at the University of California, Riverside. His work is based on collaborative projects engaging antiblackness in the United States and in Brazil. He has published, among other books, *Catching Hell in the City of Angels: Life and Meanings of Blackness in South Central Los Angeles* (2006), *Never Meant to Survive: Genocide and Utopias in Black Diaspora Communities* (2008), and *The Denial of Antiblackness: Multiracial Redemption and Black Suffering* (2018).

Ingrid Waldron is an associate professor in the Faculty of Health and the team co-lead of the Health of People of African Descent Research Cluster at the Healthy Populations Institute at Dalhousie University. Waldron's scholarship is driven by a long-standing interest in looking at the many ways in which spaces and places are organized by structures of colonialism and gendered racial capitalism. She is the director of the Environmental Noxiousness, Racial Inequities

& Community Health Project, which is investigating the socioeconomic and health effects of environmental racism in Mi'kmaq and African Nova Scotian communities. Her first book, *There's Something in the Water: Environmental Racism in Indigenous and Black Communities,* received the 2019 Atlantic Book Award for Scholarly Writing. The 2019 documentary *There's Something in the Water* is based on Waldron's book and was co-produced by Waldron, Ellen Page, Ian Daniel, and Julia Sanderson.

Kyle Powys Whyte is the Timnick Chair in the Humanities at Michigan State University, professor of philosophy and community sustainability, a faculty member of the Environmental Philosophy & Ethics graduate concentration and the Geocognition Research Lab, and a faculty affiliate of the American Indian & Indigenous Studies and Environmental Science & Policy programs. He is an enrolled member of the Citizen Potawatomi Nation. Whyte's work focuses on climate and environmental justice and Indigenous environmental studies, most recently studying issues related to Indigenous food sovereignty and Indigenous critiques of concepts of the anthropocene. His writing appears in journals such as *Climatic Change, Sustainability Science,* and *Human Ecology.* He is the recipient of several awards including the K. Patricia Cross Future Leaders Award from the Association of American Colleges and Universities in 2009, and the Bunyan Bryant Award for Academic Excellence from Detroiters Working for Environmental Justice in 2015.

Stephen Zavestoski is a professor of environmental studies at the University of San Francisco. He has co-edited books such as *Social Movements in Health* (2005), *Contested Illnesses: Citizens, Science, and Health Social Movements* (2012), and *Incomplete Streets: Processes, Practices and Problems* (2014). He is also co-editor of the Routledge book series Equity, Justice and the Sustainable City. His research areas include environmental sociology, social movements, sociology of health and illness, and urban sustainability. Zavestoski's previous research has also covered topics such as ecological identity, consumerism, and the effects of the Internet on public participation in environmental regulatory rulemaking processes.

INTRODUCTION

Michael Mascarenhas

I have wanted a book like this for some time: a book that not only provides an entry point to the field of environmental justice for those who are unfamiliar with the topic but also provides the opportunity to assemble a group of scholars who are helping to cultivate a new and vibrant wave of environmental justice scholarship, methods, and activism. Toward this effort, I invited a diverse set of scholars to contribute to this edited book. Some readers will recognize established scholars in the field. Other names are new. I wanted readers to read the words and ideas of scholars who have often been marginalized in their own fields. Justice comes in many forms, and academic writing needs to increase matters of diversity, equity, and inclusion in our research, teaching, and learning. This book project is an effort to do that.

For example, black women account for two percent of full-time professors of college and university faculty, and Latinas make up one percent or less of the total number of full-time professors. This blatant omission of women-of-color scholars, I argue, is one that the discipline of environmental sociology and efforts toward environmental justice can ignore only at their own peril. At the time of writing this book, the University of California, Berkeley, my employer, had three Native American faculty. Similarly, African Americans (men and women) occupy a paltry 4.6 percent of leadership positions in environmental organizations; Hispanics/Latinxs, 2.3 percent (Taylor, 2014). As a result, much of what we know about environmental justice and environmental racism has been written by academics who have never experienced institutional racism or sexism. I saw this book project as a means to challenge the white supremacy that lies at the heart of much of our academic writing.

My motivation for this book project was also inspired by recent developments in environmental justice theory and methods. For example, in addition to conceptualizing the environment as a place where we work, live, and play, scholars in this book have considered the conditions of how and where we eat, where people are incarcerated, and the emotional costs of being subjected to environmental injustice and environmental racism. The heinous crimes of Dylann Roof and Gregory Bush, who together murdered 11 African Americans, reveals not only the rise of white rage and antiblack genocide in the United States, discussed by João Costa Vargas, but also the need for us to expand our thinking about environmental racism in this county and around the globe.

Similarly, scholars in this book have given much thought to how to engage in research with vulnerable communities. Many of the scholars in this book have advanced a collaborative style of community engagement that works *with* and not *on* environmental justice communities. This participatory style of research can be difficult, lengthy, and complicated; and it stands in stark contradiction to the publish-or-perish mantra of the academy. Yet, the contributions of scholars in this book illustrate that social science research can be both academic and directed toward local empowerment and meaningful social and environmental change. In a post–Flint water crisis era, having community-based health monitoring research and programs that are built on the idea that ordinary residents are knowledgeable about their environment and the status of their health is now, literally, a matter of life or death.

The Flint water crisis, antiblack genocide, unremitting settler colonialism, and the possessive investment in whiteness, more generally, underscore a continuing theme in American history—the betrayal of environmental justice for nonwhite communities. This enduring racial formation also raises some important questions about the role of the state in normalizing environmental justice. And although scholars are generally critical of the abuse of state and corporate power with respect to the perpetration of environmental injustices, they generally do not question the existence of those institutions. This reformist approach, David Pellow argues, is an existential threat to social and environmental justice and ecological sustainability for all beings. How do we reconcile, Cristina Faiver-Serna asks, a society that simultaneously permits the poisoning of particular (racialized and marginalized) communities yet also sets aside funds to pay for the medical treatment of members of that same community? Thinking through environmental justice as *structure,* I suggest, can help explain this seeming contradiction.

It has also been my intention to make the theory and praxis of environmental justice relevant to students and others who are struggling to make sense of a rapidly changing world, growing levels of inequality, and deepening degrees of social and environmental injustice. By foregrounding the work of environmental justice scholars, many of them also people of color, I hope to instill a sense of hope and the chance for change for you and the communities you call home.

The book is divided into five parts. Part I begins with an introduction and history of the environmental justice movement. It is fitting that Robert Bullard writes this chapter. He has written dozens of books on the subject, helping to shape theory and influence environmental justice policy. Many people have referred to Bullard as the father of environmental justice. Yet he will admit that his wife, attorney Linda McKeever Bullard, was the original motivating force behind his efforts, as he provided the sociological data that attorney McKeever Bullard required in a discriminatory facility siting in Houston, Texas. His study, conducted in 1978, was the first comprehensive account of environmental racism in the United States. In Chapter 2, Stella Čapek explains why the environmental justice frame continues to be a powerful tool for understanding social and environmental inequalities at a systemic and personal level. Environmental justice has always been about more than just "disproportionate dumping" of toxins on vulnerable groups, Čapek argues. The environmental justice frame powerfully linked the civil rights movement to the environmental movement, and over time it has been expanded to include global processes, transnational networks, Indigenous understandings of place, urban as well as rural sites, and issues like immigration, food production, and climate justice. In Chapter 3, Indigenous scholar Kyle Powys Whyte asks us to consider how processes of imperialism, colonialism, industrialization, and settlement undermine Indigenous ways of life by altering ecosystems to suit dominant societies. A key observation to cultivating environmental justice is recognizing that wherever we are in North America (Turtle Island), we are always standing on colonized land. Not recognizing this fact is a major barrier to environmental justice efforts.

Part II explores environmental justice methodology, both in terms of epistemology (how do you know) and ethics (who should you involve). In Chapter 4, Paul Mohai reflects on meeting his colleague Bunyan Bryant, "the first African American environmental studies professor I ever met." Before their collaboration, Mohai had not stopped to consider that environmental problems may be disproportionately burdening certain groups in society over others. Mohai's chapter raises many important questions about methodology, including appropriate scale and availability of longitudinal data, in the assessment of environmental justice. In Chapter 5, Alissa Cordner and Phil Brown explain why the most impactful approaches to environmental justice research and advocacy are community engaged and reflexive. They highlight four such approaches: community-based participatory research, transdisciplinary social science–environmental work, the New Political Sociology of Science (NPSS), and reflexive research ethics. In Chapter 6, Carolina Balazs and Rachel Morello-Frosch advance the debates around environmental justice measurement, suggesting that rigor, relevance, and reach of science to support environmental justice claims must include affected communities at all levels of the research process, from conceptualizing the problem to communicating the findings to the affected communities. In Chapter 7, J. M. Bacon and Kari Marie Norgaard turn the microscope inward to address emotions. Specifically, they examine how unequal mental and emotional harms shape the way communities and individuals experience environmental injustice.

In Part III, chapters focus on the intersection of state agencies, policy, and environmental justice. With exclusive access to EPA staff, Jill Harrison asks in Chapter 8 why government agencies have been so slow to bring about meaningful environmental justice change for affected communities. Harrison suggests that government agency staff engage in racialized claims and actions that defend existing regulatory practice and undermine proposed environmental justice reforms. In so doing, they bolster a regulatory system that has long disproportionately protected white middle-class and wealthy communities. Drawing from an understanding of racial capitalism, and using Actor Network Theory, Cristina Faiver-Serna explores in Chapter 9 the ways in which the state is deeply entangled with racial capitalism to reproduce environmental racism, expressed in disproportionate asthma incidents in the Latinx community of Long Beach, California. Chapter 10, by Oday Salim, focuses on the case of the Flint water crisis and the law's ability to address environmental injustice in such cases.

The chapters in Part IV examine environments of (in)justice and activism. In Chapter 11, George Lipsitz examines the intersection between housing, health, and race. Using a social ecology framework, Lipsitz connects individual ills to discrimination; collective, cumulative, and continuing ideologies; structures; and systems. In Chapter 12, Elizabeth Hoover explains how environmental justice impacting Indian Country needs to be considered through the unique colonial history and relationship that tribes have with the United States and Canada, and their status as sovereign nations. Using three case studies, Hoover explains that in order to achieve true food sovereignty for Native communities, tribes need to be able to define their own standards for health and well-being, and demand environmental conditions that support their Indigenous food systems. In Chapter 13, Sarah Rios examines the prison as an egregious environment of injustice and racism. Rios shows how the siting of prisons in the Central Valley of California has placed inmates in proximity to noxious industries such as agricultural operations, chemical waste evaporation from oil well water ponds, ammonia gases from dairies, and open dust fields. These multiple exposures have contributed to the gestation and dissemination of Valley Fever in California's prisons. In Chapter 14, Beth Rose Middleton Manning, Kaitlin Reed, and Deniss Martinez reveal how processes of settler colonialism harm both water and Indigenous peoples. For many Indigenous communities, Middleton Manning and her co-authors argue, environmental justice struggles began with the invasion and colonization of our lands. Using various examples, including the protests at Standing Rock, efforts to recognize the rights of the Klamath River, and re-storying the Sierra Nevada, Middleton Manning et al. explain how Indigenous peoples across the globe are standing up to protect water relatives and future generations.

In the final section of the book, Part V, you will find chapters that discuss new frontiers and old questions regarding environmental justice. In Chapter 15, Ingrid Waldron provides a detailed case study of the fight against environmental racism in two African Nova Scotian communities in Canada. Waldron locates her analysis of environmental racism within Canada's settler-colonial existence. She concludes that addressing environmental racism in Nova Scotia and Canada must acknowledge and address structural as well as environmental determinants of health, and it must involve the community in consultations and monitoring to right the impacts of environmental racism. Whereas the other chapters in this text focus more explicitly on environmental justice, Chapter 16, by João Costa Vargas, delves more deeply into an enduring pattern of imposed and global marginalization on black communities that precipitates and normalizes environmental injustices. Using the framing of antiblack genocide in the United States and Brazil, Costa Vargas argues that, from residential segregation, to police brutality, to blocked access to well-being and quality medical care, and to exposure to environmental toxins, black people were never meant to survive. Julian Agyeman and Stephen Zavestoski examine in Chapter 17 how urban sustainability efforts are creating new forms of environmental injustices and racial segregation in the city. They conclude that as private interests use the rhetoric of sustainable, livable, and smart cities to gain more and more influence in urban planning and development, the environmental justice movement must adapt through a combination of drawing on lessons learned from the movement's early history and innovating new ways of identifying the earliest possible signs of speculative investment and then build coalitions capable of placing social equity at the center of such projects. In Chapter 18, David Pellow frames Black Lives Matter as an environmental justice movement. Black Lives Matter has become one of the most important sources and sites of racial justice struggle in the early 21st century. It has built on the long history of anti-colonial/decolonial resistance movements among Indigenous people, people of color, and their allies for the past half millennium. Pellow develops the critical environmental justice studies framework as a way of contextualizing environmental justice studies and the problem of environmental racism, and as a tool for rethinking the Black Lives Matter movement. As with all the other contributions to this edited volume, Pellow's intersectional analysis provides a novel and generative space for thinking about new ways to engage a range of social and environmental justice movements in the 21st century.

PART I

UNDERSTANDING ENVIRONMENTAL JUSTICE: CLAIMS, FRAMES, AND COLONIALISM

1

FROM CIVIL RIGHTS TO BLACK LIVES MATTER

Robert D. Bullard

This introductory chapter lays the foundation for understanding environmental justice from its early roots in the modern civil rights movement to the current-day Black Lives Matter (BLM) movement. It also explores how the environmental justice framework redefined environmentalism and challenged institutional racism and the dominant environmental protection paradigm. The chapter uses an environmental justice framework to examine the location of polluting facilities; government response to natural and human-made disasters; and the application and enforcement of laws, policies, and regulations governing equal protection and civil rights. The environmental justice framework rests on developing tools and strategies to eliminate unfair, unjust, and inequitable conditions and decisions. It also attempts to uncover the underlying assumptions that contribute to and produce inequality, including differential exposure, unfair treatment, and unequal protection. The framework brings to the surface the ethical, moral, and political questions of "who gets what, when, why, and how much?"

Various movements over the decades have challenged structural racism that devalued blacks and other people of color and their communities. These movements challenged the underlying assumptions that contribute to and reproduce inequality. The modern civil rights movement of the 1950s and 1960s was largely an anti-racism and anti–white supremacy movement. The 1970s and 1980s ushered in a more focused era of targeting unequal and unfair pollution burdens borne by poor people, people of color, and other vulnerable populations—including children. The 1990s and 2000s expanded the equity and justice lens to include issues ranging from health equity, parks and green space, food security and healthy food access, sustainability, climate change, community resilience, racial profiling, policing, and criminal justice.

LEGACY OF THE MODERN CIVIL RIGHTS MOVEMENT: 1950s AND 1960s

The U.S. civil rights movement waged an assault on various forms of structural racism that penetrated the daily lives of African Americans and other people of color—in voting, housing, education, employment, transportation, and equal access to other public accommodations. Protesters—young and old—were beaten, hosed with water cannons, attacked with vicious police dogs, harassed, and jailed, and some were killed in their quest for equal treatment. Yet, they persisted in their intergenerational quest for justice.

It is worth noting that Dr. Martin Luther King Jr. went to Memphis, Tennessee, in April 1968 to support the environmental and economic justice struggle of 1,300 striking sanitation workers from Local 1733. The strike shut down garbage collection and sewer, water, and street maintenance. Clearly,

---

Photo 1.1: Houston protests against the Whispering Pines Sanitary Landfill that was the subject of the *Bean et al. v Southwestern Waste Management Corps.* lawsuit—the nation's first lawsuit to challenge environmental racism using civil rights laws.
Photo by Robert D. Bullard, 1978.

the Memphis struggle was much more than a garbage strike. The "I AM A MAN" signs that black workers carried reflected the larger struggle for human dignity and rights. Black sanitation workers were on strike because of unequal pay, discriminatory labor practices, and unsafe work conditions that resulted in disproportionately high rates of injuries and deaths among them. They were also striking to be treated as men—with the same dignity and respect accorded white city workers. For Memphis strikers, Black Workers Mattered. Memphis was Dr. King's "last campaign." He was assassinated on April 4, 1968, but his legacy lives on and is an integral part of anti-racism movements around jobs, environment, health, transportation, land use, smart growth, energy, climate, and criminal justice.

HOUSTON WASTE STUDY HISTORICAL BACKDROP: 1970s

The historical backdrop of the environmental justice movement has its roots in small, isolated struggles found across the United States. A decade after Dr. King's death, black homeowners in 1978 took on a fight against a municipal landfill proposal in a mostly black suburban community in Houston, Texas. The city has seen a dramatic demographic shift over the past three decades (Bullard, 1987). In 1980, it was 52.3 percent white, 27.4 percent black, 17.6 percent Hispanic, and 2.7 percent Asian and other. By 2010, Houston became a majority people-of-color city—25.6 percent Anglo, 43.8 percent Hispanic, 23.7 percent African American, and 6.0 percent Asian.

The Houston case study examined solid waste disposal in Houston from the 1970s through 2013. Houston is the nation's fourth largest city with a population of some 2.3 million persons spread over more than 600 square miles and more than 500 neighborhoods. It is the only major American city without zoning. This no-zoning policy allowed for an erratic land-use pattern. As a result, the NIMBY (Not in My Back Yard) practice was replaced with the "PIBBY" (Place in Blacks' Back Yard) policy (Bullard, 1983). The all-white, all-male Houston city government and private industry targeted landfills, incinerators, garbage dumps, and garbage transfer stations for Houston's black neighborhoods (Bullard, 1987). Five decades of this type of thinking and discriminatory land-use practices lowered black residents' property values, accelerated physical deterioration, and increased disinvestment in Houston's black neighborhoods. Houston's black neighborhoods were in fact "unofficially zoned for garbage" (Bullard, 2005). Discriminatory siting of waste facilities stigmatized black neighborhoods as "dumping grounds" for a host of other unwanted services, including salvage yards and recycling (Rosen, 1994, pp. 223–229).

Ineffective land-use regulations created a nightmare for many of Houston's neighborhoods—especially the ones that were ill equipped to fend off industrial encroachment. From the 1920s through the 1970s, the siting of nonresidential facilities heightened animosities between the black community and the local government. This is especially true in the case of solid waste disposal. It was not until 1978, with *Bean v. Southwestern Waste Management Corp.*, that Black

Houston mounted a legal assault on environmental racism in solid waste facility siting (Bullard, 1983).

The *Bean* case exposed racism and the discriminatory practices of Houston's waste facilities as well as its flawed no-zoning legacy. For example, the Whispering Pines Landfill that triggered the lawsuit was sited less than 1,500 feet from a public school and within a two-mile radius of a half-dozen other schools in the predominately black and poor North Forest Independent School District. Although the landfill was built and plaintiffs lost their legal case in 1984, the lawsuit changed the city's solid waste facility siting practices after 1979 (Bullard, 2005). From the time of the lawsuit until the present, not a single Type I municipal landfill has been sited in Houston, in contrast to the 1920s to the late 1970s, when Black Houston became the "unofficial dumping grounds" for the city's garbage (Bullard, 1983).

For decades, the city used two basic methods to dispose of its solid waste—incineration and landfill. Eleven of 13 city-owned landfills and incinerators (84.6 percent) were built in black neighborhoods. This city siting pattern set the stage for private waste disposal firms to follow. From 1970 to 1978, the Texas Department of Health (TDH) issued four sanitary Type I solid waste landfill permits, for the disposal of Houston's solid waste, and all four were located in city council districts that were majority people of color.

In 2018, the brunt of waste disposal was still borne disproportionately by low-income people of color. In 2018, two Type I landfills, McCarty Landfill and Whispering Pines Landfill, operated in Houston, and both were in council district B, which is 93 percent people of color (53 percent black and 40 percent Hispanic). As mentioned earlier, after 1979 and the *Bean* case, no other Type I landfills were built in the city. Houston instead began sending much household garbage to four landfills located outside the city limits. In 2018, three of these four landfills were located in census tracts where the majority of the population is people of color—Waste Management (76.6 percent), Atascocita (86.0 percent), and BFI Blue Ridge (85.7 percent).

BIRTH OF THE ENVIRONMENTAL JUSTICE MOVEMENT: WARREN COUNTY, NORTH CAROLINA: 1980s

The national environmental justice movement in the United States was born in mostly African American, rural, and poor Warren County, North Carolina, in the early 1980s after the state government decided to dispose of 30,000 cubic yards of soil contaminated with polychlorinated biphenyls (PCBs) in the tiny town of Afton—more than 84 percent of the community was black in 1982 (Bullard, 2000). Protests ensued, resulting in more than 500 arrests. The landfill later became the most recognized symbol in the county, and Warren County became a symbol of the environmental justice movement. By 1993, the facility was failing with 13 feet of water trapped inside it (Exchange Project 2006). For a decade, community leaders pressed the state to clean up the leaky landfill. Although the PCB landfill has been cleaned up, the county is still economically worse off than the state as a whole. More than 24.4 percent of Warren County residents in 2008–2012 were below the poverty line, compared with North

Carolina's 16.8 percent poverty rate—a 7.6 percentage point gap. The 2008–2012 median household income for Warren County residents was only $34,803, compared with $46,450 for the state, or roughly 75 percent of the state median (U.S. Census Bureau, 2018). These data reveal the cumulative burdens that impact toxic communities.

The Warren County protests provided the impetus for a 1983 U.S. General Accounting Office (GAO) study, *Siting of Hazardous Waste Landfills and Their Correlation with Racial and Economic Status of Surrounding Communities* (U.S. General Accounting Office, 1983). The GAO study found that three out of four of the off-site, commercial hazardous waste landfills in Environmental Protection Agency (EPA) Region 4 (Alabama, Florida, Georgia, Kentucky, Mississippi, North Carolina, South Carolina, and Tennessee) were located in predominantly African American communities, although African Americans made up only 20 percent of the region's population. The protesters put "environmental racism" on the map.

The disturbing findings from the GAO report led Benjamin Chavis of the United Church of Christ Commission for Racial Justice to produce the first national study on race and waste in 1987. The study report, *Toxic Wastes and Race in the United States*, found race to be the most significant variable in predicting where these waste facilities were located—more powerful than household income, the value of homes, and the estimated amount of hazardous waste generated by industry (Commission for Racial Justice, United Church of Christ, 1987). In other words, race was a more powerful predictor than class of where toxic waste facilities are located in the United States. The *Toxic Wastes and Race* study was revisited in 1994 using 1990 census data, in which it was found that people of color were 47 percent more likely than white Americans to live near a hazardous waste facility (Goldman & Fitton, 1994).

ENVIRONMENTAL JUSTICE MOVEMENT BUILDING: 1990s

In 1990, *Dumping in Dixie: Race, Class, and Environmental Quality* became the first book to chronicle environmental injustice, environmental racism, and the convergence of two major movements—the civil rights and environmental movements—into the environmental justice movement (Bullard, 2000). *Dumping in Dixie* documented racial dynamics involved in the location of municipal landfills, hazardous waste sites, incinerators, lead smelters, refineries, and chemical plants. The book provided clear examples of "environmental racism 101" and was adopted as a textbook by dozens of U.S. colleges and universities. The southern United States or "Dixie" is different from the rest of the country. The South has a unique legacy of slavery, "Jim Crow" racial segregation, and resistance to equal justice for all its citizens. The South is also the most environmentally degraded region of the country. It is no accident that the modern civil rights and environmental justice movements were born in the South.

A growing grassroots environmental justice movement began to take shape in the late 1980s and early 1990s. The impetus for this growth centered around grassroots activism; the redefinition of environmental rights and civil rights;

alliances and coalitions; community-driven research, forums, and conferences; and the First National People of Color Environmental Leadership Summit held in October 1991. The summit was attended by more than 1,000 individuals from every state in the U.S. and at least a half-dozen other countries. Summit delegates adopted the 17 Principles of Environmental Justice. By the time the June 1992 United Nations Conference on Environment and Development (the Rio Earth Summit) started, the Principles of Environmental Justice had been distributed and translated into a half-dozen languages.

Environmental justice leaders embraced *the principle that all people and communities are entitled to equal protection of our environmental, health, employment, education, housing, transportation, and civil rights laws* (Bullard, 1993a). For decades, hundreds of communities across the nation used a variety of tactics to confront environmental injustice (Agyeman, Bullard, & Evans, 2003; Bullard, 1987, 1993a, 1993b, 1993c, 1994, 2000, 2007; Bullard, Johnson, & Torres, 2011). The federal EPA took action on environmental justice concerns in 1990 only after extensive prodding from grassroots environmental justice activists, educators, and academics who called themselves the Michigan Group, named for environmental justice leaders who were successful in meeting with EPA administrator William Reilly and his senior staff. These meetings resulted in some key first steps in advancing environmental justice at the EPA, including the creation of the Office of Environmental Equity. In 1992, under the George H. W. Bush administration, the EPA produced *Environmental Equity: Reducing Risks for All Communities*, one of the first federal reports to acknowledge the fact that some populations shouldered greater environmental health risks in some areas than others (U.S. Environmental Protection Agency, 1992).

Thus, environmental justice was not something invented by the EPA. However, in 1992 the EPA arrived at its own definition of environmental justice as "fair treatment and meaningful involvement of all people regardless of race, color, national origin, or income with respect to the development, implementation, and enforcement of environmental laws, regulations and policies." For the EPA, "fair treatment means that no group of people, including racial, ethnic, or socioeconomic groups should bear a disproportionate share of the negative environmental consequences resulting from industrial, municipal, and commercial operations or the execution of federal, state, local, and tribal programs and policies" (U.S. Environmental Protection Agency, 1998, p. 2).

In 1994, President Bill Clinton issued Executive Order 12898, "Federal Actions to Address Environmental Justice in Minority Populations and Low-Income Populations" (Clinton, 1994). This executive order attempted to address environmental injustice within existing federal laws and regulations. It also reinforced the Civil Rights Act of 1964, Title VI, which prohibits discriminatory practices in programs receiving federal funds (see Chapter 10) and put the spotlight back on the National Environmental Policy Act (NEPA), a law that set policy goals for the protection, maintenance, and enhancement of the environment. NEPA's goal is to ensure for all Americans a safe, healthful, productive, and aesthetically and culturally pleasing environment. NEPA requires federal agencies to prepare a detailed statement on the environmental effects of proposed federal actions that significantly affect the quality of human health.

Executive Order 12898 called for improved methodologies for assessing and mitigating health effects from multiple and cumulative exposures as well as impacts on subsistence fishers and consumers of wild game, in addition to the collection of data on low-income and minority populations who may be disproportionately at risk. It also encouraged participation of the impacted populations in the various phases of assessing impacts—including scoping, data gathering, considering alternatives, analysis, mitigation, and monitoring.

In 1999, environmental justice leaders were able to get the national Academy of Sciences Institute of Medicine (IOM) to examine environmental justice and health in the United States. The IOM study concluded that government, public health officials, and the medical and scientific communities need to place a higher value on the problems and concerns of environmental justice communities (Institute of Medicine, 1999). The study also confirmed what most environmental justice leaders and communities have known for decades. People of color and low-income communities are exposed to higher levels of pollution than the rest of the nation; and they experience certain diseases in greater numbers than more affluent, white communities. In the United States, all communities are not created equal. If a community happens to be poor or working class or is inhabited largely by people of color, it generally receives less protection.

The environmental justice framework challenges the dominant environmental protection paradigm that largely institutionalizes unequal enforcement; trades human health for profit; places the burden of proof on the "victims" and not the polluting industry; legitimates human exposure to harmful chemicals, pesticides, and hazardous substances; promotes "risky" technologies; exploits the vulnerability of economically and politically disenfranchised communities; subsidizes ecological destruction; creates an industry around risk assessment and risk management; delays cleanup actions; and fails to develop pollution prevention as the overarching and dominant strategy (Finkel & Golding, 1994).

The EPA is mandated to enforce the nation's environmental laws and regulations equally across the board (Collins, 2005). It is also required to protect all persons who live in the United States, not just individuals or groups who can afford lawyers, lobbyists, and experts. Environmental protection is a right, not a privilege reserved for a few who can "vote with their feet" and escape or fend off environmental stressors (Bullard, 2005, p. 20). Companies operating industrial facilities are required to obtain permits from state and sometimes federal environmental agencies and conform to local land-use regulations. However, even when operated according to accepted specifications, industrial facilities can adversely impact nearby residents when leaks, explosions, and accidents occur. Indeed, industrial facilities in some cases have posed serious safety risks to health, property, and quality of life (Baibergenova et al., 2003; Bullard et al., 2008; Edelstein, 2004; Elliot et al., 1993; Fielder et al., 2000; Gerschwind et al., 1992; Nelson et al., 1992; Vrijheid et al., 2002). As a result of these threats, public opposition to industrial facility siting has been nearly universal, especially for high-profile facilities such as incinerators and landfills (Bullard et al., 2008).

The 1990s saw a record number of communities of color "racially profiled" or targeted for polluting operations. Communities all across the country resisted, but most were still vulnerable to industrial facility siting because of their limited financial, scientific, technical, and legal resources (Cole & Foster, 2000; Lerner, 2010; Taylor, 1998). People-of-color environmental justice organizations and networks were still small and underfunded, and only a handful had paid staff. Generally, communities with the greatest needs have the least resources and organization capacity to effectively fend off environmental assaults.

In 1992, the *National Law Journal* uncovered glaring inequities in the way the EPA enforced its laws related to hazardous waste cleanup (Lavelle & Coyle, 1992). The authors wrote: "There is a racial divide in the way the U.S. government cleans up toxic waste sites and punishes polluters. White communities see faster action, better results and stiffer penalties than communities where blacks, Hispanics and other minorities live. This unequal protection often occurs whether the community is wealthy or poor" (pp. S1–S2). These findings supplement the findings of earlier studies and reinforce what many grassroots leaders have been saying all along: People of color differentially are impacted by industrial pollution, and they also can expect different treatment from the government.

NEW TECHNOLOGY, RESEARCH TO ACTION, POLICY, AND ORGANIZING TOOLS: 2000s

The 2000s ushered in new technological advances that offered tremendous benefits to the environmental justice movement and frontline communities. A record number of grassroots leaders from low-wealth and people-of-color communities were able to gain access to new computing and communication technology and the Internet, which enabled them to better connect with their constituencies and allies. New funding opportunities from private foundations allowed more grassroots community groups and their leaders to have access to cell phones, geographic information systems (GIS) and other spatial mapping tools, community-based participatory research (CBPR) (see Chapter 6), and multi-stakeholder networks, including community-university partnerships and collaborations.

A 2002 study, *Air of Injustice: African Americans and Power Plant Pollution*, found that more than 68 percent of African Americans live within 30 miles of a coal-fired power plant, the distance within which the maximum effects of the smokestack plume are expected to occur, compared with 56 percent of white Americans (Clean the Air et al., 2002). In September 2005, the Associated Press (AP) released results from its analysis of an EPA research project showing that African Americans were 79 percent more likely than whites to live in neighborhoods where industrial pollution is suspected of posing the greatest health danger (Pace, 2005). The study revealed that in 19 states, blacks were more than twice as likely as whites to live in neighborhoods where air pollution seems to pose the greatest health danger. Hispanics in 12 states and Asians in seven states were also more likely to breathe dirty air than whites in some regions of the United States. The AP found that residents of at-risk neighborhoods were generally poorer and less educated, and unemployment rates in those districts were nearly 20 percent higher than the national average. The 2007 study *Toxic Wastes and Race at Twenty 1987–2007* found that people of color make up the majority (56 percent) of those living in neighborhoods within two miles of the nation's commercial hazardous waste facilities (Bullard et al., 2008).

This study reveals that where facilities are clustered together, people of color make up 69 percent of these neighborhoods. This pattern underscores the cumulative impact from discriminatory zoning and land-use practices. People of color were also overrepresented in populations living within a one-mile radius (44 percent) and a three-mile radius (46 percent) of the nation's 1,388 Superfund sites.

A 2008 study by University of Colorado researchers on race, income, and environmental inequality in the United States concluded that African Americans experience such a high air-pollution burden that black households with incomes of $50,000 to $60,000 live in neighborhoods that are, on average, more polluted than neighborhoods of white households with incomes less than $10,000 (Downey & Hawkins, 2008). In effect, research indicates that environmental inequality for African Americans could not be reduced to a "poverty thing." That same year, Hoerner and Robinson (2008), in their study of differential impacts of climate change on vulnerable populations, found that 43 percent of African Americans live in urban "heat islands," compared to only 20 percent of whites. Nationally, African Americans have a 5.3 percent higher prevalence of heat-related mortality than whites, and 64 percent of this disparity is traced to disparities in the prevalence of home air conditioning.

In 2011, a team of Duke University researchers also found significant air pollution burden borne by people of color compared to whites (Miranda et al., 2011). They found that

> non-Hispanic blacks in the United States suffer worse air quality across multiple metrics, geographic scales, and multiple pollution metrics. Hispanics also suffer worse air quality with respect to particulate matter, but not necessarily so for ozone. It also appears that environmental justice concerns are more prominent along race/ethnicity lines, rather than measures of poverty. (Miranda et al., 2011, p. 1755)

In ranking the 75 worst polluting coal-fired power plants in the United States, a NAACP (2012) study, *Coal Blooded: Putting Profits before People,* found that four million people live within three miles of these plants. Two million people live within three miles of one of the top 12 "dirtiest" coal-fired power plants. Approximately 76 percent of these residents are people of color and the average per-capita income is $14,626, compared with the national average of $21,587. People of color are severely overrepresented in communities that host the "dirty dozen" coal power plants since they made up only 37 percent of the U.S. population in 2010 (U.S. Census Bureau, 2010).

A Coming Clean (2012) report, *Who's in Danger? A Demographic Analysis of Chemical Disaster Vulnerability Zones*, found that fence-line residents who live closest to the facilities have average home values 33 percent below the national average and average incomes 22 percent below the national average. The percentage of blacks in the fence-line zones is 75 percent greater than for the United States as a whole, and the percentage of Latinos is 60 percent greater. The percentage of adults in the fence-line zones with less than a high school diploma is 46 percent greater than for the United States as a whole, but the percentage with a college or other post–high school degree is 27 percent lower;

and the poverty rate in the fence-line zones is 50 percent higher than for the United States as a whole.

Oil trains also pose special risk to people of color, who often live on the "wrong side of the tracks." The nation's oil trains are more likely to run through communities of color and expose their residents to elevated risks from explosion and derailment "blast zones." The blast zone is everything within a mile of tracks used for the oil trains (ACTION United, ForestEthics, and PennEnvironment Research and Policy Center, 2016). The *Fumes across the Fence-Line* report details the health toll the oil and gas industry has on black communities (NAACP and Clean Air Task Force, 2017). More than a million African Americans live within a half-mile of an oil and gas operation, and more than 6.7 million live in a county that is home to a refinery. Many of these communities (Manchester in Houston; Louisiana's "Cancer Alley"; North Richmond, California; Southwest Detroit; Port Arthur, Texas) are described as "sacrifice zones" because of the concentration of oil and gas pollution. And because of heightened exposure to oil and gas pollution, African American children suffer from 138,000 asthma attacks and 101,000 lost school days each year (NAACP and Clean Air Task Force, 2017).

The 150 or so U.S. oil refineries operating in 32 states emit thousands of tons of hazardous air pollutants, including substances that cause cancer. Half of the people at an increased cancer risk from refineries' pollution are people of color (Garcia, 2014). America is still segregated, and so is pollution (Bullard et al., 2011). More than 69.2 percent of Hispanic children, 61.3 percent of African American children, and 67.7 percent of Asian American children live in areas that exceed the EPA ozone standard, compared with 50.8 percent of white children. University of Minnesota researchers found that African Americans and other people of color breathe 38 percent more polluted air than whites and are exposed to 46 percent more nitrogen oxide (Clark et al., 2014). All indicators point to pollution taking a heavy health toll on Black America—especially black children.

The former vice president under Lyndon Johnson, Hubert H. Humphrey, once said, "The moral test of government is how that government treats those who are in the dawn of life, the children; those who are in the twilight of life, the elderly; those who are in the shadows of life, the sick, the needy and the handicapped." Children, especially poor children of color, form one of our most vulnerable groups in the United States when it comes to pollution. In addition to schools, many urban parks and playgrounds are located next to refineries, coal plants, chemical facilities, and highways. The adverse health effects of living or playing so close to polluting sources is elevated asthma and respiratory disease. For example, the asthma rate among African Americans is 35 percent higher than among whites; the hospitalization rate for African Americans and Latinos is three to four times the rate for whites; African Americans and Puerto Ricans are three times more likely than whites to die from asthma-related causes; and African Americans account for 13 percent of the U.S. population but 26 percent of asthma deaths.

Lack of zoning and poor land-use planning created a pollution nightmare for children living along the petrochemical corridor. The vast majority of residents and school children living along the Houston Ship Channel are Hispanic and African American. Children living within two miles of the channel had a 56 percent greater chance of developing lymphocytic leukemia (Houston Department of Health and Human Services, 2005). This pattern of over-polluting children of color also occurs in cities with zoning. A 2006 California study found that areas that suffer from increased respiratory hazards from air toxics tend to have schools with larger percentages of poor students and students of color (Pastor et al., 2006, p. 337).

The problem of schools near polluting facilities was thrust on the national stage in the *USA Today* special report *Toxic Air and America's Schools* (Morrison, Heath and Jervis 2008). In mapping the nation's 127,800 public, private, and parochial schools, the investigative reporters found that 20,000 schools—about one in every six—are within a half-mile of a major industrial plant. Nationally, one in three U.S. school children is at risk from a chemical catastrophe. In 2011, a team of University of Michigan researchers found that students of color are more likely than their white counterparts to attend schools in heavily polluted areas (Mohai et al., 2011, pp. 852–862). In Michigan, for example, whereas 44.4 percent of all white students in the state attend schools located in the top 10 percent of the most polluted locations in the state, 81.5 percent of all African American children and 62.1 percent of all Hispanic students attend schools in the most polluted zones.

The University of Michigan researchers also found that air pollution from industrial sources near Michigan public schools jeopardizes children's health and academic success. Schools located in areas with the highest air pollution levels had the lowest attendance rates and the highest proportions of students who failed to meet state educational testing standards. California researchers found a clear link between toxics near schools and student academic performance in Los Angeles (Pastor, Morello-Frosch, & Sadd, 2006). In El Paso, Texas, residential exposure to air toxics was linked to lower grade point averages among school children (Clark-Reyna, Grineski, & Collins, 2016).

Students of color are hit especially hard by transportation pollution. One in every 11 U.S. public schools, serving roughly 4.4 million students, lies within 500 feet of highways, truck routes, and other roads with significant traffic; 15 percent of schools where more than three-quarters of the students are racial or ethnic minorities are located near a busy road, compared with just 4 percent of schools where the demographics are reversed (Hopkins, 2014). White children make up almost 52 percent of U.S. public school students, yet only 28 percent attend high-risk schools. Black students, by contrast, make up just 16 percent of the total public school population, with 27 percent attending high-risk schools. Latinos constitute 24 percent of public school students, and 34 percent attend high-risk schools (Grineski & Collins, 2018).

This systematic overexposure of African Americans to air pollution was borne out by a 2018 U.S. EPA study that found race was more powerful than poverty in predicting exposure to air pollution (Mikati et al., 2018). In 46 states, people of color live with more air pollution than whites. African Americans are exposed to 1.54 times more fine particulate matter than whites, Hispanics are exposed to 1.2 times more than whites, and those below the poverty line are exposed to 1.35 times more than those above the line. The overall pattern reveals that a disproportionate share of places where people of color live, work, play, and learn are toxic "hotspots" with dangerous operations that pose elevated health threats—especially to vulnerable children of color.

A NEW GENERATION FIGHTING TO MAKE BLACK LIVES MATTER

Blacks lives matter less than white lives because of systemic racism that is baked into every institution in America—whether in voting, education, employment, health, the environment, or the criminal justice system. The environmental racism you will read about in this text, such as the Flint water crisis, is no fluke. In the United States, some people and communities have the "wrong complexion for protection" (Bullard & Wright, 2012). The Black Lives Matter (BLM) movement emerged in response to the 2012 death of unarmed black teenager Trayvon Martin and the 2013 acquittal of his vigilante killer, George Zimmerman. The unpopular verdict sparked outrage, protests, and the Black Lives Matter hashtag posting on social media by three black women—Alicia Garza, Patrisse Cullors, and Opal Tometi. BLM co-founder Alicia Garza gave a succinct overview of what the Black Lives Matter movement stands for: "Black Lives Matter is an ideological and political intervention in a world where Black lives are systematically and intentionally targeted for demise. It is an affirmation of Black folks' contributions to this society, our humanity, and our resilience in the face of deadly oppression" (Garza, 2014).

The #BlackLivesMatter hashtag used social media to effectively shine the national spotlight on racialized state-sanctioned killings of unarmed black men. Black Lives Matter activists exploited social media through videos and testimony where African Americans were recorded being shot, beaten, choked, and/or killed by police or vigilantes (Pellow, 2016). The movement gained national attention for its street demonstrations and mass protests following the 2014 police killing of two African American males: Michael Brown in Ferguson, Missouri, and Eric Garner in New York City (Day, 2015; Luibrand, 2015). BLM protests expanded into dozens of chapters in the United States with a primary focus on addressing racial profiling, racial inequality, and racism in the U.S. criminal justice system (Cullors-Brignac, 2016). The BLM network developed 13 core principles to guide their work; among them are diversity, empathy, restorative justice, being unapologetically black, and intergenerationality (Barre, 2016, p. 2).

Systemic policies of police violence and killing, racial profiling, overpolicing, overticketing, arresting, and jailing in the criminal justice system all emanate from the same systemic forces that target, overpollute, and poison black people where they live, work, play, and attend school. Just as black people are special targets of state-sanctioned police violence, black communities and their inhabitants are also targets of state-sponsored permits to pollute and of pollution violence (poisoning men, women, children, and unborn babies is a form of violence) by industries that cause premature illnesses and deaths in the black community.

Social media and videos taken on smartphones have allowed Americans to see in living color how racialized policing kills blacks with impunity. Racism in the criminal justice system kills and denies black people equal justice and equal protection under the laws guaranteed by the U.S. Constitution. Environmental racism kills more slowly (without the vantage point of videos) and harms a disparate share of black people. Racism denies them the same rights of equal protection and equal justice by targeting black communities for environmentally risky and polluting facilities—resulting in elevated rates of cancer and respiratory and cardiovascular illnesses such as heart disease and stroke. Racism is making Black America sick.

Dismantling systemic racism is a core guiding principle of both the environmental justice movement and the Black Lives Matter movement. In the final analysis, there is only one movement—the movement that fights for an American society that values black lives the same way it values white lives. Erasing American racism from our society will make us a much healthier, safer, and more just nation.

DEEPENING OUR UNDERSTANDING

1. This chapter has reviewed a voluminous quantity of evidence that people of color endure more than their fair share of environmental burden, at root due to structural racism.

   a. Where are the environmental burdens in your own community (where you are from or where you go to school)? For example, where is garbage taken? Where are toxic substances transferred, stored, or dumped? Where are industries that produce air or water pollution located?

   b. Who lives close to these facilities? Does it fit with the dominant national trend that this chapter describes?

2. Video taken on smartphones has aided awareness of Black Lives Matter in terms of police violence. In what ways do you think technologies do or could enable more significant exposure of and activism around environmental justice concerns today?

REFERENCES

ACTION United, Forest Ethics and Penn Environment Research and Policy Center. (2016). *Environmental Justice and Oil Trains in Pennsylvania*. https://pennenvironment.org/sites/environment/files/reports/OilTrainPAReport_r1.pdf

Agyeman, J., Bullard, R. D., & Evans, B. (2003). *Just sustainabilities: Development in an unequal world*. Cambridge, MA: MIT Press.

American Heart Association. (2016, December). *High blood pressure and African Americans. AHA Fact Sheet*. https://www.heart.org/idc/groups/heart-public/@wcm/@hcm/documents/downloadable/ucm_300463.pdf

Baibergenova, A., Kudyakov, R., Zdeb, M., & Carpenter, D. O. (2003). Low birth weight and residential proximity to PCB-contaminated waste sites. *Environmental Health Perspectives, 111*(10), 1352–1357.

Barre, D. (2016, October 12). 3 *ways racial and environmental justice are connected, as explained by the vision of black lives*. https://www.greenpeace.org/usa/3-ways-racial-and-environmental-justice-are-connected-as-explained-by-the-vision-for-black-lives/

Bullard, R. D. (1983). Solid waste sites and the black Houston community. *Sociological Inquiry, 53*(Spring), 273–288.

Bullard, R. D. (1987). *Invisible Houston: The black experience in boom and bust*. College Station, TX: Texas: A&M University Press.

Bullard, R. D. (Ed.). (1993a). *Confronting environmental racism: Voices from the grassroots*. Boston: South End Press.

Bullard, R. D. (1993b). Race and environmental justice in the United States. *Yale Journal of International Law, 18*(Winter), 319–335.

Bullard, R. D. (1993c). The threat of environmental racism. *Natural Resources & Environment*, 55–56. 7(Winter), 23–26.

Bullard, R. D. (1994). Unequal environmental protection: Incorporating environmental justice in decision making. In A. M. Finkel & D. Golding (Eds.), *Worst things first? The debate over risk-based national environmental priorities* (pp. 237–266). Washington, DC: Resources for the Future.

Bullard, R. D. (2000). *Dumping in Dixie: Race, class and environmental quality* (3rd ed.). Boulder, CO: Westview Press.

Bullard, R. D. (2005). *The quest for environmental justice: Human rights and the politics of pollution*. San Francisco: Sierra Club Books.

Bullard, R. D. (2007). *Growing smarter: Achieving livable communities, environmental justice and regional equity*. Cambridge, MA: MIT Press.

Bullard, R. D., Mohai, P., Saha, R., & Wright, B. (2008). Toxic wastes and race at twenty: Why race still matters after all of these years. *Lewis & Clark Environmental Law Journal, 38*(2), 371–412.

Bullard, R. D., Johnson, G. S., & Torres, A. O. (2011). *Environmental health and racial justice in the United States: Building environmentally just, sustainable, and livable communities*. Washington, DC: APHA Press. http://ajph.aphapublications.org/doi/book/10.2105/9780875530079

Bullard, R. D., & Wright, B. (2012). *The wrong complexion for protection: How the government response to disaster endangers African American communities*. New York: New York University Press.

Casey, J. A. (2018, May). Coal and oil power plant retirements in California associated with reduced preterm birth among populations nearby. *American Journal of Epidemiology*. Accessed June 2018. https://doi.org/10.1093/aje/kwy110

Clark, L. P., Millet, D. B., & Marshall, J. D. (2014, April). National patterns in environmental injustice and inequality: Outdoor NO2 air pollution in the United States. *PLOS*. http://journals.plos.org/plosone/article?id=10.1371/journal.pone.0094431

Clark-Reyna, S., Grineski, E., & Collins, T. M. (2016). Residential exposure to air toxics is linked to lower grade point averages among school children in El Paso, Texas, USA. *Population and Environment, 37*(3), 319–340. https://link.springer.com/article/10.1007/s11111-015-0241-8

Clear the Air, Black Leadership Forum, Southern Organizing Committee for Economic and Social Justice and the Georgia Coalition for the Peoples Agenda. (2002, October). *Air of injustice: African Americans and power plant pollution*. Washington, DC: Clear the Air. http://www.energyjustice.net/files/coal/Air_of_Injustice.pdf

Clinton, W. J. (1994). *Federal actions to address environmental justice in minority populations and low-income populations*. Title 3 – The President Executive Order 12898 of February 11, 1994. *Federal Register*, February 16, Vol. 59, No. 32.

Cole, L., & Foster, S. (2000). *From the ground up: Environmental racism and the rise of the environmental justice movement*. New York: New York University Press.

Collins, R. M. (2005). *The Environmental Protection Agency: Cleaning up America's act*. Westport, CT: Greenwood.

Commission for Racial Justice, United Church of Christ. (1987). *Toxic wastes and race in the United States: A national report on the racial and socioeconomic characteristics of communities with hazardous waste sites*. New York: United Church of Christ.

Cullors-Brignac, P. M. (2016, February 22). We didn't start a movement. We started a network. *Medium*. https://medium.com/@patrissemariecullorsbrignac/we-didn-t-start-a-movement-we-started-a-network-90f9b5717668

Day, E. (2015, July 19). #BlackLivesMatter: The birth of a new civil rights movement. *The Guardian.* http://www.theguardian.com/world/2015/jul/19/blacklivesmatter-birth-civil-rights-movement

Downey, L., & Hawkins, B. (2008, December). Race, income and environmental inequality in the United Sates. *Social Perspectives, 51*(4), 759–781.

Duke, L. (2007, March 20). A well of pain. *Washington Post*, C1.

Dum, Q. (2017). Air pollution and mortality in the medicare population. *New England Journal of Medicine, 376*, 2513–2522.

Edelstein, M. R. (2004). *Contaminated communities: Coping with residential toxic exposure* (2nd ed.). Cambridge, MA: Westview Press;.

Elliot, S., Taylor, S. M., Walter, S., Stieb, D., Frank, J., & Eyles, J. (1993). Modeling psychological effects of exposure to solid waste facilities. *Social Science and Medicine, 37*(6), 805–812.

Exchange Project.(2006, September). Real People, Real Stories: Warren County: Town of Afton. Chapel Hill: Department of Health and Health Education, University of North Carolina at Chapel Hill. http://exchangeproject.unc.edu/real_people/afton_overview/

Fielder, H. M. P., Poon-King, C. M., Palmer, S. R., Moss, N., & Coleman, G. (2000). Assessment of impact on health of residents living near the Nant-y-Gwyddon landfill site: Retrospective analysis. *British Medical Journal, 320*, 19–22.

Finkel, A., & Golding, D. (1994). *Worst things first? The debate over risk-based national environmental priorities.* New York: Resources for the Future.

Garcia, L. (2014, October 8). Communities near oil refineries must demand cleaner air. *Huffington Post.* http://www.huffingtonpost.com/lisa-garcia/communities-near-oil-refi_b_5662559.html

Garza, A. (2014, December). A herstory of the #blacklivesmatter movement." *The Feminist Wire.* http://www.thefeministwire.com/2014/10/blacklivesmatter-2/

Gerschwind, S. A., Stolwijk, J., Bracken, M., Fitzgerald, E., Stark, A., Olsen, C., & Melius, J. (1992). Risk of congenital malformations associated with proximity to hazardous waste sites. *American Journal of Epidemiology, 135*, 1197–1207.

Goldman, B., & Fitton, L. (1994). *Toxic wastes and race revisited.* Washington, DC: Center for Policy Alternatives.

Grineski, S. E., & Collins, T. W. (2018, February). Geographic and social disparities in exposure to air neurotoxicants at U.S. public schools. *Environmental Research, 161*, 580–587.

Hoerner, J. A., & Robinson, N. (2008, July). *A climate of change: African Americans, global warming, and a just climate policy for the U.S.* Oakland, CA: Redefining Progress.

Hopkins, J. S. (2014, February). *Invisible hazard afflicting schools.* Center for Public Integrity. https://www.publicintegrity.org/2017/02/17/20716/invisible-hazard-afflicting-thousands-schools

Houston Department of Health & Human Services. (2005). Preliminary epidemiologic investigation of the relationship between the presence of ambient hazardous air pollutants (HAPS) and cancer incidence in harris county. http://www.houstontx.gov/health/UT.html

Institute of Medicine. (1999). *Toward environmental justice: Research, education, and health policy needs.* Washington, DC: National Academies Press.

Khan, J. (2017, January). Environmental racism is a special and urgent concern. *The Root.* https://www.theroot.com/environmental-racism-is-a-special-and-urgent-concern-1791343793

Lavelle, M.,&Coyle, M. (1992, September). Unequal protection. *National Law Journal.*

Lerner, S. (2010). *Sacrifice zones: The front lines of toxic chemical exposure in the United States.* Cambridge, MA: MIT Press.

Luibrand, S. (2015, August 7). Black lives matter: How the events in Ferguson sparked a movement in America. *CBS News.* https://www.cbsnews.com/news/how-the-black-lives-matter-movement-changed-america-one-year-later/

McConnell, R. (2010, July). Childhood incident asthma and traffic-related air pollution at home and school. *Environmental Health Perspectives, 118*(7), 1021–1026. https://www.ncbi.nlm.nih.gov/pubmed/20371422

McGurty, E. (2007). *Transforming environmentalism: Warren County, PCBs, and the origins of environmental justice*. New Brunswick, NJ: Rutgers University Press.

Mikati, I. (2018, April). Disparities in distribution of particulate matter emission sources by race and poverty status. *American Journal of Public Health, 108*(4), 480–485. https://www.ncbi.nlm.nih.gov/pubmed/29470121

Miranda, M. L., Edwards, S. E., Keating, M. H., & Paul, C. J. (2011). Making the environmental justice grade: The relative burden of air pollution exposure in the United States. *International Journal of Environmental Research and Public Health, 8*, 1755–1771.

Mohai, P., Kweon, B. S., Lee,S., & Ard, K. (2011, August). Air pollution around schools is linked to poorer student health and academic performance. *Health Affairs, 30*(5), 852–862.

Morrison, B., Heath, B.., & Jervis, R. (2008, December 7). The smokestack effect: Toxic air and America's schools. *Special Report*.

NAACP and Clean Air Task Force. (2017). *Coal blooded: Putting profits before people*. Baltimore, MD: NAACP. https://www.naacp.org/wp-content/uploads/2016/04/CoalBlooded.pdf

Nelson, A. C., Genereux, M., & Genereux, J. (1992). Price effects and landfills on house values. *Land Economics, 68*(4), 359–365.

Pace, D. (2005, December). *AP: More blacks live with pollution*. Associated Press.

Pastor, M., Jr., Morello-Frosch, R., & Sadd, J. L. (2006). Breathless: Schools, air toxics, & environmental justice in California. *Polytechnic Study Journal*, 337.

Pellow, D. (2016). Toward a critical environmental justice studies: Black Lives Matter as an environmental justice challenge. *DuBois Review, 13*, 2. https://escholarship.org/uc/item/2rw7p84x

Rosen, R. (1994). Who gets pollution: The movement for environmental justice. *Dissent* (Spring), 223–230.

Taylor, D. E. (1998). Mobilizing for environmental justice in communities of color: An emerging profile of people of color environmental groups. In J. Aley, W. R. Burch, B. Canover, & D. Field (Eds.), *Ecosystem management: Adaptive strategies for natural resource organizations in the 21st century*. Philadelphia, PA: Taylor & Francis.

U.S. Census Bureau. (2018, June). Warren County, North Carolina Quickfacts. https://www.census.gov/quickfacts/fact/table/warrencountynorthcarolina,NC/PST045216

U.S. Census Bureau (2010). Quick Facts. Retrieved from https://www.census.gov/quickfacts/fact/table/US/RHI825218#RHI825218

U.S. Environmental Protection Agency. (1992, July). *Environmental equity: Reducing risk for all communities*. Washington, DC: EPA.

U.S. Environmental Protection Agency. (1998). *Guidance for incorporating environmental justice in EPA's NEPA compliance analysis*. Washington, DC: EPA.

U.S. General Accounting Office (GAO). (1983). *Siting of hazardous waste landfills and their correlation with racial and economic status of surrounding communities*. Washington, DC: GAO.

Vrijheid, M., Dolk, H., Armstrong, B., Boschi, G., Busby, A., Jorgensen, T., & Pointerm, P. (2002). Hazard potential ranking of hazardous waste landfill sites and risk of congenital anomalies. *Journal of Occupational and Environmental Medicine, 59*, 768–776.

Wernick, A. (2016, February). This professor says Flint's water crisis amounts to environmental racism. *Living on Earth*, PRI Radio. https://www.pri.org/stories/2016-02-11/professor-says-flints-water-crisis-amounts-environmental-racism

Williams, T., & Smith, M. (2015, December 29). Cleveland officer will not face charges in Tamir Rice shooting death. *New York Times*. https://www.nytimes.com/2015/12/29/us/tamir-rice-police-shootiing-cleveland.html

Zhang, M. (2018, May). Maternal exposure to ambient particulate matter <2.5 during pregnancy and the risk for high blood pressure in childhood. *Hypertension*. http://hyper.ahajournals.org/content/early/2018/05/11/HYPERTENSIONAHA.117.10944

2

THE ENVIRONMENTAL JUSTICE FRAME

Stella M. Čapek

When the Flint, Michigan, water crisis broke into the national news, and stories appeared about low-income African American children being exposed to unusually high levels of lead, the term *environmental justice* (EJ) was already somewhat familiar to the general public. But in the later part of the 20th century, the concept was still being invented. EJ emerged from a number of different directions, as a way of "framing" (or naming in a new way) a pattern of injustices that disproportionately exposed minority and low-income communities to toxic hazards. Multiple social actors, including social movement activists and scholars, have shaped its evolution. In this chapter, I'll begin by revisiting an EJ case from my research that dates back to the time when the EJ "frame" was starting to spread through local and national networks (Čapek, 1993). The Carver Terrace case reveals some of those early dynamics and provides an interesting comparison and contrast with present-day EJ activism. I'll also discuss how I was woken up to the issue of EJ, and how it shaped my research. Then, I'll focus on "framing" theory as a useful analytical tool, and I'll reflect on how the EJ frame has evolved over time. Throughout, I'll draw selectively on the broad and rich field of EJ scholarship, which investigates not only where harm has been done but also how a socially and ecologically just society can be envisioned (see Agyeman, Schlosberg, Craven, & Matthews, 2016). But first, let's imagine a community named Carver Terrace, a place that no longer exists, whose residents learned about EJ through a persistent struggle to get justice.

CARVER TERRACE

Picture this: In Texarkana, Texas, a thriving African American neighborhood called Carver Terrace is flourishing in the 1960s. Proud homeowners inhabit the neighborhood, jobs are plentiful, strong social networks connect neighbors, and the Mt. Zion Missionary Baptist Church is an active place. Residents compete to have the best lawns; children play safely outside; and gardens yield flowers, fruit, and vegetables. The neighborhood seems close to ideal. Fast forward to the early 1990s, and the neighborhood sits empty, surrounded by a chain link fence with posted "NO TRESPASSING" signs. The houses and church are boarded up, and weeds are growing in the once well-manicured yards. You'll notice the unnatural silence. There are no people—no children playing outside, no neighbors calling to each other. By 1993, the houses and church are completely gone, and only the concrete pads, driveways, and streets are still in place. Some gardens are still blooming—the last "residents" to give up on the place. How could a place so promising disappear?

Carver Terrace was built in the 1960s by a Louisiana-based developer who intentionally designed an affordable community for prospective African American homebuyers. Given the realities of racially segregated space in the South (and elsewhere), a development like this was highly desirable. As a resident told me, "It was a drawing card to us, because there had not been any houses of this quality available to us." Those who moved in had a variety of stories—

PHOTO 2.1: Patsy Oliver, Camille Brown, and Bettye Davis.
Photo by Stella Čapek.

some were middle-class professionals; others were working-class residents who had never owned their own home. All were thrilled at the opportunity offered by Carver Terrace. The houses were eagerly bought up, and for many years, the neighborhood appeared to thrive.

CARVER TERRACE ON CONTAMINATED LAND

But there was a catch. Although it didn't seem too significant when people bought their houses, the land for Carver Terrace was part of a former industrial site that the city had rezoned "residential." Starting in 1910, a number of industries operated there, using creosote to coat wood. The most recent was Koppers, Inc., beginning in the 1940s. When Koppers ceased operations in Texarkana, they left behind buried tanks and residues of creosote. Some local residents knew about the former creosoting operation, but when they expressed concerns, they were reassured by the city that it would be safe to live there. So, they anchored their lives to this place and made it flourish.

Decades went by. One day in 1984, Environmental Protection Agency (EPA) employees in "moon suits" showed up in the neighborhood, testing the soil. This is how residents found out that there was suspected contamination under and around their homes. In 1979, Congress had asked the 50 largest chemical companies in the United States to report hazardous waste sites (a reminder about the importance of passing good laws!). Koppers reported hazardous chemicals, including creosote. Coal tar creosote, used as a wood preservative, has been declared a probable human carcinogen by the International Agency for Research on Cancer (IARC) and the Agency for Toxic Substances and Disease Registry (ATSDR, 2002). Over the years, Carver Terrace residents had noticed an unusual number of illnesses in the neighborhood, from rashes to a variety of cancers, and a surprising number of miscarriages and even deaths. But without good information, it was easy to normalize and explain away such incidents. Now, as news of the testing for toxic chemicals spread through the neighborhood, the health problems and the odd materials dug up during landscaping took on a new meaning.

As residents became more aware of the threat of toxic chemicals, their relationships to their land and houses began to change. Distrust of local and federal agencies rose. While it is possible that in the 1960s less was known about the carcinogenic nature of certain chemicals, the rezoning of an industrial site like this one as "residential" was at best careless; at worst, it was negligent. It is also highly likely that a lower standard of scrutiny and safety was applied to land use for an African American and less affluent neighborhood. Residents felt deeply betrayed. The EPA declared Carver Terrace a Superfund site in 1984 (the federal Superfund was created to fund cleanups of contaminated sites) and placed it on the National Priority List. The next year, EPA and Koppers worked on a remedial plan, and Koppers did some soil removal and sod replacement in some people's yards. But the process moved slowly, and little information was shared with residents. They anxiously wondered what would become of their neighborhood, and of their lives.

Chemical contamination is sneaky, pernicious, and unsettling. It is often invisible, and its boundaries are unclear, which makes it difficult to assess and to address (for example, it is challenging to prove in a medical or legal sense). It makes its way into physical structures, land, water, and air, and into human bodies, planting seeds of doubt and fear. Just as devastatingly, residents discover that their property has suddenly lost its value when word of the contamination gets out. They are unable to sell their homes, which are typically their largest financial investment. They find themselves literally trapped in a place that is making them sick. Children, whose bodies are more vulnerable to contaminants, are at even greater risk. The almost unimaginable stress of such a situation is well documented (Edelstein, 2018).

In response, in 1985 approximately 60 Carver Terrace residents filed a lawsuit against Koppers for damages. Lawyers for the corporation played up genetic factors among African Americans to explain away the residents' illnesses and to exonerate the corporation. The trial was a major disappointment, and many residents became discouraged. But others became even more determined to fight for their right to live in a safe place. The EPA eventually proposed a cleanup technique called soil washing/filtering, claiming that contamination levels were not high enough for a buyout and relocation. But EJ scholars and grassroots groups charged that federal health studies were based on faulty designs. Carver Terrace residents began to try out some new strategies to get justice.

To understand this shift, let's "zoom out" and look at the bigger picture. Texarkana already had an environmental group, Friends United for a Safe Environment (FUSE). Predominantly white, and active in some form since the 1970s, FUSE had experience with fighting environmental problems. One of its members, Don Preston, spoke with several acquaintances in Carver Terrace about their situation, and attended an EPA meeting:

> I was absolutely appalled by the things I heard, the things that were happening to the people who lived out there. So when I told [FUSE] member Jim [Presley] about these things, I said, "We're environmentalists, here is the biggest cause in this town. We've got people with obviously catastrophic health complications as a result of the homes they're living in, and they're being stonewalled by the people who are supposed to protect them." We decided to go out and see what the people in Carver Terrace wanted.

Preston was a "conscience constituent," someone who supports a social change movement for ethical reasons, without directly benefiting (McCarthy & Zald, 1977). He knew schoolteacher Frances Shears in Carver Terrace, and they arranged a meeting in her living room. It was a small group conversation at first. Some residents were eager to meet, while others questioned the motivation of "these white men." Some had become discouraged and didn't show up at all. Several ministers in Carver Terrace were already brainstorming about how to move forward and were looking for the best strategy. Eventually the conversation grew into an extraordinarily effective, respectful collaboration between FUSE and what became the Carver Terrace Community Action Group (CTCAG). Patsy Oliver, who became one of the most outspoken (and globally oriented) CTCAG activists, told me, "It was really an inspiration to me to be part of FUSE because they were bi-state, bi-racial—and this was a first for Texarkana, you know."

Zooming out even further, the organizers in CTCAG and FUSE would benefit enormously from the support of regional and nationally networked EJ groups, some of which were beginning to use the term *environmental racism.* With their help, and through their own dedication and persistence, Carver Terrace residents eventually won a federal buyout and relocation, over the EPA's objections. I'll return to their story later, but first, I'll comment on my personal and scholarly connection to Carver Terrace. You'll also learn about an emerging network of EJ organizations that offered advice and assistance to places like Carver Terrace.

PERSONAL INTERSECTIONS AND SOCIOLOGICAL RESEARCH

You might wonder how I came to be involved with this story as a sociologist, since I didn't live in Carver Terrace myself. In graduate school I studied movements for social justice, especially affordable housing, tenants' rights, and inclusive urban design. When I moved to Arkansas to take a position as an assistant professor at Hendrix College, I soon found out about some very disturbing environmental problems in the state, and my research turned in that direction. One site that taught me important lessons was Jacksonville, Arkansas, where Agent Orange had been produced for the Vietnam War, and where the highly toxic chemical dioxin was extracted and stored in barrels near the Jacksonville Air Force Base (Čapek, 1992). There were 29,000 leaking barrels of one of the most toxic substances created by human beings located near residential neighborhoods and environmentally sensitive waterways, and enormous disagreements in the community about what to do. The "city fathers" (the Jacksonville Chamber of Commerce) wanted to hush it up, fearing damage to the business climate. Residents experiencing health issues, especially women with children, wanted the dioxin removed. Others (including some of their spouses) didn't want to publicize the problem, fearing a loss of property values or even their jobs.

This complicated situation woke me up to the difficult realities of toxic environments. I started interviewing grassroots environmental leaders in Arkansas (grassroots refers to organizations that originate locally, rather than having a national top-down structure). I also tried to learn more about dioxin. It wasn't easy—the EPA first assessed the dangers of dioxin to human health in 1984 but withdrew its document under pressure from the chemical industry. Astoundingly, it would take more than 20 years to reissue a public reassessment declaring dioxin to be a human carcinogen with other significant health effects. This lack of access to crucial health information taught me about the often politicized nature of federal scientific research. I became involved in an environmental organization, the Environmental Congress of Arkansas (ECA), and encountered some key national anti-toxics organizations that supported Jacksonville citizens who wanted the dioxin safely removed. I learned about the challenge of underfunded, reluctant, corporate-influenced federal agencies like the EPA, and how difficult it was for people without any political power to get something done about their situation. As an environmentalist and a sociologist, I could see that *justice* needed to be paired with the word *environment,* and that *environment* needed to include the people who inhabit it.

My first trip to Carver Terrace (located about three hours away) was for a national conference on environmental justice in 1989. A FUSE member and an

ECA member had attended an inspiring meeting and demonstration in Wichita, Kansas, that brought together grassroots environmental leaders from many states. They suggested Texarkana as the next site. The Texarkana EJ conference brought in national leaders from the anti-toxics movement, including Lois Gibbs, a white working-class mother who had organized support to win the first federal buyout of a contaminated community in Love Canal, New York. She went on to become director of the Citizens Clearinghouse for Hazardous Wastes (CCHW, later renamed the Center for Health, Environment & Justice), a resource base created to help other contaminated communities with scientific information and action strategies.

In addition to talks by national anti-toxic leaders and environmentalists, the program included a "citizen's public hearing," a rally with speeches by experienced activists from around the country, informal strategizing sessions, and a culminating march through Carver Terrace demanding a federal buyout. Lois Gibbs and other speakers strongly emphasized that legal strategies could go only so far and that, based on experience of grassroots anti-toxics groups around the country, political organizing and direct action tactics were more effective. She advocated putting a "face" on the problem—identifying specific politicians and others who were accountable. Environmental writers who attended publicized the story nationally, and Jim Presley of FUSE wrote an article for the *Texas Observer* titled "Toxicana, U.S.A." As in Jacksonville, the "city fathers" were not happy.

The conference provides an early snapshot of this segment of the EJ movement, and how it envisioned (framed) environmental justice. Presentations focused on toxic chemicals, critiques of corporate capitalism, ineffective state and federal agencies, the need for more democracy, and building grassroots coalitions to challenge unequal treatment. Most of the speakers were white, including Larry and Shelia Wilson from the Highlander Center, an organization that had crossed racial lines for social justice since the 1930s, cultivating social change activists for the labor movement, the civil rights movement, the environmental movement, and more (Marguerite Casey Foundation, 2015). Pat Bryant, of the Gulf Coast Tenants' Association and one of the few African American environmental leaders at the conference, underscored the emerging EJ movement's significance:

> I'm looking here at a whole community of refugees—soon to be refugees from your own community! I live in New Orleans where there are many refugees from Central America who have been on the wrong end of the foreign and military policy of the United States government, and now I'm looking at the prospect that by the year 2000 all across this country, people who *live* and *die* to make this country will be refugees in their own communities. That is a very shocking, but real, understanding of what is happening. The numbers are so staggering, brothers and sisters, that you undoubtedly are part of *the* movement of the '90s. (Presley, 1989, 9)

I came to Carver Terrace to learn, and to express solidarity through my ECA involvement. Like Don Preston, I was a conscience constituent (but with much less knowledge and experience). Later that year, I represented the ECA at a Stop Toxic Pollution (STP) workshop at the Highlander Center and heard more firsthand testimony from residents around the country who had become anti-toxics organizers. Their stories had similar ingredients: toxic sites leading to suffering through illness and devalued homes and land; residents who weren't high on the social hierarchy of power, whether they were white or people of color; local, regional, and federal authorities who were unresponsive or outright denied the problem; stigmatizing of (and sometimes violent threats against) residents for "stirring up trouble"; and overall a strong sense of injustice.

The next year, the ECA helped organize a Rally for the Environment at the state capitol in Little Rock to protest the state's decision to burn the dioxin in Jacksonville. National organizations and grassroots activists from around the country turned up, since the decision was seen as setting a terrible precedent (in effect, "a toxic landfill in the sky"). From CTCAG and FUSE, I heard the latest updates about Carver Terrace. The more I learned, the more I felt the urgency and importance of this unfolding story. Soon after that, I accepted an invitation from CTCAG/FUSE to do a sociological study of the community that would help document and analyze what was happening there.

DOCUMENTING THE CARVER TERRACE CASE

On a hot July day, Patsy Oliver drove me through the neighborhood, narrating a house-by-house story of economic and medical disasters. The catalogue of shattered dreams was disheartening. By then, some houses had been on the market for as little as $7,000, and worsening floods had invaded part of the neighborhood. Oliver herself had replaced the floor in her home for the third time and had lost her mother, Mattie Warren, to cancer in the previous year. It was one of many sudden deaths that shocked the community.

Loss came in so many forms that it would be easy to focus only on that part of the story. But the residents' resolute fight for social and environmental justice is just as remarkable. Over a period of many months, during short, intense research trips scheduled between my work obligations, I interviewed residents, attended many types of meetings, pored over documents, and thought hard about what EJ means in practice. I continued to visit the community during the transition to a buyout, and afterward I located and reinterviewed a number of families in their new homes. Using qualitative research methods, I did my best to create a holistic case study that would offer comparisons to other communities and preserve some of the unique details of this one. Like many of my colleagues, I hoped that my research, teaching, and writing about EJ would make some positive difference.

> ***My mother could outwork me two to one.... She went down weighing 98 pounds before her death.... She was telling it in church, and she was shouting to it on Sunday, she was telling them all, it's poison over here that's killing people, EPA is lying to us! When we had a march out here in the neighborhood, she was one of the first, and she marched all the way in every march. Once I lost her to the chemicals here—and I know that's what it was, you know, no one has to second guess—I made her a promise that I would never let her down, and I would never stop the fight. The more involved I am, the more that part of her is living.***

ENVIRONMENTAL JUSTICE AND ENVIRONMENTAL RACISM

In the 1980s and 1990s, different narrative strands were coming together around the concept of environmental justice, including one focused on *environmental racism*. At that time, environmental racism most often referred to the disproportionate targeting of minority communities for toxic burdens (like the siting of landfills, incinerators, or toxic industries). Warren County, North Carolina, became an iconic place symbolizing the coming together of the civil rights movement and the EJ movement when in 1982 African American residents engaged in direct action protests against the EPA-approved placing of a landfill with contaminated waste in their area despite the potential health hazards (see Chapter 1 in this volume).

In his 1990 book *Dumping in Dixie: Race, Class, and Environmental Quality*, Robert Bullard not only researched the pattern of disproportionate impact on minority communities but also engaged in active outreach to affected communities (including Carver Terrace) to make the research usable in a fight against environmental racism. As Dorceta Taylor (2000) and others have pointed out, mainstream organizations focused on reducing human damage to the environment (often construed as "wilderness") but ignored social justice issues and the everyday spaces where people live. Thus, an EJ agenda was badly needed.

Although early images associated with environmental racism often emphasize black communities, the environmental racism component of EJ had a wide umbrella that included many other people of color—among others, Latinos, Native Americans, and U.S. Asian and Pacific Islander communities. For example, the SouthWest Organizing Project (SWOP), active since the 1980s supporting the rights of communities of color in the U.S. Southwest, easily found a place under this banner. In 1990, organizers in the Native American community formed the Indigenous Environmental Network. Also in 1990, SWOP wrote a now-famous letter to the so-called Group of 10 mainstream environmental organizations (for example, the Sierra Club and the National Wildlife Federation), which were predominantly white and male, pointing out their exclusionary structure and issues. Responding to grassroots pressure, some of the Group of 10 began to diversify their organizations and issues. In a key development in 1991, the First National People of Color Environmental Justice Leadership Summit met in Washington, D.C., and adopted 17 principles (Principles of Environmental Justice, 1991). This would prove to be a transformative and radical reframing of environmental justice.

FRAMING THEORY, SOCIAL MOVEMENTS, AND ENVIRONMENTAL JUSTICE

When I first encountered framing theory, theories of social movements had become very focused on "resource mobilization," or the so-called nuts and bolts of organizing—leadership and organizational skills, fundraising, mobilizing constituents, and the like. Although these are important, the equally significant issues of symbolic meaning and identity had taken a back seat. Noticing this gap, sociologist David Snow and his collaborators developed a theory of framing that focused on the social construction of meaning and its links to social action. They defined frames as "'schemata of interpretation' that enable individuals 'to locate, perceive, identify, and label' occurrences within their life space and the world at large. By rendering events or occurrences meaningful, frames function to organize experience and guide action, whether individual or collective" (Snow, Rochford, Worden, & Benford, 1986, p. 464). In other words, interpretive frames serve the dual purpose of constructing meaning and offering strategies for action. Just like a frame around a picture, meaningful frames highlight certain elements of reality and affect how we look at (and act in) the world. A successful frame must "resonate," that is, it has to ring true and feel authentic to those who embrace it, individually and collectively.

Snow and Benford (1988) identified three types of "core framing tasks": *diagnostic* framing (analyzing a problem and identifying its causes); *prognostic* framing (envisioning plans for a solution); and *motivational* framing (providing a motive for action). Any viable social movement, they argue, needs to perform these framing tasks to mobilize supporters. To address an injustice, we try to figure out who (or what—but there is always a "who," as Lois Gibbs pointed out) is causing it and what we can do to change it. To actually change it, you have to believe that it's possible and that you should take action. As Carver Terrace residents found out, this is much easier if you aren't facing the problem alone; connecting with others provides experience, motivation, and courage to carry out all three framing tasks. You need courage when you challenge a powerful social hierarchy built around inequalities of gender, race, class, and more. Adopting a collective action frame is also often linked to a reframing of personal identity. The #MeToo movement and the Black Lives Matter (BLM) movement (BLM) provide excellent examples of powerful collective action frames that are also deeply personal. The same can be said about the EJ frame.

Importantly, frames don't just automatically "snap into place." Social movement scholar Aldon Morris (1986) reminds us that the nonviolent protest frame, for which the civil rights movement is so well known, was initially not widely embraced by African Americans in the South, who saw being unarmed in the face of armed opponents as a potential death sentence. Successful framing is always a product of meaningful social interaction. Also, framing is not a purely rational process. Rather, an injustice frame is frequently connected to a strong sense of "moral outrage," "a 'hot' cognition...that is emotionally charged" (Taylor, 2000, p. 511). A resonant frame channels emotions in a particular direction. The polarization in the United States during the Trump presidency illustrates all too well the competing frames that resonate among his supporters and opponents, amplified by social media networks.

Framing has multiple purposes: presenting issues to the public (for example, presenting climate change in a convincing way); emphasizing certain collective strategies among a movement's own participants and supporters (for example, validating direct action protest as the best choice); and influencing the social construction (reframing) of personal identity (for example, coming to see oneself as an EJ activist).

Framing theory has greatly influenced my work. I wrote that "'[e]nvironmental justice' can be understood as a conceptual construction, or interpretive 'frame' (Snow et al., 1986), fashioned simultaneously from the bottom up (local grass-roots groups discovering a pattern to their grievances) and from the top down (national organizations conveying the term to local groups)" (Čapek, 1993, p. 5). This is what I saw happening in Carver Terrace and nationwide during a time when the language of EJ was surfacing. Besides exploring the EJ frame more generally, I wanted to pay attention to the everyday experience of residents in contaminated communities and what environmental justice meant to them. Carver Terrace was a microcosm of this search for justice, with similarities to (and of course differences from) other communities. My research suggested at least five consistent EJ frame dimensions: (1) the right to accurate information; (2) the right to a prompt, respectful, and unbiased hearing; (3) the right to democratically participate in deciding the future of the contaminated community; (4) the right to compensation from those who inflicted injuries; and (5) commitment to solidarity with victims of toxic contamination in other communities. Are these dimensions still relevant? I will say a bit more about this later in the chapter, as we consider the past, present, and future of the EJ frame.

ENVIRONMENTAL JUSTICE FOR THE 21ST CENTURY

The EJ frame has expanded greatly since I first wrote about it. Consider the sheer number of contemporary EJ issues. More recent technologies like hydraulic fracking create new inequities, health risks, and environmental destruction. Unequal global "mobilities" with ecological consequences include both elite tourists inhabiting mostly white, privileged spaces, and immigrants driven from their homes by deadly violence, climate change, and global "free-trade" agreements that undercut their livelihoods (Urry & Larsen, 2011). As Pat Bryant presciently stated in Carver Terrace, many people have become climate refugees in their own communities, whether in New Orleans after Hurricane Katrina, in Puerto Rico after Hurricane Maria, or in Shishmaref, Alaska, one of a growing number of vulnerable coastal communities literally going underwater due to climate impacts disproportionately caused by others. Indigenous resistance is more visible, as reflected in the Standing Rock encampments supporting the Standing Rock Sioux protests against the Dakota Access Pipeline (DAPL), focusing on toxic pollution but also affirming Indigenous cultural values and treaty rights. Globally, Indigenous EJ activism networks have grown, although the risks are often high. For example, human rights activist Berta Cáceres in Honduras was tragically assassinated in 2016 after organizing effective opposition to destructive dams, mining, and logging on Indigenous Lenca lands. Her daughter has carried on her work. The list of other EJ issues is long: the "dumping" of toxic electronic waste from the Global North into the Global South; EJ debates about siting national parks and nature preserves (who gets access? who/what is protected or displaced?); food justice activism (for example, "food deserts," food sovereignty, food politics, farmworkers' rights, genetically modified organisms, and agricultural chemicals); sustainability and green design (sustainability for whom?); and much more. Destruction of scientific data joined the list during the Trump administration, as federal agencies were directed to delete information and websites about climate change.

Given the growing list of EJ issues, what can we say about the EJ *frame*? The frame continues to highlight what is wrong, who is responsible, and how to fix it. Contemporary EJ research helps to explain how the framing has changed. Let's consider sustainability. EJ scholars and activists have critiqued sustainability initiatives that unwittingly create bubbles of privilege. For instance, many U.S. urban planners, including those in my city, have been enthusiastic about the so-called complete street concept, which, instead of focusing on cars, includes spaces for pedestrian strolling, bicycle lanes, traffic calming elements, and green spaces. They frame this as good for small businesses and beneficial to everyone. However, some minority communities have protested because funding these projects attracts gentrification, where higher-priced retail businesses, restaurants, and housing drive up property tax and rents, pricing lower-income residents out of their own communities. A good idea in principle becomes a bad idea in practice if it contributes even unintentionally to segregation, displacement, and distrust. In the United States, green design (a good idea) is often pitched to a higher-income clientele (exclusionary), prompting EJ critiques of "green gentrification" (Checker, 2011; Gould & Lewis, 2017). For a full discussion of this phenomenon, please see Chapter 17 in this volume. Likewise, local food movements have been critiqued for being insular and supporting "white space" while ignoring a deeper history and diverse cultural perspectives (Mares & Peña, 2014; see also Chapter 12 in this volume). A growing body of research on "just sustainabilities" (Alkon & Agyeman, 2014; Agyeman et al., 2016) is contributing significantly to an expanding EJ frame.

Framing theory also reminds us that the EJ frame will look different depending on the depth of the *diagnostic* and *prognostic* process that produces it. A range of EJ scholars point out that today the EJ frame is more likely to reach deep into the systemic roots of inequalities like racism and sexism. As environmental racism became more prominent in the EJ frame, the frame expanded beyond seeking a remedy for a particular situation and targeted the racist logic prevalent in the U.S social structure and elsewhere—a deeper diagnosis, and a message that many white people either do not want to hear, or of which they are naïvely unaware. A highly visible example is how white bodies and bodies of people of color in particular spaces are treated differently (think Flint, Michigan, and the BLM movement). Who is privileged? Who is erased? Who is put under surveillance? Who is injured or killed? A deeper EJ frame also spotlights inequalities built into the global capitalist system and its political power; Laura Pulido and her co-authors question the effectiveness of merely "tinkering with policies" (Pulido, Kohl & Cotton, 2016, p. 12). The problem is multifaceted and includes the need for stronger democracy and confronting the deep roots of inequality embedded in the economic system. By understanding the framing process, we can discover who constructed a particular set of meanings, how this relates to social power, and how framing can support or undermine social justice. For example, "counterframes" spring up to challenge successful frames like EJ, targeting minorities and poor people as the problem, rather than the system that does violence to them.

WEAVING TOGETHER PAST, PRESENT, AND FUTURE

The residents of Carver Terrace endured many institutional failures, made worse by systemic racism: the way land use was(n't) regulated, racialized housing markets, the way science and law was practiced, and encounters with outside agencies and other interactions that framed people of color as problematic and undeserving (Čapek, 1999). What about the five EJ frame dimensions that I identified earlier—Are they still relevant? Yes, but they are part of a bigger picture (and EJ frame) with many more dimensions.

Communities continue to struggle to get accurate information (dimension 1) about the safety of their land and homes, whether the problem is fracking, tar sands oil pipelines, urban air and water pollution, runoff from massive poultry or hog operations, pesticide drift, Indigenous sovereignty rights, and much more. Due to the hostility to regulation built into capitalism, citizens often don't get information without a fight. They hold agencies, politicians, and corporations accountable through protests, legal suits, and political action, as well as by creating alternative resources. When FUSE/CTCAG discovered that an important federal health assessment was withheld from the community, they held a press conference but also worked with the grassroots Environmental Health Network to collect their own data—a good example of what Phil Brown (1992) calls popular epidemiology (see Chapter 5 in this volume). Today, various nonprofit organizations continue to sidestep reluctant government agencies to study environmental health impacts and to share the information. Recently, researchers at the Silent Spring Institute, a public-interest nonprofit research organization, found that black hair products contain "multiple chemicals linked to cancer, asthma, infertility, and more" (Helm, Nishioka, Brody, Rudel, & Dodson, 2018). Researchers focusing on the "environmental injustice of beauty" found that employees of nail salons (predominantly Asian American and African American women) are disproportionately exposed to toxins from the products they work with (Zota & Shamasunder, 2017). EJ enters the most intimate spheres of our lives, especially where information is lacking. The European Union has stricter regulations for many everyday items, but U.S. corporations resist labeling their products and politically frame precautions as unnecessary, as if the labels themselves were toxic. Organizations like the Environmental Working Group (EWG) have stepped into the gap to provide important health information to the public, but the struggle for accurate information continues.

The problem is not just an information deficit. People want to be heard in an unbiased, respectful way (frame dimension 2). Sometimes disrespectful treatment results simply from an overworked bureaucracy, but beyond that, community residents have often felt the sting of second-class citizenship and—in places like Carver Terrace, Flint, and many others—racism. Carver Terrace residents who traveled to Dallas to get more information were locked out of the EPA regional office building and the police were called. Beginning with their lawsuit and extending through their struggle for a fair buyout price, they were assumed to be "wanting something for nothing." The grassroots Southwest Network for Environmental and Economic Justice provided a space where they could be respected, sending a team door to door to collect residents' stories, and creating a public forum that included representatives of federal agencies and EJ groups, amplifying the residents' voices in a system that didn't want to "hear" them. More recently, the Dakota Access Pipeline protests created a space where Indigenous rights could be affirmed instead of erased from public view, despite strenuous efforts to shut down the protests. Social media, in addition to mainstream and progressive media outlets, spread the word nationally and globally, and the DAPL opposition eventually gained support from the Obama administration. The importance of such "alternative spaces" shouldn't be underestimated, even if victory isn't immediately within reach. For example, many conscience constituents have transformative experiences in such spaces, a reframing of personal and collective identity that inspires them to work for social change—like the Native American youth who organized ReZpect Our Water, creating a community of runners who publicized opposition to the pipeline, and organized a protest relay run from North Dakota to Washington, D.C. (Greene, 2017). At the same time, Indigenous Sami youth protested Norway's investments in DAPL. An energized Indigenous Caucus convened at the U.N. Climate Change conference in Bonn, Germany, to strategize (Monet, 2018). Creating alternative networks and spaces is important, given that the "state" is an unreliable ally—the Trump administration turned a deaf ear to EJ and approved the pipeline. Finding a way to be truly heard remains a creative and challenging struggle, but countless groups are mobilizing for EJ at local, regional, national, and international levels.

The other EJ frame dimensions I mentioned (the right to democratically decide the future of contaminated communities, the right to compensation, and solidarity with other contaminated communities) also continue to be relevant. I would now simply say "communities," since toxic contamination isn't the only issue (and never was). While Carver Terrace fought for a buyout and relocation, some other EJ battles are about staying in place and claiming health, dignity, and other human rights. Flint residents, where children's development was tragically compromised by lead contamination, had no choice but to stay in place and to call for accountability at all levels of government. By then, a more established and experienced EJ movement came to their assistance, including EJ scholars Paul Mohai (Mohai, Pellow, & Roberts, 2009) and Michael Mascarenhas (2007), who testified before the Michigan Civil Rights Commission (Michigan Department of Civil Rights, 2016). But the question of compensation hovers disturbingly over every EJ case. Who will pay for all the damage done to people's bodies and dreams when they encounter environmental *in*justice? Who will pay to relocate Shishmaref, a community whose suffering has been caused by others (especially when the "others" are less visible)? The answer is "no one," in the absence of strategic organizing and framing that provides an effective leverage point for justice. This makes the fifth EJ frame dimension more important than ever: building solidarity with other communities through sharing information, creating networks, joining in protests, and working at multiple levels for social justice. This means co-inventing an inclusive future that draws on EJ research and that uses imaginative organizing and framing skills to create alliances between groups that can support EJ together, even if temporarily. Coalition building is difficult but necessary work, since not only EJ, but also racism, is receiving new infusions of energy.

CONCLUSION

> *If ten thousand people camping at Standing Rock to protect the Missouri River could not stop the siege of the Dakota Access Pipeline, then what does it take?*...I know, we all know, it will take *more*. And toward this, our work continues.—Layli Long Soldier (2017)

This chapter has invited you to explore the environmental justice frame and its evolution over time, using a case study of Carver Terrace as a point of comparison with more contemporary understandings of EJ. Carver Terrace was like many other contaminated communities but also atypical—the first minority community to demand and win a federal buyout. The story is richer and more complex than what I touched on here, but the abridged version allowed us to dive more deeply into a particular time and place and to think about connections and contrasts with other situations. I showed how a sociological framing theory is useful for understanding the EJ frame at multiple levels. Framing theory applies to much more than EJ; it helps you to think about your own life, your meaningful frames, and how they connect to your actions and those of others. It equips you to be more critically aware of the many frames that drift our way through cyberspace, mediascapes, and so many other sources. It gives you tools for social change.

Here is one last image, which is a contrast with the desolate landscape we encountered earlier. Carver Terrace residents finally won a federal buyout in 1990. Grassroots leaders who told them that "political organizing matters" were right. CTCAG/FUSE persuaded one of their key allies, Texas Democratic Rep. Jim Chapman, a member of the budget appropriations committee for the EPA, to attach a provision to the EPA's budget that authorized a Carver Terrace buyout. Without CTCAG/FUSE and their mobilizing, organizing, coalition-building, and *framing* skills, there would be no buyout. Flawed as the political-economic system might be, it pays to know how it works. Yet, consistent with deep and ongoing structural inequalities, the victory was bittersweet. Imagine the residents in new homes that they bought or built, breathing a sigh of relief, enjoying a space away from the toxic contamination. Then realize that some didn't live to see the relocation, and others died too soon afterward, including Patsy Oliver. Survivors won a safer place to live, but they lost their physical community. For those who moved away, and especially for younger generations, the battle was worth the precious chance for a healthier life and a better future. Yet to truly heal what systematically produces environmental injustice, environmental racism, and global pollution on a massive scale, as Layli Long Soldier says, "it will take *more*." The EJ frame will continue to evolve, as it must, if we want to live on a planet that is sustainable, and not only for the privileged few. As researchers, we also evolve (yes, it is happening at this very moment as I write!), discovering new questions and solutions that become part of the collective stream of EJ scholarship and recipes for action. You, too, can be part of this solution.

DEEPENING OUR UNDERSTANDING

1. Search the term *environmental justice* on the Internet, and see what you turn up. How visible is the EJ frame in cyberspace, and what do you learn from this?
2. Identify an EJ issue or event (preferably local), and analyze the effects of the framing of that issue. Do you see any evidence of a counterframe?
3. The way we construct meaning shapes how (and whether) we participate in social movements. Think about your own identity. Can you "see" yourself taking part in a protest action, for example, a march? If so, why? If not, why not?
4. Framing includes how we interpret everything—"nature," our bodies, race, gender, who/what we consider to be "other," and much more. Based on your own identity and, thus, interpretations, where do you draw the line and start feeling uncomfortable?
5. Social media has become an important part of the EJ movement. A recent strategy has been to post Indigenous names on Instagram during hikes in outdoor recreational areas, calling attention to the missing Indigenous history (https://www.yesmagazine.org/issues/affordable-housing/posting-your-hike-on-instagram-now-you-can-tag-your-locations-indigenous-name-20180523). What is your response to this? What are some other ways that technology could be used creatively for EJ?

I thank Samantha Lewis for her assistance with bibliographical research on environmental justice. Any quotes are from my own interviews unless otherwise attributed. I dedicate this to all who make EJ possible.

References

Agency for Toxic Substances and Disease Registry (ATSDR). (2002). *Toxicological profile for creosote.*Atlanta, GA: U.S. Department of Health and Human Services, Public Health Service. https://www.atsdr.cdc.gov/ToxProfiles/tp85-c1-b.pdf

Agyeman, J., Schlosberg, D., Craven, L., & Matthews, C. (2016). Trends and directions in environmental justice: From inequity to everyday life, community, and just sustainabilities. *Annual Review of Environment and Resources, 41*, 321–40.

Alkon, A. H., & Agyeman, J. (2014). *Cultivating food justice: Race, class, and sustainability*. Cambridge, MA: MIT Press.

Brown, P. (1992). Popular epidemiology and toxic waste contamination: Lay and professional ways of knowing. *Journal of Health and Social Behavior, 33*, 267–281.

Čapek, S. (1992). Environmental justice, regulation, and the local community. *International Journal of Health Services, 22*(4), 729–746.

Čapek, S. (1993). The environmental justice frame: A conceptual discussion and an application. *Social Problems, 40*(1), 5–24.

Čapek, S. (1999). Erasing community: Institutional failures and the demise of Carver Terrace. *Research in Social Problems and Public Policy, 7*, 139–162.

Checker, M. (2011). Wiped out by the "greenwave": Environmental gentrification and the paradoxical politics of urban sustainability. *City and Society, 23*(2), 210–229.

Edelstein, M. (2018). *Contaminated communities: Coping with residential toxic exposure*. New York: Routledge, Taylor & Francis Group.

Gould, K. A., & Lewis, T. L. (2017). *Green gentrification: Urban sustainability and the struggle for environmental justice*. New York: Routledge.

Greene, E. (2017). ReZpect our water. *Green American, 109*, 23.

Helm, J., Nishioka, M., Brody, J., Rudel, R., & Dodson, R. (2018). Measurement of endocrine disrupting and asthma-associated chemicals in hair products used by black women. *Environmental Research, 165*, 448–458. doi:10.1016/j.envres.2018.03.030

Long Soldier, L. (2017). Words for water. *Orion, 36*(4), 52–53.

Mares, T., & Peña, D. (2014). Environmental and food justice: Toward local, slow, and deep food systems. In A. H. Alkon & J. Agyeman (Eds.), *Cultivating food justice: Race, class, and sustainability* (pp. 197–219). Cambridge: MIT Press.

Marguerite Casey Foundation. (2015). Highlander Center—training generations of change-makers. https://caseygrants.org/who-we-are/inside-mcf/highlander-center-training-generations-of-change-makers/

Mascarenhas, M. (2007). Where the waters divide: First Nations, tainted water and environmental justice in Canada. *Local Environment, 12*(6), 565–577.

McCarthy, J. D., & Zald, M. N. (1977). Resource mobilization and social movements: A partial theory. *American Journal of Sociology, 82*(6), 1212–1241.

Michigan Department of Civil Rights. (2016, September). Michigan Civil Rights Commission to hold third public hearing on Flint water crisis Thursday. https://www.michigan.gov/mdcr/0,4613,7-138-392873-,00.html

Mohai, P., Pellow, D., & Roberts, T. (2009). Environmental justice. *Annual Review of Environment and Resources, 34*, 405–430.

Monet, J. (2018). What Standing Rock gave the world. *Yes! 85*, 26–29.

Morris, A. D. (1986). *The origins of the civil rights movement: Black communities organizing for change*. New York: Free Press.

Presley, J. (1989, March). Toxicana USA—the growing drive to clean up Texarkana—and the nation. *The Texas Observer*.

Principles of Environmental Justice. (1991). Retrieved from https://www.ejnet.org/ej/principles.html

Pulido, L., Kohl, E., & Cotton, N. (2016). State regulation and environmental justice: The need for strategy reassessment. *Capitalism Nature Socialism, 27*(2), 1–20.

Snow, D., & Benford, R. D. (1988). Ideology, frame resonance and participant mobilization. *International Social Movement Research, 1*, 197–217.

Snow, D. A., Rochford, E. B., Jr., Worden, S. K., & Benford, R. D. (1986). Frame alignment processes, micromobilization, and movement participation. *American Sociological Review, 51*(4), 464–481.

Taylor, D. E. (2000). The rise of the environmental justice paradigm: Injustice framing and the social construction of environmental discourses. *American Behavioral Scientist, 43*(4), 508–580.

Urry, J., & Larsen, J. (2011). *The tourist gaze 3.0*. London: SAGE.

Zota, A. R., & Shamasunder, B. (2017). The environmental injustice of beauty: Framing chemical exposures from beauty products as a health disparities concern. *American Journal of Obstetrics and Gynecology, 217*(4), 418.e1–418.e6. doi:10.1016/j.ajog.2017.07.020

3

ENVIRONMENTAL JUSTICE, INDIGENOUS PEOPLES, AND CONSENT

Kyle Powys Whyte

Indigenous peoples are living societies who continue to exercise their own political and cultural self-determination despite facing conditions of invasion, exploitation, and colonization (Anaya, 2004; Sanders, 1977). Self-determination refers to a society's capacity to pursue freely its own plans and future in ways that support the aspirations and needs of its members. Conditions of invasion, exploitation, and colonization are caused by groups from other societies. The groups include nations and for-profit and nonprofit organizations, such as multinational corporations, local industries, and conservationist groups. In very simple terms, invasion occurs when one society (or certain groups from it) forcefully seizes the lands and waters that another society lives on and flourishes from. The latter society is the Indigenous people. Seizure is likely aimed at several goals, including exploitation and colonization. Exploitation occurs when the invaders seek to earn economic profits at the expense of harming the Indigenous peoples. Colonization occurs when the invaders seek to create strategies to undermine the Indigenous peoples' self-determination in preventing themselves from being exploited.

Colonial strategies for denying the colonized society's self-determination often involve military protection of people who seek to engage in industries such as mining that take resources from Indigenous lands. Indigenous peoples do not profit from these industries and are often harmed by environmental consequences, such as pollution. Or colonial strategies involve the invading society actually forcing the creation of conditions for its members to live permanently in the new lands. In North America, the United States and Canada, as well as the European nations that preceded them, invaded Indigenous peoples' lands and continue to exploit and colonize Indigenous peoples today. Corporations, operating with the sanction of these countries, have profited from dirty environments at the expense of Indigenous peoples' health, cultural integrity, and economic well-being. Economic exploitation, stealing of resources, and polluting the land are all strategies to stop people from pursing their own plans and aspirations, that is, the self-determination of their societies.

For example, General Motors and Reynolds Metals dumped unacceptably high levels of polychlorinated biphenyls (PCBs), dioxin, and mercury in the territories of the St. Regis Mohawk Tribe (New York) and the Mohawk Nation at Akwesasne (Ontario) (Hoover, 2017; Tarbell & Arquette, 2000). In the 1980s and 1990s, numerous studies showed Mohawk persons suffered health problems, threats to their cultural practices, and decline of their fishing economy (Hoover, 2017). Moreover, U.S. and Canadian laws and policies restricted the capacity of the Mohawks to prevent their communities from suffering harms and living with unacceptable risks. Alice Tarbell and Mary Arquette (2000), who have played critical roles in this issue, have written that the laws and policies involved "relax[ed] treatment standards and promot[ed] substandard, temporary cleanups at Superfund sites." The governments' responses were much too slow to address "environmental problems [at Akwesasne] and [lacked] the will and support to enforce their own decisions." In failing to "[respect] the decisions of governments like those of Akwesasne," the U.S. and Canadian

---

PHOTO 3.1
Diego G. Diaz / Shutterstock

governments have been "biased toward industry and the local economy at the expense of Native peoples" (Tarbell & Arquette, 2000, p. 99).

Indigenous peoples are among the leading groups that are working to address environmental injustice. At Akwesasne, Mohawk peoples founded projects, such as the Akwesasne Task Force on the Environment, the Environmental Division of the St. Regis Mohawk Tribe, and the Mother's Milk Project, that revived their own traditions in order to exercise self-determination in how they cleaned up the environment and protected their community members' safety. The Mother's Milk project, for example, led by Katsi Cook, involved creating strategies for women affected by pollution to study their own exposure and implement their own solutions (Tarbell & Arquette, 2000). The Akwesasne Task Force, among other key roles, offers alternatives that protect health, cultural integrity, and economic vitality, such as aquaculture projects. The Traditional Mohawk Nation Council of Chiefs has advocated at the level of the United Nations, presenting in 1995 a document called *Haudenosaunee Environmental Restoration: An Indigenous Strategy for Human Sustainability* (Tarbell & Arquette, 2000).

Indigenous peoples have never *consented* to the pollution of their lands. That is, they have never consented to invasion, exploitation, and colonization. When someone consents to an action by someone else that affects them, it means they have willingly accepted that action. An action can be acceptable when those affected by it understand and are "okay" with the risks, see the action as having impacts that are in their best interests, and are secure in knowing that they have the chance to influence and monitor how the action is carried out and completed. When someone dissents to an action, it means they neither understand the risks sufficiently (to consent confidently) nor wish to shoulder them, do not see any converging interests, and have reason to believe they will have no role in influencing or monitoring how the action is carried out. For the people of Akwesasne, they clearly did not consent to the presence of and pollution caused by the dirty industries in their region. The U.S. and Canada, and the corporations and communities benefiting from the industries, exploited the health, cultural integrity, and economic vitality of Mohawk peoples. They colonized Mohawk peoples by limiting Indigenous self-determination to prevent these harms and risks.

Denial of consent is *one* important factor causing environmental injustice across a number of cases. Broadly, *environmental injustice* refers to the problem that there are some groups or societies who suffer more harms and shoulder greater risks, such as the health problems of pollution. What really makes a particular situation an *injustice* is when the prevalence of harm or risk is a product of another group's or society's seeking their own benefits by taking advantage of others. Hence, the Mohawks, because they are Indigenous, suffered more harms and shouldered greater risks for the benefit of corporations, communities, and individuals of U.S. and Canadian societies. Colonialism, then, is an environmental injustice since it undermines self-determination through land dispossession, pollution, and other environmental threats. A key strategy of colonialism is the denial of consent, which can produce environmental injustice. Nations and corporations often consider groups of people, such as Indigenous peoples, as not worthy of consent.

This chapter examines Indigenous environmental justice issues through the lens of consent. In it, I seek to show why environmental injustices against Indigenous peoples are problems of consent. I compare the current situation of consent today with Indigenous traditions that privilege consent in terms of how a society is organized. Part of colonialism in contexts like the United States and Canada has been the reorganizing of societies in North America to undermine Indigenous consent. The dismantling of traditions of consent is one way to understand how colonialism attacks self-determination. Although the United States is a unique context, many of the consent issues in relation to environmental justice arise in other contexts around the world. Readers should come away from this chapter with a good sense of why consent matters in relation to injustice, and why affirming consent is a strategy for achieving environmental justice for the sake of future generations.

CONSENT, INDIGENOUS JUSTICE, AND THE DAKOTA ACCESS PIPELINE

Denial of consent is a key reason why environmental injustice occurs in the case of the Standing Rock Sioux Tribe's resistance to the construction of the Dakota Access Pipeline in its territory. The story of the Tribe's resistance to the pipeline was covered widely in national and international media in 2016 and 2017, and was considered a major issue in the U.S. presidential election during that time. Understanding why consent matters in Indigenous struggles against oil and gas pipelines and coal terminals involves learning about the culture, history, and self-determination of Indigenous peoples who are involved and the motivations behind the projects that threaten to cause pollution, land dispossession, and other environmentally related harms and risks.

The Lakota and Dakota peoples of the Standing Rock Sioux Tribe, who are part of the larger Oceti Sakowin Nation, are Indigenous peoples in North America. Oceti Sakowin peoples have a rich cultural heritage that involves respect and care for water, including *Mni Sose* (the Missouri River). The river flows through their territories. Jaskiran Dhillon and Nick Estes (2016) write that "Mni Sose (the Missouri River) is not a thing that is quantifiable according to possessive logics. Mni Sose is a relative: the Mni Oyate, the Water Nation. She is alive. Nothing owns her." For Oceti Sakowin peoples, their lives and the waters of the river are interdependent. According to Craig Howe and Tyler Young (2016), "*Mnisose*, the Missouri River, is a living being.... Throughout her life, Mni Sose has nurtured the adjacent fertile bottomlands by intermittently inundating them with upriver nutrients, and she serves as a transportation corridor for peoples and their nonhuman relatives. Her waters and riparian areas provide sustenance to countless living beings."

Nick Estes (2019) and Dina Gilio-Whitaker (2019) document decades of the Standing Rock Sioux Tribe's resistance to nonconsensual actions of the government, corporations, and private businesses of U.S. settlers. Let's consider some examples. In the 19th century, U.S. settlers overharvested bison, an animal depended upon by the tribe, without having ever attempted to make legitimate agreements or diplomatic arrangements over sharing the bison harvest. They overran the region with mining speculation in the late 1840s as gold was discovered in California and the tribe's sacred Black Hills region. While treaty agreements were made that protected some Oceti Sakowin lands, such as in 1851 and 1868, continuous settlement and prospecting violated the terms of those treaties. The effects ultimately reduced the tribe's land area from more than 100 million acres to 2 million acres by the late 19th century. The lack of Indigenous consent to these actions and outcomes led the U.S. Supreme Court in 1980 to claim that "[a] more ripe and rank case of dishonorable dealings will never, in all probability, be found in our history" (*United States v. Sioux Nation of Indians*, 1980, 448 U.S. 371).

The end of the 19th century was by no means the end of nonconsensual U.S. actions. The United States imposed the Dawes Act or General Allotment Act (1887) on Indigenous peoples to break up their lands into private property, which would open up lands not owned by Indigenous persons to settlers. Indigenous extended kinship networks were forcibly broken up during allotment to make way for nuclear families who would be sedentary farmers. The United States literally endorsed actions that separated families and discontinued relationships that Indigenous persons had with lands outside their private property. The United States brokered leases of Indigenous private property, in many cases to settler farmers and ranchers and, later, to extractive industries, such as mining. Well into the 20th century, the United States, via its government or supported by churches, forcibly sent many Dakota and Lakota children to boarding schools, some as far away as Virginia and Pennsylvania. The schools divested students of their language, cultures, and knowledges, replacing them with technical skills for settler occupations. The children were sometimes physically abused or even murdered. In an effort to get Tribes like the Standing Rock Sioux to adopt a government that would endorse extractive industries, the United States imposed boilerplate constitutions, fashioned after corporate charters, that tribes could refuse to adopt at the peril of losing potential programmatic support from the U.S. government.

By the time we get to the national and international media stories about the Dakota Access Pipeline (DAPL), there already had been decades of resistance to nonconsensual actions by the U.S. government (Estes, 2019). Of course, consent is a major issue regarding the construction of the pipeline. DAPL is a 1,172-mile pipeline, in operation since 2017, that transports crude oil from the Bakken Formation in Williston, North Dakota, to refineries and terminals in Illinois. The Bakken is one of the largest repositories of oil in North America. DAPL's investors profit by offering a cheaper and safer alternative to rail or truck that also increases employment, state tax revenues, and energy independence. DAPL's advocates claim the pipeline will meet the highest environmental safety standards. To avoid having to navigate multiple government permits for construction, the pipeline builders designed a route that passes mostly through private lands, with several exceptions. One exception is an area below the Missouri River, which required a U.S. Army Corps of Engineers permit and affects *Mni Sose*.

Part of the concerns expressed by Standing Rock Tribal members and many other Indigenous advocates is that neither the Tribe nor the water itself consented to the construction of the pipeline and to the future risks of leaks. James Grijalva (2017) discusses some of the concerns the Tribe had:

> The Army Corps of Engineers did not address the Tribes' expert reports documenting numerous EA [environmental assessment] flaws and gaps, including: dismissing impacts on Indian treaty rights without analysis; violating NEPA [National Environmental Policy Act]

> regulations for actions with impacts that are "highly controversial" and "highly uncertain"; understating the risk of significant pipeline leaks; ignoring the inability of detection systems to identify slow leaks that could result in large oil discharges over time; inadequately analyzing spill risks; and depriving the public of comment by keeping the underlying spill modeling data secret.

Grijalva documents how an initial draft of the environmental assessment did not mention the Standing Rock Reservation less than a mile downstream from the proposed Lake Oahe crossing. Moreover,

> The final EA "recognized" that fact, but rather than confront the health and cultural impacts of contaminating the Reservation's largest water body and its shorelines, the EA instead re-emphasized the pipeline's off-Reservation location, the expensive and high-tech nature of the horizontal drilling technique for putting the pipeline below the lake, and the very low likelihood of spills. Illogically, that same low risk of spills justified rejecting an alternate route upstream of the State Capitol of Bismarck, whose racial composition is overwhelmingly White. (Grijalva, 2017)

The Bakken region, where the oil going down the DAPL comes from, resonates with Indigenous persons as a place where Indigenous women and children are sex trafficked at the "man camps" for construction workers. Sarah Deer and Mary Kathryn Nagle (2017) describe how "the trafficking of Native women and children is not a new phenomenon.... Sexual exploitation of Native women and children, dating back to the times of the Spanish Conquistadors, often times accompanies the colonial conquest of tribal lands" (p. 36). Yet "the Bakken oil boom has created a renewed sense of urgency in areas that have recently experienced a rapid increase in oil extraction" (p. 36).

Decades of land dispossession, cultural assimilation, and violence against Indigenous peoples' bodies are attacks on Oceti Sakowin self-determination. In all these cases, the lack of Indigenous consent was not accidental. In 2016, thousands of people, led by Standing Rock Tribal members, gathered at camps near the crossing of the Missouri and Cannon Ball Rivers to stop the construction. Allard (2017) writes,

> This movement is not just about a pipeline. We are not fighting for a reroute, or a better process in the white man's courts. We are fighting for our rights as the Indigenous peoples of this land; we are fighting for our liberation, and the liberation of Unci Maka, Mother Earth. We want every last oil and gas pipe removed from her body. We want healing. We want clean water. We want to determine our own future.

Self-determination means consent.

VIOLATIONS OF CONSENT: ALL TOO COMMON

That thousands of Indigenous persons from different nations and communities supported the Standing Rock Sioux Tribe is not surprising. Violations of Indigenous consent are all too common in the U.S. and Canadian energy sectors. The United States and Canada both have millions of miles of energy pipelines that connect to many hundreds of shipping terminals, such as major ports in Seattle and New Orleans, and fossil fuel basins, such as the Bakken, Powder River, or Athabasca Oil Sands regions. Energy is also transported via rail and truck, which enlarges the network of fossil fuel–related transportation infrastructure. Maps of the U.S. and Canadian energy sectors show how lands above and waters below are covered with this massive infrastructure.

Scholars, including Deborah Cowen and Anne Spice, are examining the relationship between infrastructures and colonialism. Cowen (2017) writes that "[i]n colonial...contexts, infrastructure is often the means of dispossession." Indigenous peoples never consented to being completely surrounded and enmeshed by pipelines, terminals, and processing/refining centers. Nor did Indigenous peoples consent to being surrounded by the numerous other industries, technologies, and populations whose activities and lifestyles are fueled by the burning of fossil fuels. Hence, Spice calls such infrastructure "invasive infrastructure," since its construction and use were not consented to by Indigenous peoples, among other populations who face risks and harms (Spice, 2018). Such invasive infrastructure expresses the self-determination of various settler groups in U.S. and Canadian societies at the expense of Indigenous self-determination.

Indigenous peoples in North America are fighting against harms and risks associated with numerous pipelines, basins, and shipping terminals. In North America alone, resistance actions are taking place in response to hundreds of such situations. The Enbridge Line 3 pipeline, which crosses through Anishinaabe territories neighboring Minnesota and Wisconsin, threatens treaty-protected wild rice–harvesting areas. The Keystone XL oil pipeline crosses multiple Indigenous lands in North America, threatening to leak dangerous chemicals into waters and soils. The Atlantic Coast Pipeline in North Carolina is being built along a route that includes many Indigenous territories, and Indigenous persons were left out of the consultation process and environmental justice analysis in all of them (Sadasivam, 2020). The Enbridge Line 5 pipeline crosses Lake Michigan waters that are culturally and economically important for the Anishinaabe people, and the state of disrepair of the pipeline poses serious risks to water quality. The TransCanada Coastal GasLink threatens the lands and sovereignty of the the Wet'suwet'en people. The resistance movement of the Unist'ot'en Camp has been vocal about the lack of consent by Canada and the corporate backers of the project (see http://unistoten.camp).

The Gwich'in Tribe has been engaged in a long resistance to oil drilling and transportation in the Arctic National Wildlife Refuge (ANWR). The refuge is home to the tribe's history and relations with the Porcupine caribou herd. Luci Beach, a representative of the Gwich'in Steering Committee, said in a press conference that it is "unacceptable that another nation is allowed to be destroyed [for oil]." The Gwich'in Nation's homeland spreads across northeastern Alaska and northwestern Canada, and includes the ANWR. Gwich'in have occupied this land for more than 20,000 years and subsist primarily on hunting caribou. This land is sacred to the Gwich'in people, who call it "the sacred place where life begins" (quoted in *Cultural Survival*, 2005).

For decades, the Northern Cheyenne Tribe has struggled with the fact that its territory is in one of the largest coal basins in the United States, the Powder River Basin. The U.S. Bureau of Indian Affairs leased close to half the reservation for coal strip-mining, paying only 17 cents per ton to the tribe. "The extensive strip mining of the Northern Great Plains will cause considerable changes in the amount, distribution and quality of water in mined areas" (Woessner, Andrews, & Osborne, 1979). Gail Small (1994) writes, "Like many Cheyenne, I feel as if I have already lived a lifetime fighting [coal] strip-mining. We live with fear, anger, and urgency. And we long for a better life for our tribe."

The Lummi Nation, one of the same group of Northwest Treaty Tribes, has taken action to block the establishment of a coal shipment terminal and train railway near its treaty-protected sacred area Xwc'chi'eXen. In addition to environmental protection, the Lummi reject the industrial capitalist values and colonial policies that ignore treaties for the sake of expanding carbon-intensive industries such as coal. The tribal chair, Tim Ballew, says: "We're taking a united stand against corporate interests that interfere with our treaty-protected rights" (Connelly, 2015).

At the same time, Indigenous peoples are not always of the same mind on pipelines and extraction basins. Tribes such as the Navajo Nation, the Crow Nation, and the Three Affiliated Tribes have substantial investments in coal, gas, and oil. Yet, in each of these cases, the United States had used political and economic pressure to reengineer the governments of each of these tribes to be dependent on extractive industries, often forcing them to make hard choices between having access to resources or being recognized as sovereign. Scholars such as Melanie Yazzie (2018) are retelling some of the histories of Indigenous persons who for decades were resisting these pressures on tribal governance. For example, John Redhouse had researched, going back many decades, how the Navajo Nation was trapped in a network of fossil fuel infrastructure that pressured them to become dependent on extraction. While some people interpret the conflicts regarding fossil fuels between tribes to prove that tribes are advocates of extraction, that is too simple an interpretation. Tribal governments that advocate for extraction must be put into historical context.

In all these cases, the advocacy movements themselves emphasize the denial of consent as a critical reason why Indigenous peoples are facing the infrastructure of extractive industries in the first place. Or, in cases where tribes are co-opted by powers such as the U.S. government and corporations, the historical record clearly shows that these tribes were put in a position where they had to make hard choices that were not conducive to their self-determination.

TRADITIONS OF CONSENT

There is an irony when we reflect on how the United States has failed to value and respect Indigenous consent in situations such as those facing the Akwesasne Mohawks or the Standing Rock Sioux Tribe. The irony is that many Indigenous peoples' own traditional philosophies for how to organize their societies place great emphasis on the importance of consent. These philosophies translate into how Indigenous peoples understood the best ways to engage in diplomacy with other societies. Consent also mattered to Indigenous peoples in their interactions with the nonhuman world and featured prominently in traditions of how humans should interact with plants, animals, entities and flows (such as water and wind), and particular landscapes. Consider some examples of these Indigenous philosophies that privilege consent in their very understanding of how to organize a society.

Haudenosaunee and Anishinaabe peoples are well known for traditions of treaty making that prioritized the idea that all parties to the agreement should be able to consent or dissent. The Haudenosaunee *Kaswentha* refers to a philosophy that political agreements between two parties are like two vessels navigating parallel running rivers in a shared ecosystem. In the agreement, each party should maintain its independence and way of life, yet both parties should find beneficial ways to cooperate. In this way of thinking of political agreement, the core of treaty making is respect for each party's independence, or consent. Haudenosaunee people today continue to use the *Kaswentha* philosophy as the basis for environmental protection and justice. For example, the Akwesasne Task Force has created protocols for how environmental scientists from outside the tribe can collaborate with the tribe in ways that respect each other's mutual independence and consent (Hill, 2008; Ransom & Ettenger, 2001).

The Dish with One Spoon refers to another treaty-making tradition that connects the Anishinaabe and Haudenosaunee peoples. In one interpretation, the philosophy is that both parties live in a common bowl or dish (ecosystem) and have just one spoon to share together in order to eat from the dish. Every time someone seeks to take from the ecosystem in order to satisfy their survival and sustenance, they have to think about the implications for the other party who shares the same dish and spoon. In this way, parties in the agreement have to respect each other's consent to the actions they take because of how they impact one another. The Dish with One Spoon philosophy is being used today in one of the peoples' ancestral homelands, in what is currently Toronto, Ontario, Canada. When I travel there, I witness how a lot of Indigenous persons are making it apparent that the territory can be called "the Dish," which is intended to show the settler population the importance of working to protect the environment and taking responsibility for the dispossession of Indigenous lands, languages, and sovereignty in the region (King, 2015; Simpson, 2008).

Within particular Indigenous peoples, consensus is also privileged as a best practice for how to organize a society. The Haudenosaunee people actually made major political decisions through local leaders coming to a consensus about the wisest action. The goal of making decisions by consensus was to ensure that local leaders actually had a dialogue and learned from each other before deciding what to do. Those who ultimately did not consent to the decision that the majority made were not bound by it, yet they were nonetheless supported in their decision making (Ransom & Ettenger, 2001). In the Navajo Nation, local leaders were selected by informal consensus. Robert Yazzie (1996–1997) writes that this ensures "everyone can have their say, and when someone is out of line, they get a 'talking to' by a naat'aanii [peacemaker/mediator]" (p. 122). Yazzie describes this process "as a circle, where everyone (including a naat'aanii) is an equal. No person is above the other. In this 'horizontal' system, decisions and plans are made through consensus" (p. 122). Similar to the Haudenosaunee consensus process, the Navajo process encourages discussion (long, when needed), the sharing of perspectives, and in-depth learning about the nature of the problem being looked at (p. 122).

Indigenous justice traditions are also consensual in nature. Yazzie describes the Navajo restorative justice process:

> For example, to Navajos, the thought that one person has the power to tell another person what to do is alien. The Navajo legal maxim is "it's up to him," [*sic*] meaning that every person is responsible for

> his or her own actions, and not those of another. As another example, Navajos do not believe in coercion. Coercion is an undeniable aspect of a vertical justice system. However, because coercion tends to be authoritarian, it is thus alien to the Navajo egalitarian system.... It is illustrated as a circle where everyone is equal. (p. 120)

For the Lakota, James Zion (2005) writes that the justice system involves the relatives of the people who are affected by the wrongdoing: "They use a talking-out process among relatives to reach a practical consensus about what to do about a problem. They use ceremony and prayer to bond people to the process and to involve the spirits both in forming the path to a solution and in making the decision binding" (p. 70).

Consent also plays a role in some Indigenous cultural and intellectual traditions in terms of consenting to environmental expertise and leadership. Coash Salish societies, for example, are well known for their giveaway traditions. Ronald Trosper (2002) discuss this in his work. In the case of salmon stewardship, leaders of houses had to go through educational processes, widely understood by society, that would give them the basis for expertise in managing salmon habitats. Given that the ecosystems were interconnected, a giveaway ceremony meant that a titleholder in a house had to show to others that they had done a good job harvesting. If one's harvest that one gave away was inadequate or inappropriate for some reason, then one's position as a title holder could be challenged. Title holders, who played roles as both leaders and experts, were accountable to the consent of those who were affected by their decisions (Trosper, 2002).

Indigenous consent traditions go beyond just human relationships with other humans. They also include the dynamics of the nonhuman world. Rosalyn LaPier, in her work on Blackfeet cosmology, discusses how people respect the importance of the nonhuman world consenting to any knowledge or powers that would be passed on to humans. In one story she shares, "a beaver and his family invite a human to live an entire winter in their lodge." The beaver shared critical knowledge with the human, such as the role of the tobacco plant and an understanding of water (power of the waters). LaPier describes, in one account, how the knowledge exchange occurs only when the human makes a respectful request and the beaver "consents" to give the power. "The being...offers or consents upon request to give power [to a human].... The being conferring power...transfers it to the recipient.... This is regard[ed] as a compact between the recipient and the being...and each is expected to fulfill faithfully his own obligations. The compact is a continuous relationship" (LaPier, 2017, pp. 69–72). In this way, a consent relationship is part of human allyship with the nonhuman world.

Melissa Nelson (2013) discusses the Anishinaabe mythical creature *Mishipizhu*, an underwater panther and powerful Manitou. Manitou means, as Nelson quotes Basil Johnston, "spirit...property, essence, transcendental, mystical, muse, patron, and divine" (p. 213). Nelson writes that "Mishipizhu is a protector of natural resources and a mediator between the water, land and sky beings" (p. 213). Nelson quotes Victoria Brehm in discussing how "[i]n his role as guardian of resources, he is immortal, reappearing to punish anyone who attempts to upset the balance of eco-social relations" (p. 213). Nelson shows how the storytelling tradition looks at Mishipizhu as relating to larger environmental, even earth systems, to which humans do not themselves consent to. She describes this tradition as a "[m]oral lens for water relations," which invokes the capacity of water to treat us with "malevalence or compassion" (p. 222). Nelson's work shows how the challenge of understanding how to treat such an entity that we cannot consent to is similar to an environmental planning process in which we go through multiple possibilities in order to be best prepared for unknown futures.

Indigenous philosophies of consent suggest the importance of ensuring that all our relationships are consensual and that people can freely dissent. Susan Hill (2008) writes, "We [Haudenosaunee] govern ourselves by consensus and endeavor to build consensus in all aspects of our society." Consent is just one example of how Indigenous philosophies work in terms of how they understand the organization of society. I raise this issue to put in stark contrast what it means for a society that places great value on consent to suddenly be confronted with a nation like the United States that does not value consent, whether the consent of humans or nonhumans.

COLONIALISM AND CONSENT

Given the importance of consent within many Indigenous traditions, what is the best way to characterize the lack of consent in the relationship between nations like the United States and Indigenous peoples?

One way in which environmental injustices are made possible is when societies and relationships between societies do not value consent. For many Indigenous peoples, I would argue, one way of understanding environmental injustice is as the systematic undermining of the consent relationships of one society by another society. The undermining is "systematic" when the perpetrating party, such as the U.S. government or a corporation, fulfills two conditions. First, a key reason why the perpetrator harms certain people is because of the afflicted persons' identities, such as their ethnicity, indigeneity, race, class, gender, sexual orientation, and age, among other identities. Second, the perpetrator creates a false reality in its own mind that such harms are (a) negligible concerns, (b) morally acceptable situations, or (c) even deserved by those who are harmed. In the cases of environmental injustice against Indigenous peoples that have been discussed so far, nations like the United States, and many of their citizens, see the denial of Indigenous consent as (depending on the case) a negligible concern, a morally acceptable situation, or deserved (by Indigenous peoples). The actual cultural, social, and political systems of the United States and many other nations have as fundamental features the disrespect of Indigenous consent and dissent. Consider some examples.

Bruce Duthu's study of U.S. Indian law shows that the United States justified its own sovereignty in North America by the *doctrine of discovery* (Duthu, 2008). The doctrine refers to the idea that by virtue of landing in North America, Europeans and eventually the United States are sovereign over the continent (and for no other reason). In the 1830s, U.S. Chief Justice John Marshall established that, by virtue of "possession," that "discovery gave title to the government by whose subjects, or by those authority, it was made." Interestingly, Duthu writes, "Only Christian colonizers in their encounters with non-Christian peoples could invoke the discovery doctrine. An Indigenous seafaring tribe, by contrast, could not plant a flag in the British Isles or on the beaches of Normandy and make comparable claims to England or France under the doctrine" (pp. 70–71). Today, the United States believes that it is able to choose which tribes it recognizes as Indigenous peoples. The U.S. Congress exercises "plenary power" over tribes, which means that the U.S. Congress believes it can even decide who counts as a member of a tribe. In these ways, Indigenous consent to invasion, colonialism, and U.S. sovereignty was considered to be negligible.

Another illustrative case is how the United States endorsed the breakup of Indigenous territories into private property. The allotment policy sanctioned by the 1887 Dawes Act made individual Indians own land in fee simple as a way to get them to organize into nuclear families that made income from farming. Religious groups, such as the Quakers, saw this as a great moral mission. Of course, the great moral mission of doing this made advocates of the policy feel it was legitimate to break up Indigenous kinship systems to force nuclear family formations. A second way is to cast violent military, containment, and other colonial practices as ways of saving Indigenous peoples. For example, the liquidation of Indigenous lands into private property in the 19th century, which also involved the breaking up of Indigenous kinship networks and families, was cast as helping Indigenous persons become independent "competent" farmers and was advocated by settler organizations that often called themselves "friends" (Royster, 1995; Stremlau, 2005). Even institutions as assimilative and violent as boarding schools were also seen as civilizing Indigenous peoples (Archuleta, Child, & Lomawaima, 2000). The dispossession of Indigenous lands was considered to be a morally acceptable situation (Mascarenhas, 2012).

Violent confrontations and forced land dispossession are also often cast as inevitabilities of progress. Thomas Morgan (1889), a U.S. commissioner of Indian affairs, wrote in the late 19th century,

> The Indians must conform to the "white man's ways" peaceably if they will, forcibly if they must. They must adjust themselves to their environment, and conform their mode of living substantially to our civilization. This civilization may not be the best possible but it is the best the Indians can get. They cannot escape it, and must either conform to it or be crushed by it.

Across these examples is an important truth about U.S. imperialism, namely, that integral to its formation as a modern nation-state, the United States set up an entire system and tradition of denying Indigenous consent in an effort to justify the violence of its colonial legacy.

CONCLUSION: RESTORING CONSENT FOR THE FUTURE

After hundreds of years of colonialism, we are certainly living in times that many of our ancestors would have seen as dystopian. Often in dystopian science fiction stories, the main characters find themselves in a situation where dominant groups of people exercise nearly absolute power over them. For example, the popular movie *The Matrix* is about a time in the future when artificial intelligence (AI) has dispossessed humans of their land. Decades of abuse by humans and AI have heavily polluted the environment. AI now uses humans for energy it needs to survive. It created a system in which humans who are being tapped for energy have the illusion that they are living free and fulfilling lives. In reality, humans live under nearly absolute domination and do not consent to the way they are treated by AI machines.

Were our ancestors several hundred years ago to have heard a story about what life is like today for Indigenous peoples, it would likely have sounded to them like the story of *The Matrix*, where the United States is the AI and Indigenous peoples, among other oppressed groups, are the humans. Like the characters in *The Matrix*, our situation involves living under nearly total domination, and our possibilities for liberation are greatly limited because of the lack of consent. The educational, social, cultural, and political institutions of the United States seek to create the false reality that Indigenous peoples were not violently dispossessed of land and that U.S. colonialism is largely over. In this false reality, Indigenous peoples are seen as either tragically eliminated or romantically clinging to a holistic environmental ethic.

In terms of environmental injustice, our ancestors would have likely believed that today's times are ones in which environmental destruction has occurred at the same time that relationships of consent have been diminished. Today, through actions such as the resistance to the Dakota Access Pipeline and the Unist'ot'en Camp, Indigenous peoples largely do not have the right to consent to the actions of other groups that affect environments in which Indigenous peoples live, work, and play. There is a lack of concern, broadly, with whether our nonhuman relatives consent to how they are treated by industries such as oil and gas, commercial agriculture, and mining. The landscape-scale environmental change caused by these industries demonstrates little concern for the well-being of nonhuman lives and ecosystems. Our ancestors would have also noticed a marked lack of respect for those entities for whom we cannot consent, such as the climate system. It seems like some people will not be sufficiently concerned about climate change until it is too late.

As a solution, when Indigenous peoples invoke their own cosmologies and philosophies, they seek to bring attention both to the maintenance of colonial domination and to Indigenous traditions of environmental stewardship that offer underrepresented visions for environmentally just futures. Indigenous cosmologies and philosophies call for concrete reforms, such as land reclamation, at the same time that they suggest shifts in consciousness of human relationships to the nonhuman world. When Indigenous peoples advocate for their own philosophies and cosmologies, they should not be mistaken for expressions of spirituality reminiscent of the conceptions of faith, spirituality, or transcendence that are common in some types of Christianity. Rather, Indigenous philosophies and cosmologies often refer to moral relationships, especially moral qualities such as consent, that are connected systematically with human relationships to the environment.

Environmental justice, then, is about consent, but not just any type of consent. It is about understanding the ways in which consent and dissent are part of our daily lives, engagement with politics and economics, and our connections to the land and nonhuman worlds around us. Indigenous environmental justice pushes us to be aware of the different dimensions of consent around us. But, again, consent is just one moral quality, and further study and practice of Indigenous environmental traditions speaks to many other moral relationships, such as responsibility and accountability, and moral qualities, such as trust and reciprocity.

DEEPENING OUR UNDERSTANDING

1. On a daily basis, people are affected by the food they eat, the air and water they are exposed to, and materials and chemicals they interact with in buildings and the land.
   a. On a piece of paper, place yourself in the center. Using lines, draw out how many different types of exposures as you can think of that you are exposed to every day from food, air, water, buildings, and the land. These exposures may be risky or not.
   b. For each exposure, guess what industry or government entity controls it.
   c. For each one, describe whether you feel that you have consented or would have a chance to dissent to it.
2. Think about other systems of oppression that result in environmental injustice (imperialism, settler colonialism, patriarchy, capitalism, democracy, and the like). For a particular system, list the ways in which your participation, or consent, is integral to this environmental injustice.

References

Allard, L. (2017, February 4). How powerful could we be if we agree to stand our ground on our treaty land. Yes! Magazine. https://www.yesmagazine.org/orphan/2017/02/04/to-save-the-water-we-must-break-the-cycle-of-colonial-trauma/

Anaya, J. (2004). *Indigenous peoples in international law*. New York: Oxford University Press.

Archuleta, M., Child, B. J., & Lomawaima, K. T. (2000). *Away from home: American Indian boarding school experiences, 1879–2000*. Phoenix, AZ: Heard Museum.

Cultural Survival. (2005, October 24). Gwich'in human rights threated by ANWR drilling. *Cultural Survival News & Articles*.

Dhillon, J., & Estes, N. (2016). Introduction: Standing rock, #NoDAPL, and Mni Wiconi cultural anthropology, Hot Spots. https://culanth.org/fieldsights/1007-introduction-standing-rock-nodapl-and-mni-wiconi

Duthu, N. B. (2008). *American Indians and the law*. New York: Viking.

Estes, N. (2019). *Our history is the future: Standing Rock versus the Dakota Access Pipeline, and the long tradition of Indigenous resistance*. Brooklyn, NY: Verso.

Gilio-Whitaker, D. (2019). *As long as grass grows: The Indigenous fight for environmental justice, from colonization to Standing Rock*. Boston, MA: Beacon Press.

Grijalva, J. M. (2017, April 12). Resistance, resilience and reconciliation. *JURIST - Academic Commentary*. http://jurist.org/forum/2017/4/James-Grijalva-resistance-resilience-and-reconciliation-indigenous-human-rights-to-environmental-protection-in-afrenzy.php

Hill, S. (2008). "Travelling down the river of life together in peace and friendship, forever": Haudenosaunee land ethics and treaty agreements as the basis for restructuring the relationship with the British Crown. In L. Simpson (Ed.), *Lighting the eighth fire: The liberation, resurgence, and protection of Indigenous nations* (pp. 23–45). Winnipeg, Manitoba, Canada: Arbeiter Ring Publishing.

Hoover, E. (2017). *The river is in US.* Minneapolis: University of Minnesota Press.

Howe, C., & Young, T. (2016). Mnisose. Cultural anthropology, Hot Spots. https://culanth.org/fieldsights/1016-mnisose

LaPier, R. R. (2017). *Invisible reality: Storytellers, storytakers, and the supernatural world of the Blackfeet.* Lincoln: University of Nebraska Press.

Mascarenhas, M. (2012). *Where the waters divide: Neoliberalism, white privilege, and environmental racism in Canada.* Lanham, MD: Lexington Books.

Morgan, T. (1889). The Indian Commissioner's report. *Friends Intelligencer and Journal, 46,* V.

Ransom, J. W., & Ettenger, K. T. (2001). "Polishing the Kaswentha": A Haudenosaunee view of environmental cooperation. *Environmental Science & Policy*, *4*(4), 219–228.

Royster, J. V. (1995). The legacy of allotment. *Arizona State Law Journal*, *27*, 1–77.

Sanders, D. E. (1977). *The formation of the World Council of Indigenous Peoples* Copenhagen, Denmark: International Secretariat of International Work Group for Indigenous Affairs.

Simpson, L. (2008). Looking after Gdoo-naaganinaa: Precolonial Nishnaabeg diplomatic and treaty relationships. *Wicazo Sa Review*, *23*(2), 29–42.

Small, G. (1994). War stories: Environmental justice in Indian country. *Daybreak*, *4*(2).

Spice, A. (2018). Fighting invasive infrastructures: Indigenous relations against pipelines. *Environment and Society*, *9*(1), 40–56.

Stremlau, R. (2005). "To domesticate and civilize wild Indians": Allotment and the campaign to reform Indian families, 1875–1887. *Journal of Family History*, *30*(3), 265–286.

Tarbell, A., & Arquette, M. (2000). Akwesasne: A Native American community's resistance to cultural and environmental damage. In R. Hofrichter (Ed.), *Reclaiming the environmental debate: The politics of health in a toxic culture* (pp. 93–111). Cambridge, MA: MIT Press.

Trosper, R. L. (2002). Northwest coast indigenous institutions that supported resilience and sustainability. *Ecological Economics*, *41*, 329–344.

Woessner, W. W., Andrews, C. B., & Osborne, T. J. (1979). The impacts of coal strip mining on the hydrogeologic system of the Northern Great Plains: Case study of potential impacts on the Northern Cheyenne Reservation. *Journal of Hydrology*, *43*(1–4), 445–467.

Yazzie, R. (1996–1997). Hozho-Nahasdlii–We are now in good relations: Navajo restorative justice. *St. Thomas Law Review*, *9*, 117–124.

Zion, J. (2005). Punishment versus healing. In W. McCaslin (Ed.), *Justice as healing: Indigenous ways* (pp. 68–72). St. Paul, MN: Living Justice Press.

PART II

ENVIRONMENTAL JUSTICE METHODOLOGY

4

MEASURING ENVIRONMENTAL INJUSTICE

Paul Mohai

The environmental justice movement has spurred much academic interest in and policy debates about the existence of, causes for, and solutions to environmental inequalities based on racial and socioeconomic factors. The earliest research attempted to determine the existence and magnitude of such disparities. Evidence of the existence of such disparities has been enough to spur government action. Although some researchers have questioned the existence and seriousness of such disparities, systematic reviews have shown that the weight of the evidence supports the claims of the movement. Nevertheless, challenges to the claims of environmental justice activists, supported at times by contrary research evidence, have stimulated a great deal of attention to questions about the validity of the methodologies for assessing racial and socioeconomic disparities around environmentally hazardous sites. There has also been much interest in understanding how racial and socioeconomic disparities in the distribution of environmental hazards come about, and in probing the economic, health, and other quality-of-life impacts associated with living near environmentally burdened sites. I have been and am currently involved in all aspects of the research seeking answers to these questions.

RACE AND CONCERN FOR THE ENVIRONMENT: DISPELLING OLD MYTHS

Initially I became involved as a researcher in the area of what is now termed *environmental justice* because of my interest in better understanding the environmental concerns and attitudes of African Americans and other people of color. This interest emerged in the late 1980s. Earlier research found that there was little empirical evidence to show that working-class people or poor people are less concerned about the environment than others (Buttel & Flinn, 1978; Mitchell, 1979; Mohai, 1985; Van Liere & Dunlap, 1980). I began wondering whether the conventional wisdom of the day, that African Americans are unconcerned, or at least less concerned about the environment than white Americans, was also unfounded. This conventional wisdom was already being questioned in the 1980s by African American scholars (Bullard, 1983; Bullard & Wright, 1986, 1987a, 1987b; Taylor, 1989). I had at my disposal data from a large national survey of environmental attitudes, based on more than 7,000 face-to-face interviews conducted by Louis Harris and Associates. Almost 600 African Americans were interviewed in this survey, making it the largest national survey of African American environmental attitudes ever conducted. In addition, I saw the potential of the annual (now biannual) General Social Survey (GSS) for comparing African American and white American concerns about the environment. From these national datasets, I demonstrated that, contrary to the conventional wisdom, African Americans express levels of concern about the environment that are as great as—if not greater than—those of their white counterparts (Mohai, 1990). Indeed, data from the GSS demonstrate that this has been a long-held trend (Figure 4.1). In most years since 1973, larger

---

Photo 4.1
istockphoto.com/EvanTravels

FIGURE 4.1 ■ Percentages of African Americans and whites indicating the nation spends too little money protecting and improving the environment, 1973–2018

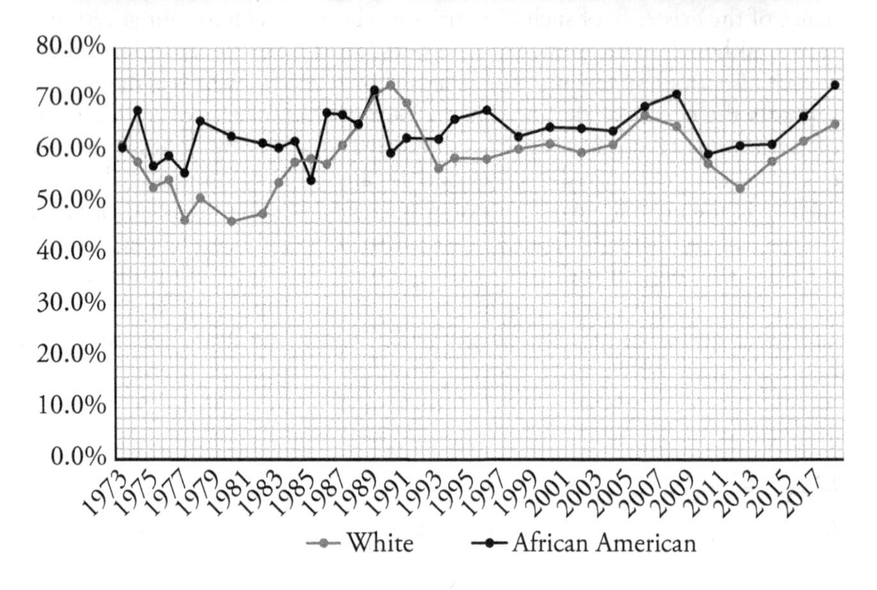

Source: 1973–2018 General Social Surveys

percentages of African Americans than whites say that we as a nation spend too little money protecting and improving the environment. See also Mohai and Kershner (2002), Mohai (2003), and Mohai and Bryant (1998).

In 1987 I began as an assistant professor at the University of Michigan, where I met Bunyan Bryant, the first African American environmental studies professor I came to know. I told him about my research interests, and he pointed me to the recently released report *Toxic Wastes and Race in the United States,* published by the Commission for Racial Justice of the United Church of Christ (1987). This report had a major influence on me. Until that time, I had not stopped to consider that environmental problems may be disproportionately burdening certain groups in society over others. Public discourse at that time generally held that environmental problems are everyone's problems and that everyone has an equal stake in solving them. However, the United Church of Christ (UCC) report suggested that racial and ethnic minorities and the poor may be at significantly greater risk and, thus, that they may have a greater stake than others in solving environmental problems. The most compelling aspects of the report were the findings that, nationally, the minority percentages of ZIP code areas with hazardous waste facilities were twice as great as those of ZIP code areas without such facilities—with the percentages being three times greater for ZIP code areas with two or more facilities. Furthermore, multivariate statistical analyses revealed that, among a variety of variables related to facility location, including mean property values, mean incomes, and the presence of abandoned and uncontrolled hazardous waste, the minority percentage of the ZIP code areas was found to be the best predictor of where hazardous waste facilities are located. I was also struck that the reason this report existed was in response to a growing movement by people of color to protest the environmental burdens in their communities and to seek remedies to their perceptions of unfair and unequal treatment by industry and government.

I joined forces in 1987 with Bryant in an effort to understand these issues better. We began with three objectives. One was to determine whether other quantitative studies like *Toxic Wastes and Race in the United States* existed, and whether these studies produced similar findings. This effort resulted, to our knowledge, in the first systematic review of such quantitative studies (see Mohai & Bryant, 1992). A second objective was to conduct a comprehensive study of the distribution of environmental hazards in the Detroit metropolitan area and to examine how such a distribution affected African American environmental attitudes (see Mohai & Bryant, 1992, 1998). Our third objective was to organize a conference bringing together all the researchers around the nation that we could identify who were examining racial and socioeconomic disparities in the distribution of environmental hazards to present and discuss their recent research. In doing so, we sought to increase the visibility of this issue among other academic researchers as well as to get the attention of state and national policymakers, especially that of the U.S. Environmental Protection Agency (EPA). We named our conference the Michigan Conference on Race and the Incidence of Environmental Hazards, which convened on the University of Michigan campus in January 1990 (Bryant & Mohai, 1992a, 1992b).

THE 1990 MICHIGAN CONFERENCE

One of the outcomes of the conference was the decision by the participants to draft a letter to then–EPA administrator William Reilly requesting a meeting with him and his staff to talk about the evidence pertaining to environmental inequalities, and to explore what the agency could do to address this problem. One of the most important outcomes of our meeting was Reilly's decision to create an internal EPA working group (the Environmental Equity Workgroup) to investigate the evidence and draft a set of proposals to address environmental inequalities. (The EPA later created an Office of Environmental Equity, which was renamed the Office of Environmental Justice under the Clinton administration.) In 1992, the EPA's Environmental Equity Workgroup produced a report entitled *Environmental Equity: Reducing Risk for All Communities* (U.S. Environmental Protection Agency, 1992). Although the report's draft proposals were heavily criticized for not going far enough in addressing the problem of environmental injustice, the report was nevertheless significant: As the first official acknowledgment of the problem by the agency and federal government, the report raised the visibility of this issue and affirmed environmental injustice as an issue warranting further federal attention. Indeed, subsequent to the report, a series of hearings on environmental justice were held in the U.S. Congress and a number of bills were introduced. Although none passed, the increasing attention and efforts by activists, academics, and the Congressional Black Caucus and other supporters eventually led President Clinton in 1994 to issue Executive Order 12898 ("Federal Actions to Address Environmental Justice in Minority Populations and Low-Income Populations"), calling upon on all the agencies in the federal government, not just the EPA, to make environmental justice a priority in their rulemaking.

EXAMINING THE EVIDENCE

At the same time Bryant and I were planning the 1990 Michigan Conference, we wanted to see if there was corroborating evidence to that presented in the UCC report. With the help of Robert D. Bullard, the University of Michigan Library, and other sources, I found 15 empirical studies over a 20-year period (1971–1992) that had examined racial and socioeconomic disparities in the distribution of environmental hazards. As mentioned earlier, I was specifically interested in determining whether the findings of other empirical studies pointed in the same direction as that of the UCC report. Specifically, I wanted to know (a) whether other quantitative studies found significant racial and socioeconomic disparities in the distribution of environmental hazards and (b) whether racial disparities tended to be more important than income and other socioeconomic disparities. All 15 studies I reviewed found significant racial or socioeconomic disparities. Additionally, where it was possible to weigh the relative importance of race or income in predicting where environmental hazards were located, in five out of eight studies race was found to be the most important predictor (Mohai & Bryant, 1992). Given that these studies varied in their geographic scope (some studied specific metropolitan areas, others were regional, while still others were national in scope), examined a wide variety of hazards (including air pollution, garbage dumps, and hazardous waste sites), and varied in their methodological approaches, the robustness of the pattern appeared to offer solid confirmation of the UCC report's findings (Table 4.1).

TABLE 4.1 ■ Empirical Studies, 1992 and Earlier, Providing Systematic Evidence Regarding the Burden of Environmental Hazards by Income and Race

Study	Hazard	Focus of Study	Distribution Inequitable by Income?	Distribution Inequitable by Race?	Income or Race More Important?
Council on Environmental Qualtiy (1971)	Air pollution	Urban area	Yes	NA	NA[a]
Freeman (1972)	Air pollution	Urban areas	Yes	Yes	Race
Harrison (1975)	Air pollution	Urban areas	Yes	NA	NA
		Nation	No	Yes	NA
Kruvant (1975)	Air pollution	Urban area	Yes	Yes	Income
Zupan (1973)	Air pollution	Urban area	Yes	NA	NA

(*Continued*)

TABLE 4.1 ■ Empirical Studies, 1992 and Earlier, Providing Systematic Evidence Regarding the Burden of Environmental Hazards by Income and Race (*Continued*)

Study	Hazard	Focus of Study	Distribution Inequitable by Income?	Distribution Inequitable by Race?	Income or Race More Important?
Burch (1976)	Air pollution	Urban area	Yes	No	Income
Berry et al. (1977)	Air pollution	Urban areas	Yes	Yes	NA
	Solid waste	Urban areas	Yes	Yes	NA
	Noise	Urban areas	Yes	Yes	NA
	Pesticide poisoning	Urban areas	Yes	Yes	NA
	Rat bite risk	Urban areas	Yes	Yes	NA
Handy (1977)	Air pollution	Urban area	Yes	NA	NA
Asch & Seneca (1978)	Air pollution	Urban areas	Yes	Yes	Income
Gianessi et al. (1979)	Air pollution	Nation	No	Yes	Race
Bullard (1983)	Solid waste	Urban area	NA	Yes	NA
U.S. General Accounting Office (1983)	Hazardous waste	Southern region	Yes	Yes	NA
Commission for Justice, United Church of Christ (1987)	Hazardous waste	Nation	Yes	Yes	Race
Gelobter (1988, 1992)	Air pollution	Urban areas	Yes	Yes	Race
		Nation	No	Yes	NA
West et al. (1992)	Toxic fish consumption	State	No	Yes	Race

[a]NA = Not Applicable

Source: Mohai & Bryant (1992).

In addition to conducting a review of existing evidence, Bryant and I wanted to conduct a new study, examining environmental disparities in the Detroit metropolitan area. We also wanted to know how such disparities affect racial differences in environmental awareness and concerns. Although prior environmental inequality studies employed census data at that time, it was apparent that survey data would be especially ideal for achieving our two objectives. We also realized that the University of Michigan's annual Detroit Area Study (DAS) would be an ideal vehicle, so we submitted our proposal for consideration for the 1990 DAS and were successful. We not only developed a comprehensive survey to compare African American and white American environmental attitudes, but we also mapped the locations of the homes of the 793 DAS respondents as well as the locations of the hundreds of polluting industrial facilities, commercial hazardous waste facilities, and abandoned hazardous waste sites in the Detroit area. We measured the distances between respondents and the hazardous sites. Because of the UCC report's focus on commercial hazardous waste facilities, we were specifically interested in examining the proportion of people of color living near such facilities in the Detroit area. We found that, with increasing proximity to the hazardous waste facilities, the proportions of African American residents relative to white residents increased (Mohai & Bryant 1992). Through multivariate statistical analyses, we also found that the race of the respondents was more important than their incomes in predicting their proximity to these facilities. Subsequently, with the help of graduate students in the School of Natural Resources (now the School for Environment and Sustainability), we completed our analyses of racial disparities in the distribution of a wide range of environmentally hazardous sites (for example, polluting industrial facilities and abandoned hazardous waste sites) and found the patterns consistent with what we initially found for commercial hazardous waste facilities (see Table 4.2 and Mohai & Byrant, 1998). That is, larger proportions of the African American population in the Detroit metropolitan area relative to the white population were found living near hazardous sites and areas of poor environmental quality. The results of our studies thus appeared to corroborate the findings of the UCC report and earlier environmental inequality studies.

TABLE 4.2 ■ Percentages of African Americans and Whites in the Detroit Metropolitan Area Living in Proximity to a Potential Environmental Hazard or Poor Environmental Quality

		African Americans	Whites	Difference
Proximity to Polluting Industrial Facility		*n* = 130	*n* = 629	
(1)	More than 1.5 miles	26.7	50.1	-23.4
(2)	Between 1.0 and 1.5 miles	22.3	18.4	3.9
(3)	Within 1.0 mile	51.0	31.5	19.5
	Chi-square = 25.63***			

(*Continued*)

TABLE 4.2 ■ Percentages of African Americans and Whites in the Detroit Metropolitan Area Living in Proximity to a Potential Environmental Hazard or Poor Environmental Quality (*Continued*)

	African Americans	Whites	Difference
Proximity to Commercial Hazardous Waste Facility	*n* = 130	*n* = 629	
(1) More than 1.5 miles	64.5	90.3	-25.8
(2) Between 1.0 and 1.5 miles	19.6	6.7	12.9
(3) Within 1.0 mile	15.9	3.0	12.9
Chi-square = 63.18***			
Proximity to Uncontrolled Hazardous Waste Site	*n* = 130	*n* = 629	
(1) More than 1.5 miles	26.1	46.3	-20.3
(2) Between 1.0 and 1.5 miles	26.7	14.0	12.7
(3) Within 1.0 mile	47.2	39.7	7.5
Chi-square = 22.45***			
Commercial or Industrial Structure on Resident's Block	*n* = 130	*n* = 624	
(0) No	91.7	93.0	-1.3
(1) Yes	8.3	7.0	1.3
Chi-Square = 0.28			
Vacant Buildings on Resident's Block	*n* = 130	*n* = 624	
(0) No	75.7	97.0	-21.3
(1) Yes	24.3	3.0	21.3
Chi-square = 79.25***			
Upkeep of Structures on Resident's Block	*n* = 130	*n* = 619	
(1) Very well	22.6	66.2	-43.6
(2) Mixed	66.6	32.4	34.2
(3) Poorly	7.4	1.0	6.4
(4) Very poorly	3.4	0.3	3.1
Chi-square = 100.03***			

(Continued)

TABLE 4.2 ■ Percentages of African Americans and Whites in the Detroit Metropolitan Area Living in Proximity to a Potential Environmental Hazard or Poor Environmental Quality (*Continued*)

		African Americans	Whites	Difference
Upkeep of Sidewalks and Yards on Resident's Block		n = 130	n = 621	
(1)	Very well	24.3	55.3	-31.0
(2)	Mixed	60.5	41.4	19.1
(3)	Poorly	11.0	3.2	7.8
(4)	Very poorly	4.1	0.0	4.1
	Chi-square = 67.54***			

***$p < .001$

[a]Number represents weighted sample size.

Source: Based on Mohai & Bryant (1998). Reprinted with permission.

REASSESSING RACIAL AND SOCIOECONOMIC DISPARITIES IN ENVIRONMENTAL JUSTICE RESEARCH: QUESTIONS ABOUT METHODOLOGY

The UCC report (Commission for Racial Justice, 1987); the 1990 Michigan Conference; the book *Dumping in Dixie: Race, Class, and Environmental Quality* (Bullard, 1990); the volume *Race and the Incidence of Environmental Hazards: A Time for Discourse* (containing the published versions of the papers presented at the 1990 Michigan Conference; Bryant & Mohai, 1992a); the EPA report *Environmental Equity: Reducing Risk for All Communities* (U.S. Environmental Protection Agency, 1992); as well as many other articles, hearings, and conferences on environmental justice led in the 1990s to a flurry of new academic research in a wide variety of disciplines. Most of this new research tended to support the findings of the UCC report and earlier environmental inequality studies and the claims of environmental justice activists. However, a serious challenge to the claims that environmental hazards are distributed disproportionately by race and socioeconomic status emerged in 1994 with the publication of an article in the prestigious journal *Demography*, by researchers at the Social and Demographic Research Institute (SADRI) at the University of Massachusetts (Anderton et al., 1994).

In this article, the researchers claimed to have replicated the UCC study but instead found that no statistically significant racial disparities existed in the distribution of commercial hazardous waste facilities. They concluded that the reason for the contrary findings was their use of census tracts rather than ZIP codes. Census tracts are generally smaller and thus less likely to lead to "reaching conclusions from a larger unit of analysis that do not hold true in analyses of smaller, more refined units" (Anderton et al., 1994, p. 232). This article and subsequent articles by SADRI researchers (Anderton et al., 1997; Davidson & Anderton, 2000; Oakes et al., 1996) motivated a hard look at the methodologies used in conducting environmental inequality analyses (Mohai, 1995; Mohai & Saha, 2006). Given that both the SADRI article and the UCC study claimed similar objectives, were of national scope, focused on commercial hazardous waste facilities, and used the same census decade (1980), how did they come to such contrasting findings and conclusions? Was the only difference the use of census tracts instead of ZIP code areas?

In reviewing both studies, I noticed they also differed in the way they identified the comparison groups, known as nonhost units (Mohai, 1995). The UCC study used all nonhost ZIP code areas in the United States, whether in metropolitan or nonmetropolitan areas, for the comparison group; in the SADRI study, only nonhost census tracts in metropolitan areas already containing a hazardous waste facility were considered for the comparison group. For the SADRI study, in short—unlike the UCC study—nonmetropolitan areas were unrepresented, as were all metropolitan areas currently without hazardous waste facilities.

Given that the comparison groups of nonhost units were constructed differently, I compared the percentages of people of color in the host units and nonhost units of the respective studies. I found the discrepancy in the results was in the estimates of the minority percentages of the comparison group of nonhost units. The average minority percentage of the nonhost ZIP code areas in the UCC study was half that of the nonhost census tracts in the SADRI study. Thus, differences in the construction of the control (comparison) populations used in the two respective studies, rather than differences in the selection of their units of analysis, appeared to be the key to understanding the differences in the two studies' findings.

Ironically, although the SADRI researchers had defended their approach of excluding from their study metropolitan areas not containing a hazardous waste treatment, storage, and disposal facility (TSDF) on the assumption that such areas likely do not serve "the markets that TSDFs currently serve" (Anderton et al., 1994, p. 246), I found that the mean values of their chief indicator of industrial activity—the proportion of people employed in precision manufacturing and labor occupations—to be nearly identical for included and excluded metropolitan areas (Table 4.4). Furthermore, it was the variable least able to predict which metropolitan areas contained a TSDF and which did not (significance level of 0.8935; Table 4.5).

TABLE 4.3 ■ Average Percentages of People of Color in Areas Hosting Commercial Hazardous Waste Facilities and the Respective Control Populations Used in the UCC and SADRI Studies

	UCC Study (ZIP Code Areas)	SADRI Study (Census Tracts)	SADRI Study (Host Tracts and Tracts within 2.5- Mile Radius Aggregated)	SADRI Study–25 Largest SMSAs (Host Tracts and Tracts within 2.5 Mile Radius Aggregated)
Average percentages of people of color in host areas[1]	25%	24%	35%	42%
Average percentages of people of color in respective control populations	12%	23%	21%	24%

[1]Average percentages of people of color are for areas containing one or more hazardous waste facilities. Percentages are computed from information provided in the UCC (1987, p. 41) and SADRI studies (Anderton et al., 1994, pp. 235, 238; Anderton et al., 1994, p. 134). The SADRI study included only African Americans and Hispanics in the analyses. These percentages were summed in order to obtain the percentages of people of color in this table.

Although the SADRI study included only African Americans and Hispanics in the analyses, other groups of people of color made up only 2.3 percent of the U.S. population in 1980 (U.S. Census Bureau, 1980: Table 233). Thus, the results of summing the percentages of the African American and Hispanic categories is unlikely to be appreciably different if the SADRI study included other groups of people of color in the analyses. Also not likely to appreciably affect the results is the slight overlap between the African American and Hispanic categories employed in the SADRI study. This is because the overlap is so small. Only 390,492 out of the 40,695,540 people (or 0.96 percent) classified in the 1980 Census as either "Black" or "Person of Spanish Origin" are indicated as both (U.S. Census Bureau, 1980: Table 233).

Source: Adapted from Mohai (1995). Demographic values in both the UCC and SADRI studies are based on the 1990 Census.

TABLE 4.4 ■ Comparison of Mean Demographics of SMSAs Containing a TSDF with Mean Demographics of SMSAs Not Containing a TSDF

Variable	Means for SMSAs Containing a TSDF (and hence included in SADRI study)	Means for SMSAs Not Containing a TSDF (and hence excluded from SADRI study)
Percentage black	11.55%	7.79%****
Percentage Hispanic	5.69%	5.36%

(Continued)

TABLE 4.4 ■ Comparison of Mean Demographics of SMSAs Containing a TSDF with Mean Demographics of SMSAs Not Containing a TSDF (*Continued*)

Variable	Means for SMSAs Containing a TSDF (and hence included in SADRI study)	Means for SMSAs Not Containing a TSDF (and hence excluded from SADRI study)
Percentage of families below poverty line	12.34%	12.80%
Percentage of households receiving public assistance	7.50%	7.01%
Percentage of males employed in civilian labor force	57.18%	57.13%
Percentage employed in precision occupations	31.16%	31.03%
Mean value of housing stock	$54,551.17	$51,449.15

****$p < .0000$

Source: William R. Freudenburg and Robert Wilkinson (2008). Equity and the Environment.

TABLE 4.5 ■ Results of Logistic Regression Model Predicting Whether or Not SMSA Contains a TSDF (and hence whether or not SMSA was included in the SADRI study)

Variable	Coefficient	Standard Error	Significance
Percentage black	.0804	.0167	.0000
Percentage Hispanic	.0304	.0140	.0297
Percentage of families below poverty line	−.2321	.0600	.0001
Percentage of households receiving public assistance	.1729	.0658	.0086
Percentage of males employed in civilian labor force	.0487	.0677	.4724
Percentage employed in precision occupations	−.0031	.0234	.8935
Mean value of housing stock	−.0000	.0000	.8158
Constant	−1.9352	3.6733	.5983
Chi-square (7 df)	37.559		.0000

Source: Research in Social Problems and Public Policy, Volume 15, pp. 35. Copyright 2008 by Elsevier Ltd.

In addition to the discrepant way in which the SADRI study constructed its control population, I was further struck that when the SADRI researchers went on to examine the minority percentages of nearby tracts within 2.5 miles of the geographic centers (or "centroids") of the host tracts, the average minority percentage increased from 24 percent to 35 percent. When tracts within 2.5 miles of the host for the 25 largest metropolitan areas were examined, the minority percentage increased again to 42 percent (Table 4.3). Based on these outcomes, it seemed ironic to me that, rather than overestimating the percentage of people of color around hazardous waste facilities, as the SADRI researchers had implied, the UCC study may have underestimated them.

I wondered further what would happen to these demographics if the 2.5-mile circular buffers were centered *at the facilities* rather than at the host tract centroids. My work with the 1990 Detroit Area Study, in which we took into account the precise locations of hazardous sites and their proximity to households, and my comparative analyses of the UCC and SADRI studies caused me to wonder whether the then-standard approach of using census data in environmental justice analyses, such as that used by the UCC and SADRI, led to an underestimation of the racial and socioeconomic disparities around hazardous sites. In this approach, also known as the "unit-hazard coincidence method," the researcher (a) selects a predefined geographic unit, such as census tracts or ZIP code areas; (b) determines which of the units "host" the hazard of interest and which do not; and then (c) compares the demographic characteristics of the host and nonhost units to see if there are statistically significant differences. Although the precise location of the hazard within the unit is not taken into account, it is assumed that people residing in the host units live closer to the hazard under investigation than people living in the nonhost units. But do they?

I noticed that maps of hazardous waste facility locations in the Detroit metropolitan area showed them often to be located near the boundaries of their host tracts (for example, see Facility A and other facilities in Figure 4.2). Were the bordering (nonhost) tracts then also being affected by the facility, and were they also disproportionately composed of poor and minority residents?

Through a series of grants, I worked with my then doctoral student and current co-author and colleague, Robin Saha, to map the precise geographic locations of all the hazardous waste treatment and storage facilities in the United States. We found that the pattern we observed in the Detroit metropolitan area extended elsewhere. That facilities are located near the boundaries of their host tracts is not a rare occurrence; nationally, 49 percent of hazardous waste TSDFs are located within 0.25 miles of their host tract boundaries, while 71 percent are located within 0.5 miles (Mohai & Saha, 2006). If people of color and poor people are located disproportionately where environmental hazards are located, then would not the racial and socioeconomic characteristics of such adjacent tracts be more similar to the host tracts proper than nonhost tracts much farther away?

We also found, as we had observed for the Detroit metropolitan area, that there is a great deal of variation in the size of the tracts hosting hazardous waste facilities. Nationally the smallest host tract is less than 0.1 square miles, while the largest is more than 7,500 square miles (Mohai & Saha, 2006). Whereas it is a reasonable assumption that anyone living in a tract as small as 0.1 square

FIGURE 4.2 ■ 1990 Census Tracts and Hazardous Waste Facility Locations in Wayne County, Michigan

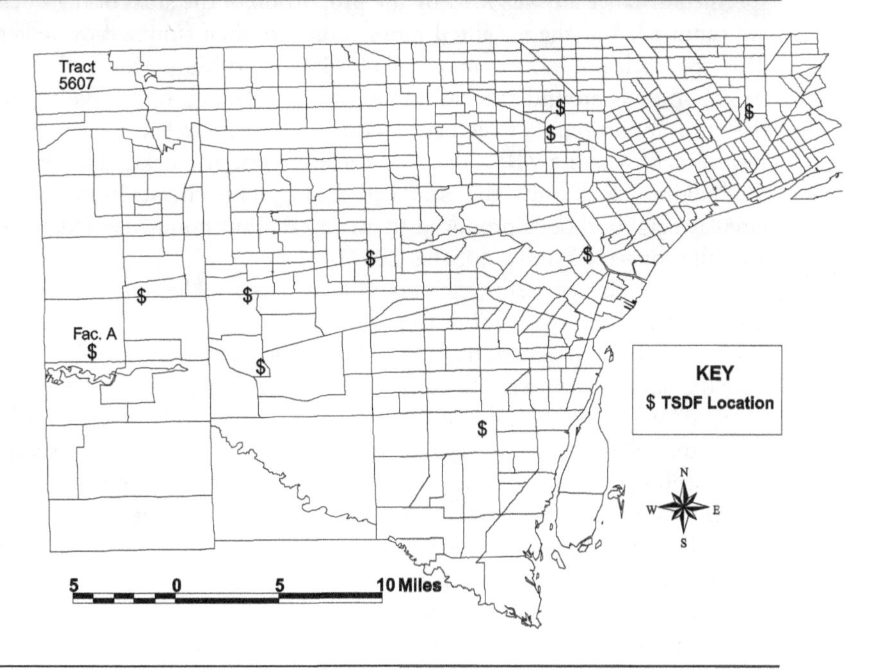

Source: William R. Freudenburg and Robert Wilkinson (2008). Equity and the Environment.

miles lives close to the facility in it, the same cannot be said for the very large tract. If most residents in a large tract live far from the facility, is there any reason to expect an association between the presence of the facility and the racial and socioeconomic characteristics of the tract? It seemed to us that the unit-hazard coincidence method failed to control adequately for proximity, and we began to experiment with various ways of controlling for proximity by using distance-based methods, similar to what was employed in the 1990 Detroit Area Study.

There was a catch. In the 1990 DAS the respondents in the survey could be treated as geographic points; thus, there was little ambiguity about whether or not they were located inside or outside specified distances to the hazardous sites. However, when circular buffers of a specified distance from the hazardous sites are drawn, the units, such as census tracts, don't always lie completely inside or outside the buffers. Often, these units are only partially captured by the circular buffers. How should these partially captured units be assessed?

One promising approach, modeled after that employed by the SADRI researchers (Anderton et al., 1994; Davidson & Anderton, 2000), is to include as part of the host neighborhood all tracts in which at least 50 percent of their areas were captured by circular buffers of specified distances from the hazards. We termed this the "50% areal containment method." A critical difference in our approach from that of the SADRI studies, however, is that we centered the radii at the facilities rather than at the centroids of the host tracts. We applied radii of various distances (up to three miles) around the nation's hazardous waste facilities to define the host neighborhoods (see Figure 4.3A). An even more promising approach is the "areal apportionment" method (see, for example, Chakraborty & Armstrong, 1997; Glickman, 1994; Hamilton

& Viscusi, 1999; and Sheppard et al., 1999). In applying this method, the populations of all census tracts wholly or partially captured by a radii of a specified distance are weighted by the proportion of the areas of the tracts that are captured, and the weighted populations are then summed or aggregated. This approach produces demographic estimates within perfectly circular host neighborhoods around the hazardous sites (see Figure 4.3B). We found that the areal apportionment method produces especially reliable results. That is, regardless of whether ZIP code areas, census tracts, or block groups are used as the building block units, the estimated demographic characteristics within the circular neighborhoods produced by the areal apportionment method remain virtually constant (Mohai & Saha, 2007).

If we are able to define neighborhoods that are closer to the hazards of interest than is possible from using the unit-hazard coincidence method, will we find that the proportions of people of color and poor people are greater near the hazards than what the unit-hazard coincidence method has previously revealed? We indeed found that this was the case (Mohai & Saha, 2006, 2007). Using the 1990 Census, we found the racial and socioeconomic disparities in the distribution of hazardous waste facilities to be much greater when either of the two distance-based methods was employed than when the unit-hazard coincidence method was applied (Table 4.6). For example, the difference in the non-white percentages between host and nonhost tracts was only 1.2 percent using the unit-hazard coincidence method, but the difference in the non-white percentages of host and nonhost neighborhoods defined by a one-mile radius using either the 50% areal containment or the areal apportionment method was more than 20.0 percent (Table 4.6). Similarly, the difference in the poverty percentages between host and nonhost tracts was only 0.5 percent, while the difference in the poverty percentages of host and nonhost neighborhoods defined by a one-mile radius using either distance-based method was more than 6.0 percent. Furthermore, racial and socioeconomic disparities between host and nonhost neighborhoods defined by a three-mile radius using either distance-based method were also found to be greater than disparities found between host and nonhost tracts (not shown in Table 4.6).

Subsequently, Saha and I teamed with environmental justice scholars Bullard and Wright to update the 1987 UCC report, applying distance-based methods to an updated list of hazardous waste treatment, storage, and disposal facilities and the 2000 Census data (Bullard, Mohai, Saha, & Wright 2007, 2008). We found that the concentration of people of color near hazardous waste TSDFs was greater than what the 1987 study and other previous studies had found. Of the population who live within 3.0 kilometers of a TSDF, 56 percent are people of color. Where TSDFs are clustered, this concentration increases to 68 percent. Furthermore, the racial composition of the neighborhoods remains an independent and statistically significant predictor of the location of TSDFs, even after controlling for socioeconomic variables.

FIGURE 4.3 ■ Host neighborhoods surrounding hazardous waste facility defined by 50% areal containment and areal apportionment methods

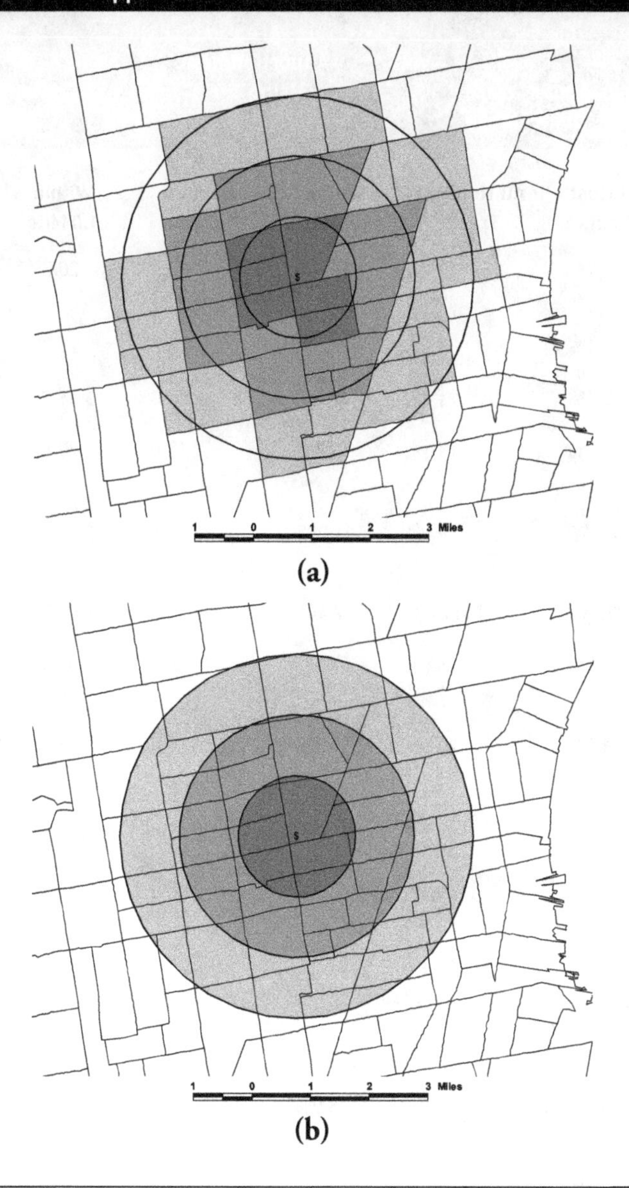

Source: Adapted from Mohai & Saha, 2006.

TABLE 4.6 ■ Comparisons of population characteristics in host and nonhost neighborhoods defined by areal apportionment and 50% areal containment methods versus unit-hazard coincidence method

	Unit-Hazard Coincidence		Areal Apportionment Using 1.0-Mile Radius		50% Areal Containment Using 1.0-Mile Radius	
	All Host Tracts	All Nonhost Tracts	Within 1.0 Mile	Beyond 1.0 Mile	Within 1.0 Mile	Beyond 1.0 Mile
Percentage African American	12.7%	12.0%	18.8%	11.9%	20.2%	11.9%
Percentage Hispanic	10.1%	8.8%	20.1%	8.6%	21.8%	8.7%
Percentage non-white	25.4%	24.2%	42.8%	24.0%	46.2%	24.0%
Percentage below the poverty line	13.6%	13.1%	19.1%	13.0%	20.6%	13.0%
Mean household income	$34,526	$38,491	$31,977	$38,545	$30,598	$38,543
Mean property value	$88,892	$111,883	$92,442	$110,320	$89,747	$111,856
Percentage without a high school diploma	28.5%	24.7%	34.4%	24.6%	36.5%	24.6%
Percentage with a college degree	14.4%	20.4%	14.1%	20.4%	13.1%	20.4%
Percentage employed in executive, management, or professional occupation	21.4%	26.4%	20.7%	26.4%	19.7%	26.4%
Percentage employed in precision production, transportation, or labor occupation	31.4%	26.1%	31.3%	26.1%	32.2%	26.1%
Percentage unemployed	6.7%	6.3%	8.9%	6.3%	9.6%	6.3%

Source: Adapted from Mohai & Saha (2007). Demographic values are based on the 1990 census.

THE RACE VERSUS CLASS DEBATE

There has been much interest in environmental justice research to determine whether racial disparities in the distribution of environmental hazards are independent of socioeconomic disparities. There are several reasons why such a determination is of interest. From an academic standpoint, knowing whether or not racial disparities in the distribution of environmental hazards are independent of socioeconomic disparities is an important step in narrowing down explanations of why the environmental disparities exist. If in a multivariate statistical analysis controlling for the average incomes or average property values in a community eliminates the significance of the race variables (for example, the African American or Hispanic percentages of the community) in predicting the presence of environmental hazards, this suggests that economic explanations of the distribution of hazardous waste and industrial sites may be more salient than racial explanations. For example, such facilities may tend to be located where racial and ethnic minorities live because industries desire to minimize production costs by siting new facilities in places where land values are low (Boone & Modarres, 1999; Hamilton, 1995; Hird & Reese, 1998; Rhodes, 2003). These locations may be coincidentally where minorities live. Alternatively, the facilities, once sited, may cause a decline in property values and quality of life, motivating affluent whites to move away and the poor and minorities to move in because of increased affordability of housing. However, if race variables remain significant after controlling for economic variables, it may suggest that factors uniquely related to race play a role. For example, housing discrimination may limit the ability of people of color to move away from such hazardous sites, beyond the limitations imposed by constrained incomes (Mohai & Bryant, 1992; Szasz & Meuser, 2000). Or perhaps minority communities are targeted for new facilities because minority communities over time have come to be recognized as the "paths of least resistance" by government and industry (Bullard, 1990; Bullard & Wright, 1987b; Hurley, 1995; Pellow, 2002; Pulido et al., 1996).

Supporting environmental justice activists' claims, systematic reviews of the quantitative research have tended to find that race effects are stronger than economic effects (Goldman, 1994; Mohai & Bryant, 1992; Ringquist 2005). However, there is nevertheless a range in the magnitude of racial and socioeconomic disparities found, with some studies finding none. In a recent meta-analysis of 49 quantitative studies, Ringquist (2005) found that race effects are stronger than economic effects but cautioned that these effects are nevertheless modest. An argument we made in our article in *Demography* (Mohai & Saha, 2006), however, is that if the unit-hazard coincidence method does a poor job of controlling for proximity, and hence leads to an underestimation of the magnitude of both racial and socioeconomic disparities in the distribution of environmentally hazardous sites, then the results of multivariate analyses involving the unit-hazard coincidence method aimed at assessing the relative importance of race and socioeconomic variables will also be unreliable. We demonstrated that when distance-based methods are employed, the outcomes of multivariate analyses are different and, hence, assessments about the relative importance of race and socioeconomic variables (and racial and socioeconomic explanations of environmental inequality) are changed (Mohai & Saha, 2006).

Results of the logistic regression analyses in Table 4.7 illustrate this point. In the analysis involving the unit-hazard coincidence method, the dependent variable takes a value of one if a tract hosts a hazardous waste facility and a value of zero if it does not. In the analysis involving one of the distance-based methods (specifically, 50% areal containment), the dependent variable takes a value of one if a tract is within one mile of a hazardous waste facility and a value of zero if it is not. Results for the two regression analyses are dramatically different. In the logistic regression involving the unit-hazard coincidence method, neither the African American percentage nor the Hispanic percentage is a statistically significant (at the 0.05 level) predictor of hazardous waste facility location, whereas the occupation variables are. Specifically, the larger the proportion of residents employed in manufacturing or labor jobs, and the smaller the proportion employed in executive, management, or professional jobs, the greater the likelihood there will be a hazardous waste facility nearby. Such an outcome might be interpreted to mean that any racial disparities found in the bivariate analyses are explained by the disproportionate number of African Americans in manufacturing jobs versus professional and management jobs, and that hazardous waste facilities tend to be located where manufacturing labor pools are nearby. In contrast, in the logistic regression associated with the 50% areal containment method, the race variables remain highly statistically significant, indicating that the racial disparities are not purely a function of the occupational and other socioeconomic variables. Some other factors associated with race appear to account for the disparities. In this case, housing discrimination or the possible targeting of minority communities by government and industry have not been ruled out. Whereas the results of the first regression motivate further exploration of economic explanations of environmental inequality and not racial ones, the results of the second regression clearly motivate further exploration of the latter.

TABLE 4.7 ■ Logistic regression results comparing unit-hazard coincidence and 50% areal containment methods

	Unit-Hazard Coincidence		50% Areal Containment (1.0-mile radius)	
Variables	Coefficient	Significance	Coefficient	Significance
Percentage African American	−.003	.986	.698	.000
Percentage Hispanic	.431	.066	1.482	.000
Mean household income ($1,000s)	.012	.000	−.025	.000
Mean property value ($1,000s)	−.002	.058	.005	.000
Percentage with a college degree	.338	.673	−1.704	.012

(Continued)

TABLE 4.7 ■ Logistic regression results comparing unit-hazard coincidence and 50% areal containment methods (*Continued*)

	Unit-Hazard Coincidence		50% Areal Containment (1.0-mile radius)	
Percentage employed in executive, management, or professional occupation	2.323	.000	1.787	.000
Percentage employed in precision production, transportation, or labor occupation	2.323	.000	1.787	.000
Constant	-5.052	.000	-4.197	.000
-2 Log likelihood	6010.2		8077.3	
Model chi-square	153.743	.000	548.233	.000
Sample size	59,050		59,050	

Source: Adapted from Mohai & Saha (2006).

CURRENT AND FUTURE RESEARCH

Better methods in environmental justice research should lead to more accurate assessments about the magnitude of racial and socioeconomic disparities in the distribution of environmental hazards as well as to more accurate hypothesis tests regarding cause and effect relationships in hazard distribution. However, most quantitative research to date has been based on analyzing present-day disparities from cross-sectional data, and to determine how present-day disparities have come about requires analysis of longitudinal data. This is because present-day disparities may have resulted from two processes: (a) disproportionate siting of hazardous waste facilities and other locally unwanted land uses in people-of-color communities at the time of siting, or (b) demographic changes after facility siting. Of course, both processes working together could account for present-day disparities. "Which came first, people or pollution?" is the chicken-or-the-egg question in environmental justice research (Mohai & Saha, 2015).

An answer to this question is important because it can help researchers to understand the social, economic, and political processes by which environmental disparities occur. It can also help in crafting policies to prevent environmental disparities from occurring. For example, if present-day environmental disparities occur largely because of a pattern of disproportionately placing facilities in poor and people-of-color communities at the time of siting, then policies that focus on managing the siting and permitting process to avoid such disparities could be adopted. If, however, the present-day disparities occur because of demographic changes after siting, then policies managing the siting process may not be enough. Policies that ensure newcomers have adequate access to information about the potential health risks in a neighborhood (so that they can weigh the risks against the desirability and affordability of the housing) and policies that make certain that people of color are not intentionally steered into contaminated neighborhoods should be implemented (Burby & Strong, 1997; Pastor et al., 2001).

Currently, there are relatively few longitudinal studies addressing the "chicken or egg" question. This is partly the result of the difficulty of conducting such studies. First, researchers need to obtain reliable information about when facilities were sited so that they can check census data at or near the time of siting, not the most recent census, to see if there has been a pattern of disproportionately placing the facilities in people-of-color communities. For example, if a facility was sited in 1970, one needs to examine the demographics around 1970. A second major difficulty in conducting longitudinal studies is the fact that census tract boundaries can shift from decade to decade. This makes it difficult to know whether demographic changes in an area are the result of population shifts or simply tract boundary shifts.

Another difficulty is that the earliest longitudinal environmental justice studies relied on the unit-hazard coincidence method. Not surprisingly, because this method tends to underestimate racial and socioeconomic disparities around environmentally hazardous sites, longitudinal studies using this method have led to inconclusive findings as to whether disproportionate siting or postsiting demographic change explains present-day disparities (Been & Gupta, 1997; Oakes et al., 1996).

In contrast, more recent studies employing distance-based methods have provided clear evidence of patterns of racially disproportionate exposure at the time of siting (Mohai & Saha, 2015; Pastor, Sadd, & Hipp, 2001; Saha & Mohai, 2005). In the first national-level longitudinal environmental justice study to employ distance-based methods (Mohai & Saha, 2015), we found clear evidence of this pattern in the siting of hazardous waste treatment, storage, and disposal facilities in the United States. Furthermore, when we reanalyzed our data applying the unit-hazard coincidence method, these patterns were no longer evident.

In the future, more longitudinal environmental justice studies employing distance-based methods are needed to determine whether these studies would reveal similar patterns of disproportionately placing (at the time of siting) a wide variety of pollution sources in poor and people-of-color communities, including industrial facilities, major highways, airports, municipal landfills, power plants, and others. In addition to more work being needed to address the environmental justice "chicken or egg" question and to test hypotheses about the causes of present-day environmental disparities, more work is needed to model pollution exposures and assess the role such exposures play in accounting for racial and socioeconomic disparities in health and mortality (Ash & Fetter, 2004; Mohai et al., 2009; Zwickl et al., 2014). Quantitative studies of environmental injustice and its consequences need to be extended to countries outside the United States as well (Glatter-Götz et al., 2019; Laurent, 2011).

DEEPENING OUR UNDERSTANDING

1. Explain how two studies examining the same question and using the same or similar data can come up with different conclusions. You can use the SADRI and UCC studies as an example if you like. Can you find other examples of environmental justice research with different conclusions? What caused those differences?

2. (a) Find a study that uses the unit-hazard coincidence method, and choose a subset of the data (for example, from a particular area of interest to your group). Use GIS or other methods to locate the environmental hazards (such as TSDFs) in question. (b) Can you locate data about race and class in a one- and/or three-mile radius from the hazard sites (that is, use a distance-based method)? How easy is it to do this? If you are able to locate the data, how do the results using the distance-based method differ from the unit-hazard method?

References

Anderton, D. L., Anderson, A. B., Oakes, J. M., & Fraser, M. R. (1994). Environmental equity: The demographics of dumping. *Demography, 31*, 229–248.

Anderton, D. L., Oakes, J. M., & Egan, K. L. (1997). Environmental equity in Superfund: Demographics of the discovery and prioritization of abandoned toxic sites. *Evaluation Review, 21*, 3–26.

Asch, P., & Seneca, J. J. (1978). Some evidence on the distribution of air quality. *Land Economics, 54*, 278–297.

Ash, M., & Robert Fetter, T. (2004). Who lives on the wrong side of the environmental tracks? Evidence from the EPA's risk-screening environmental indicators model. *SocialScience Quarterly, 85*(2), 441–462.

Been, V., & Gupta, F. (1997). Coming to the nuisance or going to the barrios? A longitudinal analysis of environmental justice claims. *Ecology Law Quarterly, 24*, 1–56.

Berry, B. J. L., Caris, S., Gaskill, D., Kaplan, C. P., Piccinini, J., Planert, N., ... de Ste Phalle, A. (1977). *The social burdens of environmental pollution: A comparative metropolitan data source*. Cambridge MA: Ballinger.

Boone, C. G., & Modarres, A. (1999). Creating a toxic neighborhood in Los Angeles County: A historical examination of environmental inequality. *Urban Affairs Review, 35*, 163–87.

Bryant, B., & Mohai, P (Eds.). (1992a). *Race and the incidence of environmental hazards: A time for discourse*. Boulder CO: Westview Press.

Bryant, B., & Mohai, P. (1992b). The Michigan Conference: A turning point. *EPA Journal, 18*, 9–10.

Bullard, R. D. (1983). Solid waste sites and the Houston black community. *Sociological Inquiry, 53*, 273–288.

Bullard, R. D. (1990). *Dumping in Dixie: Race, class, and environmental quality*. Boulder, CO: Westview Press.

Bullard, R. D., & Wright, B. H. (1986). The politics of pollution: Implications for the black community. *Phylon, 47*, 71–78.

Bullard, R. D., & Wright, B. H. (1987a). Blacks and the environment. *Humboldt Journal of Social Relations, 14*, 165–184.

Bullard, R. D., & Wright, B. H. (1987b). Environmentalism and the politics of equity: Emergent trends in the black community. *Mid-American Review of Sociology, 12*, 21–38.

Bullard, R. D., Mohai, P., Saha, R., & Wright, B. (2007). *Toxic wastes and race at twenty 1987–2007: Grassroots struggles to dismantle environmental racism in the United States*. Cleveland, OH: Justice and Witness Ministries, United Church of Christ.

Bullard, R. D., Mohai, P., Saha, R., & Wright, B. (2008). Toxic wastes and race at twenty: Why race still matters after all of these years. *Environmental Law, 38*, 371–411.

Burby, R. J., & Strong, D. E. (1997). Coping with chemicals: Blacks, white, planners, and industrial pollution. *Journal of the American Planning Association, 63*, 469–480.

Burch, W. R. (1976). The Peregrine falcon and the urban poor: Some sociological interrelations. In P. Richerson and J. McEvoy (Eds.), *Human ecology, an environmental approach* (pp. 308–316). Belmont CA: Duxbury Press.

Buttel, F. H., & Flinn, W. L. (1978). Social class and mass environmental beliefs: A reconsideration. *Environment and Behavior, 10*, 433–450.

Chakraborty, J., & Armstrong, M. P. (1997). Exploring the use of buffer analysis for the identification of impacted areas in environmental equity assessment. *Cartography and Geographic Information Systems, 24*, 145–157.

Commission for Justice, United Church of Christ.(1987).*Toxic wastes and race in the United States: A national report on the racial and socioeconomic characteristics of communities with hazardous waste sites*.New York:United Church of Christ.

Council on Environmental Quality. (1971). *The second annual report of the Council on Environmental Quality*. Washington, DC: U. S. Government Printing Office.

Davidson, P., & Anderton, D. L. (2000). The demographics of dumping II: Survey of the distribution of hazardous materials handlers. *Demography, 37*, 461–466.

Freeman, A. M. (1972). The distribution of environmental quality. In A. V. Kneese and B. T. Bower (Eds.), *Environmental quality analysis*. Baltimore, MD: Johns Hopkins University Press for Resources for the Future.

Gelobter, M. (1988, June). The distribution of air pollution by income and race. Paper presented at the Second Symposium on Social Science in Resource Management, Urbana, Illinois.

Gelobter, M. (1992). Toward a model of environmental discrimination. In B. Bryant & P. Mohai (Eds.), *Race and the incidence of environmental hazards: A time for discourse* (pp. 64–81). Boulder, CO: Westview Press.

Gianessi, L., Peskin, H. M., & Wolff, E. (1979). The distributional effects of uniform air pollution policy in the U.S. *Quarterly Journal of Economics, 93*, 281–301.

Glatter-Götz, H., Mohai, P., Haas, W., & Plutzar, C. (2019). Environmental inequality in Austria: Do inhabitants' socioeconomic characteristics differ depending on their proximity to industrial polluters? *Environmental Research Letters, 14*, 074007. https://iopscience.iop.org/article/10.1088/1748-9326/ab1611

Glickman, T. (1994). Measuring environmental equity with GIS. *Renewable Resources Journal, 12*, 17–21.

Goldman, B. A. (1994). *Not just prosperity: Achieving sustainability with environmental justice*. Washington, DC: National Wildlife Federation.

Hamilton, J. T. (1995). Testing for environmental racism: Prejudice, profits, political power? *Journal of Policy Analysis and Management, 14*, 107–132.

Hamilton, J. T., & Kip Viscusi, W. (1999). *Calculating risks? The spatial and political dimensions of hazardous waste policy*. Cambridge, MA: MIT Press.

Handy, F. (1977). Income and air quality in Hamilton, Ontario. *Alternatives, 6*, 18–24.

Harrison, D., Jr. (1975). *Who pays for clean air: The cost and benefit distribution of automobile emission standards*. Cambridge, MA: Ballinger.

Hird, J. A., & Reese, M. (1998). The distribution of environmental quality: An empirical analysis. *Social Science Quarterly, 79*, 694–716.

Hurley, A. (1995). *Environmental inequities: Class, race and industrial pollution in Gary Indiana 1945–1990*. Chapel HIll: University of North Carolina Press.

Kruvant, W. J. (1975). People, energy, and pollution. In D. K. Newman & D. Day (Eds.), *The American energy consumer* (pp. 125–167). Cambridge, MA: Ballinger.

Laurent, E. (2011). Issues in environmental justice within the European Union. *Ecological Economics*, *70*(11), 1846–1853.

Mitchell, R. C. (1979). Silent springs/solid majorities. *Public Opinion*, *55*, 16–20.

Mohai, P. (1985). Public concern and elite involvement in environmental-conservation issues. *Social Science Quarterly*, *66*, 820–838.

Mohai, P. (1990). Black environmentalism. *Social Science Quarterly*, *71*, 744–765.

Mohai, P. (1995). The demographics of dumping revisited: Examining the impact of alternate methodologies in environmental justice research. *Virginia Environmental Law Journal*, *13*, 615–653.

Mohai, P. (2003). Dispelling old myths: African American concern for the environment. *Environment*, *45*, 10–26.

Mohai, P. (2008). Equity and the environmental justice debate. *Research in Social Problems and Public Policy*, *15*, 21–49. https://doi.org/10.1016/S0196-1152(07)15010-9

Mohai, P., & Bunyan, B. (1992). Environmental racism: Reviewing the evidence. In B. Bryant & P. Mohai (Eds.), *Race and the incidence of environmental hazards: A time for discourse* (pp. 163–176). Boulder, CO: Westview Press.

Mohai, P., & Bunyan, B. (1998). Is there a "race" effect on concern for environmental quality? *Public Opinion Quarterly*, *62*, 475–505.

Mohai, P., & Kershner, D. (2002). Race and environmental voting in the U.S. Congress. *Social Science Quarterly*, *83*, 167–189.

Mohai, P., Lantz, P. M., Morenoff, J., House, J. S., & Mero, R. P. (2009). Racial and socioeconomic disparities in residential proximity to polluting industrial facilities: Evidence from the Americans' Changing Lives Study. *American Journal of Public Health*, *99*(S3), S649–S656.

Mohai, P., & Saha, R. (2006). Reassessing racial and socioeconomic disparities in environmental justice research. *Demography*, *43*, 383–399.

Mohai, P., & Saha, R. (2007). Racial inequality in the distribution of hazardous waste: A national-level reassessment. *Social Problems*, *54*(3), 343–370.

Mohai, P., & Saha, R. (2015). Which came first, people or pollution? Assessing the disparate siting and post-siting demographic change hypotheses of environmental injustice. *Environmental Research Letters*, *1*(0.11), 115008. https://iopscience.iop.org/article/10.1088/1748-9326/10/11/115008

Oakes, J. M., Anderton, D. L., & Anderson, A. B. (1996). A longitudinal analysis of environmental equity in communities with hazardous waste facilities. *Social Science Research*, *25*, 125–148.

Pastor, M., Sadd, J., & Hipp, J. (2001). Which came first? Toxic facilities, minority move-in, and environmental justice. *Journal of Urban Affairs*, *23*, 1–21.

Pellow, D. N. (2002). *Garbage wars: The struggle for environmental justice in Chicago*. Cambridge, MA: MIT Press.

Pulido, L., Sidawi, S., & Vos, R. O. (1996). An archaeology of environmental racism in Los Angeles. *Urban Geography*, *17*, 419–39.

Rhodes, E. L. (2003). *Environmental justice in America: A new paradigm*. Bloomington and Indianapolis: Indiana University Press.

Ringquist, E. J. (2005). Assessing evidence of environmental inequities: A meta-analysis. *Journal of Policy Analysis and Management*, *24*, 223–247.

Saha, R., & Mohai, P. (2005). Historical context and hazardous waste facility siting: Understanding temporal patterns in Michigan. *Social Problems*, *52*, 618–648.

Sheppard, E., Leitner, H., McMaster, R. B., & Tian, H. (1999). GIS-based measures of environmental equity: Exploring their sensitivity and significance.

Journal of Exposure Analysis and Environmental Epidemiology, 9, 18–28.

Szasz, A., & Meuser, M. (2000). Unintended, inexorable: The production of environmental inequalities in Santa Clara County, California. *American Behavioral Scientist, 43*, 602–632.

Taylor, D. E. (1989). Blacks and the environment: Toward explanation of the concern and action gap between blacks and whites. *Environment and Behavior, 21*, 175–205.

U.S. Census Bureau. (1980). Census of U.S. population: General social and economic characteristics summary, Table 233. Washington, DC: U.S. Census Bureau.

U.S. Environmental Protection Agency. (1992, June). *Environmental equity: Reducing risks for all communities.* EPA 230-R-92-008. Washington, DC: U.S. Environmental Protection Agency.

U.S. General Accounting Office. (1983). *Siting of hazardous waste landfills and their correlation with racial and economic status of surrounding communities.* Washington, DC: U. S. General Accounting Office.

Van Liere., D, K., & Dunlap, R. E. (1980). The social bases of environmental concern: A review of hypotheses, explanations and empirical evidence. *Public Opinion Quarterly, 44*, 181–197.

West, P. C., Fly, J. M., Larkin., F., & Marans, R. (1992). Minority anglers and toxic fish consumption: Evidence from a state-wide survey of Michigan. In B. Bryant & P. Mohai (Eds.), *Race and the incidence of environmental hazards: A time for discourse* (pp. 100–113). Boulder, CO: Westview Press.

Zupan, J. M. (1973). *The distribution of air quality in the New York region.* Baltimore, MD: Johns Hopkins University Press for Resources for the Future.

Zwickl, K., Ash, M., & Boyce, J. K. (2014). Regional variation in environmental inequality: Industrial air toxics exposure in US cities. *Ecological Economics, 107*, 494–509.

5

SCIENCE, EXPERTISE, AND ENVIRONMENTAL JUSTICE

Alissa Cordner

Phil Brown

For individuals impacted by environmental health problems, whether localized sites of contamination or broader exposures through daily life and consumer products, science is a necessary tool to uncover and reduce toxic exposures, identify and alleviate associated health effects, and prevent future exposures. Science and expertise are fundamental environmental justice (EJ) issues because exposures and effects usually are felt disproportionately by low-income people and people of color, and marginalized populations are often excluded from scientific decision-making processes.

In this chapter, we argue that the most impactful approaches to EJ research and advocacy are *community engaged*, working with EJ communities as participants in the research process rather than simply research subjects; and *reflexive*, continuously and deliberately reflecting on the research goals, processes, and outcomes. Although not all environmental research is conducted in or with EJ communities, all researchers can be guided by environmental health and justice principles, and can advance the tools and goals sought by EJ activists. This involves employing a political-economic perspective, advancing democratic science governance, and seeking actions that benefit the public health and equity over corporate profits and the failures of many regulatory agencies.

In this chapter, we highlight key theoretical and empirical challenges for research conducted with and in EJ communities. We will demonstrate this approach with examples of research and activism on two classes of contaminants that have inspired significant research, regulation, and activism in recent years: flame retardants and per- and poly-fluoroaklyl substances (PFASs). Though these examples come from the field of environmental health research, our arguments are equally relevant for other areas of environmental sociology, such as natural resource extraction and use, climate change, or food and agriculture.

FRAMEWORK AND BACKGROUND

Issues of science and expertise are central to the work of EJ researchers and communities because science is often a primary guide and justification for environmental policy decisions (National Research Council, 2009, p. 4). Though there are many possible definitions and conceptualizations of *science*, we use the term to refer to the systematic collection of evidence and observations to describe and explain something about the world. Extensive social science research has interrogated supposed boundaries between science and other areas of society, including politics, economics, and policymaking, demonstrating that there is no definitive line separating scientific practices and findings from their social contexts (Gieryn, 1983, 1999; Jasanoff, 1987). Although scientific authority rests on science being seen as "value-free and politically neutral" (Kinchy & Kleinman, 2003, p. 380), most sociologists argue that science

---

PHOTO 5.1 Audience at the PFAS Conference at Northeastern University, Boston, June 2019. Photo courtesy of Toxics Action Center.

is as much socially constructed as it is empirically based. This is, in part, because science is done by people with diverse social positions and because science takes place within a social context (Jasanoff et al., 1995). Scientific arguments are also required for participation in scientized fields like science policy, chemical product development, or environmental activism. The process of *scientization* refers to how scientific authority is increasingly valued and required for regulatory, legal, and social movement activities (Kinchy, 2010; Michaels & Monforton, 2005; Morello-Frosch et al., 2006). Participation in these scientized fields typically depends on *expertise*, which we define as in-depth and appropriately credentialed technical knowledge and experience that is particular to a topic, sector, or discipline. Highly scientized fields routinely exclude lay voices and the experiences of those directly affected by risks, such as workers or residents who live near polluting facilities (Morello-Frosch et al., 2006). These questions of expertise and participation in policy and scientific debates are extremely relevant for EJ scholarship.

Several methodological and theoretical approaches offer guidance for engaged and reflexive EJ researchers and advocates who are attentive to these issues of science and expertise. We highlight four such approaches: community-based participatory research, transdisciplinary social science–environmental work, the New Political Sociology of Science (NPSS), and reflexive research ethics.

Community-Based Participatory Research

Community-based participatory research (CBPR) is a research practice that shares control over all aspects of the research process with community partners. Goals of CBPR projects include increased community engagement in research to generate more accurate scientific knowledge, improved public trust and understanding of environmental health science, utilization of culturally and socially appropriate interventions, improved public health decisions, policy changes, and reductions in environmental injustice (O'Fallon & Dearry, 2002; Wallerstein et al., 2017). Building off earlier work in participatory action research (Whyte, 1991), the practice of CBPR has greatly expanded since its origination in the 1990s, though some would say the goals and practices of CBPR have been diluted as it has become more institutionalized and as funding sources have grown in number and size (Wallerstein et al., 2017). The EJ focus of CBPR is particularly important given historical power imbalances between researchers and low-income communities of color.

The National Institute of Environmental Health Sciences (NIEHS) has been instrumental in facilitating CBPR collaborations between the social and health sciences and community-based organizations (CBOs) in the past two decades (Hoover et al., 2015). NIEHS introduced new programing in 1995 to address connections between race, class, and environmental exposures. This new programing—the first of its kind in any National Institutes of Health agency—was prompted by grassroots organizing; the First National People of Color Environmental Leadership Summit (1991) Principles of Environmental Justice; and growing recognition that people of color and marginalized communities are most exposed to health threats from toxic contamination, toxic waste disposal, and other industry externalities. The NIEHS programming focused on the ethical, legal, and social implications of science, and featured new funding specifically for CBPR and EJ research. It provided crucial and, at the time, made available federal funding for social and environmental scientists to collaborate with community groups; incorporate local knowledge in research; and put residents at the forefront of problem definition, data collection, analysis, and the reporting of results (Baron et al., 2009; Israel et al., 1998). These grants were unique in that they required researchers to work with CBOs and affected residents as *partners* who contributed to the scientific process, not just as research participants.

Between 1994 and 2007, NIEHS, the National Institute for Occupational Safety and Health, and the U.S. Environmental Protection Agency (EPA) funded 54 projects, including 27 led by CBOs (Baron et al., 2009). Many of the funded projects included co-principal investigators from CBOs and academic institutions, such as the partnership between West Harlem Environmental Action and Columbia University. Other projects were led by a CBO with scientists on their staff, as was the case for toxic contamination research done by Alaska Community Action on Toxics. Although the CBPR program ended in 2007, NIEHS continues to fund and promote interdisciplinary collaboration within multi-institutional, university-based projects, many of which involve collaborations with EJ organizations and tribal groups (Hoover et al., 2015; National Institute of Environmental Health Sciences, 2012). Brown University's Superfund Research Program, for example, has worked with the Environmental Justice League of Rhode Island on state legislation for polluter fines, school siting issues, and a Community Environmental College serving mostly high school students of color (Senier et al., 2008).

Transdisciplinary Social Science–Environmental Work

Sociologists and other social scientists are increasingly engaged in transdisciplinary collaborations in environmental health fields (Finn & Collman, 2016; Hoover et al., 2015; Matz et al., 2016). Research has moved away from isolated disciplinary silos toward engaged, transdisciplinary work in partnership with impacted communities to investigate exposures and health effects, mitigate hazards, influence environmental policy, and prevent new exposures. In such collaborations, sociologists and other EJ researchers become active members of environmental health research teams rather than just observers.

Social scientists have longed critiqued environmental health research itself through what Hoover et al. (2015) call "social science *of* environmental health." This research investigates environmental health crises, exposures, contamination, and disasters from a political-economic or traditional EJ perspective (see Chapters 1, 12, and 18 in this volume). Although this legacy remains strong, social scientists are increasingly moving from the social science *of* environmental health to "social science *with* environmental health" by directly collaborating with health scientists, residents, and CBOs (Hoover et al., 2015). Hoover et al. illustrate this transition through a case study of collaboration in the Awkwesasne Mohawk Nation. Katsi Cook, women's health activist, CBO director, and Awkwesasne midwife, initiated a biomonitoring research collaboration between members of the Mohawk Community at Akwesasne and scientists at the State University of New York at Albany that identified polychlorinated biphenyl

(PCB) contamination of community waters from nearby industry. After an initial New York Department of Health assessment recommended that residents refrain from eating fish from the river to limit the ingestion of PCBs, social scientists working with Cook argued that the loss of fishing was itself a cultural and social impact that would worsen health problems in the community (Hoover, 2013). This led to a revised fishing advisory that recommended avoiding only certain types of fish. This case demonstrates the inseparability of social impacts and environmental health impacts from toxic exposures, and the necessity of social science research to identify the full range of harms—environmental, health, and social—from toxic exposures.

Public participation in environmental decision-making, collaboration between all stakeholders, and risk communication to communities contribute additional layers of complexity, which researchers may be ill equipped to handle. Few environmental health scientists receive formal training in the nuances of environmental communication and risk surrounding contaminated sites. Likewise, few social scientists studying environmental health issues receive formal training in the environmental health sciences they are studying and must learn to communicate with environmental health scientists. This cross-training is especially important when researchers are working with EJ groups and communities. Feagin (2001) advocates for an explicitly *social justice–oriented sociology* that celebrates a commitment "to social justice in ideals and practice" (p.10).

Social scientists, environmental health scientists, and CBOs have created a new generation of social science–environmental health collaborations, largely guided by EJ principles (Bullard, 2008). The challenges of environmental health also create a number of opportunities for social scientists that go far beyond merely translating scientific jargon into publicly accessible text or studying residents' risk perceptions. Social scientists taking part in transdisciplinary projects can facilitate research translation, community engagement, or evaluation; document the initiatives through the lens of the social sciences; and serve as useful liaisons between researchers, policymakers, and the community. They can also facilitate investigation and teaching of social, legal, and ethical implications of scientific research projects in the academy.

New Political Sociology of Science and Undone Science

The New Political Sociology of Science (NPSS) studies the networks, institutions, and power structures that affect the production and consumption of scientific knowledge (Frickel & Moore, 2006). This theoretical approach argues that the practice, interpretation, and use of science is inherently and inevitably influenced by distributions of power, and is attentive to disciplinary norms and practices, institutional structures, distributions of resources, and inequality. In particular, NPSS assumes that science is inherently political, with a focus on how power is unevenly contained and expressed through networks and institutions (Kleinman & Suryanarayanan, 2012; Moore et al., 2011; Woodhouse, 2006). This makes it a particularly relevant theoretical framework for thinking about science as an EJ issue.

NPSS work has shown that gaps in scientific knowledge, common in all environmental fields, are socially produced—through both deliberate actions as well as unintentional but still influential institutional logics; disciplinary practices; and social and cultural understandings of risk, safety, and scientific proof (Frickel & Edwards, 2014; Frickel et al., 2010; Hess, 2009; Kempler et al., 2011). *Undone science* refers to research gaps and areas where data are not developed because of institutional, political, or economic factors (Frickel et al., 2010; Hess, 2009). Funding priorities, for example, are often set by federal agencies or the military and thus reflect elite priorities (Moore, 2008), disciplines compete for intellectual territory and scarce grant dollars (Frickel & Gross, 2005), and the research questions of interest to government or industry often receive greater attention than those of interest to communities and non-elites (Frickel et al., 2010; Hess, 2009).

Another important category of research is *unseen science*, research that is conducted but never shared because the findings are never published or shared beyond institutional boundaries (Richter, Cordner, & Brown 2018). Some unseen science is produced by regulated industries, never made public, and shared only selectively with regulatory agencies, or is made public only through the legal discovery process. The uneven ability to access or withhold information that creates and perpetuates areas of unseen science results from unequal distributions of power and resources, and from benefits regarding knowledge access and protection that are codified in rules, regulations, and laws. Unpublished findings are also common because academics are institutionally rewarded for novel discoveries rather than nondiscoveries or repetitions of prior research, creating unseen science within research institutions (Hopewell et al., 2009).

EJ communities often face significant data gaps that prevent them from making claims to reduce exposure and protect their health (Allen, Ferrier, & Cohen, 2016). Some uncertainty is inherent in the environmental health research process. This uncertainty might be related to limited research questions or methods, or it might be due to the interpretation of scientific results, or particular communication strategies (Cordner & Brown, 2013). Other areas of uncertainty are more direct examples of undone science. For example, fracking companies may refuse to fully disclose the contents of fracking fluid, leaving residents in the dark about potential sources and types of water contamination; chemical manufacturers are not required to fully test new chemical compounds for a broad range of toxicological endpoints, limiting what is known about chemicals' health effects; and regulatory agencies may design health studies with such narrow population or exposure parameters that no exposure-disease links can be identified. All of these issues matter greatly for scholars working with impacted EJ communities.

Reflexive Research Ethics

We argue that EJ scholarship and engaged advocacy is most effective when it is guided by *reflexive research ethics*, the self-conscious, interactive, and iterative reflection upon researchers' relationships with research participants, relevant communities, and principles of professional and scientific conduct (Cordner et al., 2012). Reflexive research ethics do not assert a set of preconceived principles of how a particular moment of ethical uncertainty should be addressed. Rather, this approach entails the continued evaluation and adjustment of research practices according to more relational and reflexive understandings of what might be beneficent or harmful. This is particularly important when working with EJ communities because of the power inequalities between researchers and marginalized communities.

As an example, Brown and other academic collaborators conducted an air and dust sampling project with the EJ group Communities for a Better Environment in Richmond, California (Brown et al., 2012). The community group was concerned with expansion of a refinery on the fence-line of two low-income communities with a majority of people of color. The collaborative project involved close engagement with local CBOs and an advisory board made up of breast cancer organizations, other EJ groups, state health scientists, and community residents. Before research commenced, community collaborators and the advisory board sought research design changes that proved very beneficial, including adding a control "clean" community far from the local pollution and allowing volunteer participants in addition to randomly selected ones (Brown et al., 2012).

The project deliberately incorporated reflexive ethics into all phases of the process. To gauge overall community support and reactions at each community meeting, the research team took detailed notes of verbal questions and responses, collected written evaluation forms, and held frequent team debriefings. To assess participants' understanding of the overall study and their data, social scientists interviewed them before sampling and after results were returned. They held monthly team calls that included a deliberate examination of how the research process was progressing as well as advisory board meetings at which they recorded notes, interviewed board members, conducted team interviews of collaborating organization, and analyzed government policies, court decisions, and media coverage (Brown et al., 2012).

ENVIRONMENTAL HEALTH MULTISECTOR ALLIANCES AND SOCIOLOGICAL RESEARCH

Sociologists working with community groups must be attentive to how science is mobilized, criticized, and developed in pursuit of stakeholder goals. A key feature of many environmental social movements today is the *multisector alliance*, which involves a broad coalition of traditional and unexpected participants and multiple levels of state and nonstate targets aimed at influencing environmental governance (Cordner & Brown, 2015). Environmental health campaigns are no longer targeted solely or even primarily at the state but, rather, focus their organizing energies on a broad range of targets including local, state, and federal agencies; retailers; manufacturers and supply chain companies; code- and standard-setting organizations; and other nonprofit organizations.

To show what the multisector alliance looks like and describe its relevance for engaged and participatory EJ research, we now describe case studies of research and activism on two classes of emerging contaminants from our own work: flame retardants and per- and poly-fluoroaklyl substances (PFASs). For both case studies, we describe the activism that emerged around the class of chemicals, the EJ connections with exposure and activism, and challenges to the production and use of science around each class of chemicals. We also describe our own research as engaged sociologists to work directly with impacted communities.

Flame Retardant Chemicals

Flame retardants are a broad class of chemicals added to reduce the flammability of consumer goods, homes, workplaces, cars, and airplanes. We began studying the full field of stakeholders who were working on flame retardant chemicals in 2009 (Brown & Cordner, 2011; Cordner, 2016; Cordner & Brown, 2015). We had learned about this class of chemicals through a prior CBPR partnership with Silent Spring Institute, a nonprofit research organization that had detected several flame retardants in indoor air and dust on Cape Cod (Rudel et al., 2003). In our research, we were interested in the science, regulation, activism, and industry decision-making around this contested class of chemicals. With funding from the National Science Foundation, we engaged in multimethod qualitative research—including more than 110 in-depth interviews; a year of participant observation at six sites; and detailed analysis of regulatory, media, and historical documents. Though this project did not meet a formal definition of CBPR work, we engaged closely with community groups and environmental health nonprofits to ensure that our research was useful to environmental health advocacy and to other social scientists.

Concerns about possible health risks from flame retardant exposure first emerged in the 1970s, after contaminated animal feed poisoned more than one million livestock, and then after mutagenic flame retardants were shown to be used in children's pajamas (Blum et al., 1978; Egginton, 2009). Starting in the 2000s, polybrominated diphenyl ether (PBDE) flame retardants received much scientific and regulatory attention. European biomonitoring research showed the presence and rapid accumulation of PBDEs in women's breast milk, and subsequent work in the United States found even higher concentrations of PBDEs in house dust and people (Meironyte, Noren, & Bergman, 1999; Zota et al., 2008). In animal studies, the chemicals have been shown to be neurological and reproductive toxics and potentially carcinogenic (Birnbaum & Staskal, 2004). Epidemiological studies have connected PBDE exposure to numerous developmental and reproductive effects in people, ranging from reduced fecundity to decreased IQ in children (Wikoff & Birnbaum, 2011).

The first PBDE ban was adopted in the European Union in 2002, and the chemicals were added to the Stockholm Convention's list of Persistent Organic Pollutants in 2009 (U.S. Environmental Protection Agency, 2006; Stockholm Convention, 2014). Because of the weaknesses of federal regulations in the United States, state governments enacted state-level bans on certain flame retardants (Cordner, Mulcahy, & Brown, 2013). Eleven states passed bans on PBDE formulations, the chemical manufacturers announced voluntary phase-outs of the chemicals starting in 2003, and the EPA proposed several regulations to permanently restrict their use and production (U.S. Environmental Protection Agency, 2017a). These actions have successfully reduced levels of PBDEs in household dust and serum, though the use of replacement chemicals has increased (Dodson et al., 2012; Zota et al., 2013).

From spring 2012 to fall 2013, a series of significant policy changes sprang rapidly from a relatively new multisector environmental health alliance around flame retardant chemicals, involving environmental groups, public health organizations, manufacturing companies and industry groups, hospital and building sustainability representatives, journalists, film makers, firefighters, and fire safety professionals (Cordner & Brown, 2015). Due to the rise in public attention and supported by the multisector alliance, the state of California revised an influential furniture flammability standard so that it can be met without the addition of flame retardants. Other states subsequently adopted the revised flammability standard, and a number of product manufacturers rapidly reformulated furniture and baby products to be free of flame retardants.

This multisector alliance is distinct from what has been observed in most prior research on environmental social movements and environmental governance research in two ways (Cordner & Brown, 2015). First, this multisector alliance includes both traditional allies, such as environmental social movement organizations, and unexpected participants, including furniture manufacturers and firefighters (Cordner et al., 2015). Second, the multisector alliance engages with a broad range of targets. In addition to environmental governance processes such as federal chemicals reform, the flame retardant advocacy coalition targeted unexpected sites of decision making including the media, fire prevention groups, and code- and standard-development processes.

Flame retardants are an EJ concern in important but somewhat surprising ways. Although exposure to these chemicals is ubiquitous across the U.S. population, levels of the chemicals are higher in the bodies of low-income and minority populations, possibly because more chemicals are released from older consumer goods like furniture and electronics (Zota et al., 2010). The flame retardant activists we interviewed were well aware of the importance of EJ for their work and of past and current tensions between mainstream environmental organizations over exclusion, lack of representation, and racism (Gottlieb, 1993). Flame retardant activists noted that minorities and the poor are disproportionately affected by exposures to flame retardants, and they deliberately sought to include relevant groups in their coalitions (for example, the International Association of Black Professional Fire Fighters).

Per- and Poly-Fluoroalkyl Substances

PFASs are another broad class of toxic chemicals that are used in a wide range of consumer and industrial applications because of their water- and stain-resistant properties. We became interested in PFASs following collaborations (built through our flame retardant project) with the California-based nonprofit Green Science Policy Institute, which identified PFASs as one of their "six classes" of chemicals known to be harmful to human health and the environment (Green Science Policy Institute, 2018). After we completed our research on flame retardants, we learned more about PFASs and identified several features of this class of chemicals worthy of study: their broad and varied use in consumer and industrial products; the ubiquitous population exposure; numerous communities whose drinking water had been seriously contaminated by industrial or military sites; and a long history of questionable practices by PFAS manufacturers. We received grants to investigate the social and scientific discovery of this class of chemicals and to organize the first national conference dedicated to research and community activism on PFASs. For this project, we partnered more directly with community groups and residents whose drinking water was contaminated, while maintaining our ongoing collaborations with other researchers and our multimethod qualitative research approach (Cordner, Richter, & Brown, 2016; Richter et al., 2018).

PFASs are a class of nearly 5,000 human-made chemicals containing chains of carbon and fluorine atoms widely used in industrial processes and consumer goods (Organization for Economic Co-operation and Development, 2018). Two PFAS compounds are most widely known: perfluorooctanoic acid (PFOA), which was used in the manufacture of Teflon cookware coatings and is a by-product of many other chemical processes, and perfluorooctane sulfonate (PFOS), which was used in Scotchgard fabric protectors, firefighting foam, and semiconductor devices. Although these compounds have been in production since the 1940s and studied by manufacturers for toxicological and exposure concerns since the 1960s (Lyons, 2007), significant awareness of PFASs within the regulatory and academic science communities did not occur until decades later (Richter et al., 2018).

The general public's exposure to multiple PFAS compounds is ubiquitous: Research by the Centers for Disease Control and Prevention (2018) national biomonitoring program found multiple PFASs in the blood of nearly all people tested. Academic, nonprofit, and regulatory studies documenting widespread exposure have brought PFASs to the attention of a new audience of environmental health scientists and involved residents whose drinking water is contaminated with PFASs (Environmental Working Group, 2017; U.S. Environmental Protection Agency, 2017b).

In addition to widespread human exposure, this class of chemicals is particularly concerning because they have low-dose or hormone-disrupting effects, and they do not naturally degrade in the environment (Post et al., 2012). A large epidemiological study of communities in Ohio and West Virginia with PFOA-contaminated drinking water linked exposure to high cholesterol, ulcerative colitis, thyroid disease, testicular and kidney cancers, and pregnancy-induced hypertension (C8 Science Panel, 2011). Other suspected health impacts of exposure to certain PFASs include endocrine disruption, obesity, reproductive problems, birth defects, additional types of cancer, stroke, and developmental problems in children (Lau, 2015). Although PFOA and PFOS are no longer produced in the United States, replacement compounds are used extensively in spite of growing concerns about widespread exposures and toxicity (Danish Ministry of the Environment, 2015; Rosenmai et al., 2016; Sun et al., 2016).

High-level exposure is concentrated at known contaminated sites. A database of contaminated sites that we maintain currently includes 93 sites in the United States, including many linked to manufacturing facilities (www.pfasproject.com). The U.S. Department of Defense has identified more than 400 current or former military sites with known or suspected PFAS contamination, including 126 sites with PFOA or PFAS levels above the EPA's health advisory levels, mostly due to use of PFAS-containing firefighting foam for training or fire suppression (Sullivan, 2018). Potential EJ issues are significant since African Americans are overrepresented in the U.S. military (Reynolds & Shendruk, 2018). Also, as with flame retardants, working-class firefighters are extremely exposed through their use of PFAS-containing firefighting foams during training exercises.

When we began our PFAS research in 2015, social movement organizations largely ignored these chemicals, and the advocacy that did exist involved mainly professional, scientific advocacy organizations. We were particularly surprised at the lack of organized, grassroots opposition to PFAS drinking water contamination, especially given the outcomes of class-action suits against DuPont in the Mid-Ohio Valley (Judge et al., 2016). Today, however, a broad field of stakeholders are working on PFASs, including scientists at regulatory, academic, and independent institutions; industry advocates and scientists along the supply chain; regulators at the local, municipal, state, and federal levels; military scientists and policymakers; legislators at the state and federal levels; journalists; lawyers and other legal experts; residents of impacted communities; and community and social movement groups at the local, state, regional, national, and international levels.

Long-standing CBPR and place-based relationships between academics and CBOs facilitated a rapid response to the emerging PFAS contamination crisis, and we have been very involved in this work. Sociologist Phil Brown has been working for more than 15 years with Toxics Action Center (TAC), a New England anti-toxics organization. Starting in 2015, as communities around New England learned that their drinking water was contaminated with PFAS from industrial and military sites, TAC worked with community groups coalescing around PFAS contamination, and our team organized PFAS-related panels at their 2016 and 2017 conferences.

In 2015, we connected with activists from Testing for Pease (TFP), an active and highly organized CBO that formed after residents learned that the drinking water supplying the Pease Tradeport, an industrial park on the site of the former Pease Air Force Base, was contaminated with high levels of PFASs. We have collaborated extensively with TFP, writing several grant proposals, supporting scientific presentations and community events, attending community meetings, and interviewing community members to document their own accounts of dealing with PFASs. We continue to work with TFP to pursue funding and regulatory support for additional research on chemical exposure and toxicity.

A second major collaboration has been with Environmental Working Group (EWG), a national environmental nonprofit that was one of the first organizations to look at PFAS contamination. We worked with EWG to develop an interactive map displaying information from our Contamination Site Database, which was publicly released in June 2017 and updated in April 2018 with an additional 42 sites. Combining our contamination tracking data with the EPA's national sample of PFAS contamination, EWG (2018) estimates the number of Americans drinking highly contaminated water to be 16 million in 33 states and Puerto Rico.

Our engagement with community groups and nongovernmental organizations has been an intentional and important part of our research program, but we also regularly engage with regulatory processes and with government science. These efforts are carried out largely in an official advisory capacity, providing feedback to ongoing government assessments and advocating in favor of greater government-funded research on the health effects of PFAS exposure. For example, Brown has provided feedback to a New York Department of Health report on contamination in Hoosick Falls, and Cordner serves on the Steering Committee for the Washington State Department of Ecology's PFAS

Chemical Action Plan. Finally, we have engaged with regulatory and scientific colleagues to advocate for stronger research on the health effects of PFAS exposure, signing the Green Science Policy Institute's commentary calling for strong national health studies of PFASs published in the influential journal *Environmental Health* (Bruton & Blum, 2017) and engaging with government researchers at the EPA.

Finally, we organized the first national conference to bring together diverse stakeholders working on PFASs, an issue that often feels invisible and ignored for impacted communities (Social Science Environmental Health Research Institute, 2017). The conference addressed advocacy goals of advancing PFAS regulation discussions, connected scientists across disciplines, and raised the public and media profiles of the issue. Most significantly, a network of activists working on PFAS issues, the National PFAS Contamination Coalition, developed out of the conference and now includes 16 CBOs and more than 40 individuals from around the country.

One of the most significant things that academics can do in their research partnerships is to maximize university resources in terms of personnel, expertise, grant-writing capacity, and other resources. With the national conference, for example, we leveraged resources by writing and executing an NIEHS conference grant, seeking additional institutional support, and ensuring reduced cost access to university facilities. We continue to seek additional grant funding for our own engaged sociology and for the work of our CBO partners, and we have written multiple grant proposals with CBO and academic partners for funding of projects aligning with their priorities.

CONCLUSION

In this chapter we have provided an overview of science and expertise issues that are relevant for environmental justice scholars and advocates, and have identified theoretical perspectives and methodological tools that may be useful for any sociologist working with EJ communities or engaging with EJ scholarship. Our primary argument is that EJ research should be participatory in terms of deep and meaningful engagement with affected communities, and reflexive regarding researchers' relationships with activists, communities, and social movements.

The two examples described in this chapter center on multisector alliances related to specific environmental health threats from chemical exposure. Given their ubiquitous exposure, neither flame retardants nor PFASs represent typical EJ topics. Neither project worked with just one geographically distinct community affected by a particular environmental threat, and population-level exposure to some (though not all) PFASs may actually be slightly higher for white U.S. residents than for African Americans (Calafat et al., 2007). But EJ considerations are clearly relevant for both classes of chemicals. Flame retardant exposure is higher in low-income populations and communities of color (Zota et al., 2010), and the number of communities whose drinking water has been contaminated by PFASs indicates a lack of distributive justice as well as procedural and recognition injustice when residents' requests for testing and remediation are denied. On the hundreds of military bases contaminated with PFASs, military policy and secrecy make it harder for affected people to mobilize.

Furthermore, our research orientation is greatly informed by EJ principles of deep and meaningful involvement of marginalized populations, the protection of all populations from environmental hazards, an emphasis on prevention and precautionary approaches as the best risk mitigation, and redress of disproportionate exposures (Bullard, 2008). We partner and collaborate with EJ organizations, and also with other environmental health groups that, though not explicitly oriented toward EJ activism, work actively on EJ-related issues. We consistently advocate for the *precautionary principle*, which advocates erring on the side of caution when risks are suspected and shifting the burden of proof from the public to those who benefit from risk activities (Kriebel et al., 2001; Raffensperger & Tickner, 1999).

Our two case studies demonstrate the utility of the four pillars of our theoretical perspective. Both projects built off relationships developed through previous CBPR collaborations and aligned with CBPR principles in terms of allowing social movement and CBO goals to influence and guide the research process. We have continually integrated deliberate and continuous reflexivity into our research process and relationships with all research partners.

Both projects epitomize transdisciplinary social science–environmental health research collaborations. On the flame retardant project, we partnered with toxicologists and exposure scientists at Silent Spring Institute to write about the importance of firefighters in the flame retardant movement (Cordner et al., 2015). We have maintained and strengthened these collaborations in the PFAS project, with active projects addressing PFAS drinking water advisories, water contamination, and immunotoxicity in children. Finally, both projects followed the NPSS orientation toward science, looking critically at how research on these chemicals has been developed, interpreted, and used by various stakeholders. Our attention to the broad field of stakeholders involved in advocacy around both classes of chemicals aligns with the NPSS attention to networks of power and how stakeholders interact across disciplinary boundaries. Our work consistently embraces a political-economic critique of the chemical industry, highlighting how profit motivates firm behavior and the obfuscation of knowledge production and dissemination.

DEEPENING OUR UNDERSTANDING

Find product labels in your immediate environment, in stores, and online. These can be on food products, personal care products, furniture, electronics, or on any number of other products.

1. Read the whole label carefully. What information is being conveyed, including ingredients or product contents, messages about health and safety, and suggested uses and unintended uses? What potentially relevant information isn't included on the label?

2. Think broadly about the product's lifestyle and how it will be used or consumed. What are some environmental, health, and justice concerns associated with this product? Do additional research on websites of nonprofit groups such as Silent Spring Institute and Environmental Working Group to see how they talk about these products and what suggestions they make for personal and social change.

3. Reflect on the industry and regulatory context of the product and do some background research. Which people, companies, organizations, and agencies determine what information must be included on the label? Do additional research to ensure you have identified all responsible parties. For U.S. products, consider federal agencies including the Environmental Protection Agency, the Occupational Safety and Health Administration, the Food and Drug Administration, and the Consumer Products Safety Commission. Also research whether there are any relevant industry standards that apply to your product.
4. Does your product contain any logos or certifications (e.g., Rainforest Alliance, USDA Certified Organic)? Do additional research on these logos or certifications. What purpose do they serve? What information is the logo designed to convey? What information is missing?
5. Develop a plan for avoiding harmful ingredients or products, and for preventing harmful ingredients in productsin the first place. For each responsible party (individuals, companies, organizations, and agencies), develop a plan to create change and improve environmental and/or health factors associated with the product. Consider using Silent Spring Institute's Detox Me smartphone app to learn how to learn about toxics in products and track changes.
6. Pick an area of regulation or a statute (e.g., Clean Water Act or Toxic Substances Control Act), and identify changes that would improve the regulation's ability to protect public and environmental health for everyone. Consider the role that science would have to play in improving the regulation.

References

Allen, B. L., Ferrier, Y., & Cohen, A. K. (2016). Through a maze of studies: Health questions and "undone science" in a French industrial region. *Environmental Sociology, 3*(2), 134–144.

Baron, S., Sinclair, R., Payne-Sturges, D., Phelps, J., Zenick, H., & Collman, G. W. (2009). Partnerships for environmental and occupational justice: Contributions to research, capacity and public health. *American Journal of Public Health, 99*, S517–S525.

Birnbaum, L. S., & Staskal, D. F. (2004). Brominated flame retardants: Cause for concern? *Environmental Health Perspectives, 112*(1), 9–17.

Blum, A., Gold, M. D., Ames, B. N., Kenyon, C., Jones, F. R., Hett, E. A., & Thenot, J. P. (1978). Children absorb tris-BP flame retardant from sleepwear: Urine contains the mutagenic metabolite, 2,3-dibromopropanol. *Science, 201*(4360), 1020–1023.

Brown, P., Brody, J. G., Morello-Frosch, R., Tovar, J., Zota, A. R., & Rudel, R. A. (2012). Measuring the success of community science: The Northern California Household Exposure Study. *Environmental health perspectives, 120*(3), 326–331.

Brown, P., & Cordner, A. (2011). Lessons learned from flame retardant use and regulation could enhance future control of potentially hazardous chemicals. *Health Affairs, 30*(5), 906–914.

Bruton, T. A., & Blum, A. (2017). Proposal for coordinated health research in PFAS-contaminated communities in the United States. *Environmental Health, 16*, 120.

Bullard, R. (2008). *The quest for environmental justice: Human rights and the politics of pollution.* San Francisco: Sierra Club Books.

C8 Science Panel. (2011). *Probable link findings pregnancy related diseases.* http://www.c8sciencepanel.org/pdfs/Probable_Link_C8_PIH_5Dec2011.pdf

Calafat, A. M., Wong, L. W., Kiklenyik, Z., Reidy, J. A., & Needham, L. N. (2007). Polyfluoroalkyl chemicals in the U.S. population: Data from the National Health and Nutrition Examination Survey (NHANES) 2003–2004 and comparisons with NHANES 1999–2000. *Environmental Health Perspectives, 115*(11), 1596–1602.

Centers for Disease Control and Prevention. (2018). *Fourth national report on human exposure to environmental chemicals.* Updated tables, March, Volume 1C. Atlanta, GA: Centers for Disease Control and Prevention.

Cordner, A. (2016). *Toxic safety: Flame retardants, chemical controversies, and environmental health.* New York: Columbia University Press.

Cordner, A., & Brown, P. (2013). Moments of uncertainty: Ethical considerations and emerging contaminants. *Sociological Forum, 28*(3), 469–494.

Cordner, A., & Brown, P. (2015). A multisector alliance approach to environmental social movements: Flame retardants and chemical reform in the United States. *Environmental Sociology, 1*(1), 69–79.

Cordner, A., Ciplet, D., Brown, P., & Morello-Frosch, R. (2012). Reflexive research ethics for environmental health and justice: Academics and movement-building. *Social Movement Studies, 11*(2), 161–176.

Cordner, A., Mulcahy, M., & Brown, P. (2013). Chemical regulation on fire: Rapid policy advances on flame retardants. *Environmental Science & Technology, 47*, 7067–7076.

Cordner, A., Richter, L., & Brown, P. (2016). Can chemical-class based approaches replace chemical-by-chemical strategies? Lessons from recent U.S. FDA regulatory action on perfluorinated compounds. *Environmental Science & Technology, 50*(23), 12584–12591.

Cordner, A., Rodgers, K. M., Brown, P., & Morello-Frosch, R. (2015). Firefighters and flame retardant activism. *New Solutions, 24*(4), 511–534.

Danish Ministry of the Environment. (2015). *Short-chain polyfluoroalkyl substances (PFAS). A literature review of information on human health effects and environmental fate and effect aspects of short-chain PFAS.* https://www2.mst.dk/Udgiv/publications/2015/05/978-87-93352-15-5.pdf

Dodson, R., Perovich, L. J., Covaci, A., Eede, Van Den., N, Ionas., C, A., ... Rudel, R. A. (2012). After the PBDE phase-out: A broad suite of flame retardants in repeat house dust samples from California. *Environmental Science & Technology, 46*(24), 13056–13066.

Egginton, J. (2009). *The poisoning of Michigan.* New York: Norton.

Environmental Working Group. (2017). *Mapping a contamination crisis.* https://www.ewg.org/research/mapping-contamination-crisis

Feagin, J. (2001). Social justice and sociology: Agendas for the twenty-first century. *American Sociological Review, 66*(1), 1–20.

Finn, S., & Collman, G. (2016). The pivotal role of the social sciences in environmental health sciences research. *New Solutions, 26*(3), 389–411.

Frickel, S., & Edwards, M. (2014). Untangling ignorance in environmental risk assessment. In S. Boudia & N. Jas (Eds.), *Powerless science? Science and politics in a toxic world* (pp. 215–233). New York: Berghahn Books.

Frickel, S., Gibbon, S., Howard, J., Kempner, J., Ottinger, G., & Hess, D. (2010). Undone science: Charting social movement and civil society challenges to research agenda setting. *Science, Technology, & Human Values, 35*(4), 444–476.

Frickel, S., & Gross, N. (2005). A general theory of scientific/intellectual movements. *American Sociological Review, 70*(2), 204–232.

Frickel, S., & Moore, K. (2006). *The new political sociology of science.* Madison: University of Wisconsin Press.

Gieryn, T. (1983). Boundary-work and the demarcation of science from non-science: Strains and interests in professional ideologies of scientists. *American Sociological Review*, *48*(6), 781–795.

Gieryn, T. (1999). *Cultural boundaries of science*. Chicago: University of Chicago Press.

Gottlieb, R. (1993). *Forcing the spring: The transformation of the American environmental movement*. Washington, DC: Island Press.

Green Science Policy Institute. (2018). *The six classes approach to reducing chemical harm*. www.sixclasses.org

Hess, D. (2009). The potentials and limitations of civil society research: Getting undone science done. *Sociological Inquiry*, *79*(3), 306–327.

Hoover, E. (2013). Cultural and health implications of fish advisories in a Native American community. *Ecological Process*, *2*, 4.

Hoover, E., Renauld, M., Edelstein, M., & Brown, P. (2015). Social science contributions to transdisciplinary environmental health. *Environmental Health Perspectives*, *123*, 1100–1106.

Hopewell, S., Loudon, K., Clarke, M. J., Oxman, A. D., & Dickersin, K. (2009). Publication bias in clinical trials due to statistical significance or direction of trial results. *Cochrane Database of Systematic Reviews*, *21*(1), MR000006.

Israel, B. A., Schulz, A. J., Parker, E. A., & Becker, A. B. (1998). Review of community-based research: Assessing partnership approaches to improve public health. *Annual Review of Public Health*, *19*, 173–202.

Jasanoff, S. (1987). Contested boundaries in policy-relevant science. *Social Studies of Science*, *17*(2), 195–230.

Jasanoff, S., Markle, G., Peterson, J., & Pinch, T. (1995). *Handbook of science and technology studies*. Thousand Oaks, CA: Sage.

Judge, M. J., Brown, P., Brody, J. G., & Ryan, S. (2016). The exposure experience: Ohio River valley residents respond to local perfluorooctanoic acid (PFOA) contamination. *Journal of Health and Social Behavior*, *57*(3), 333–350.

Kempler, J., Merz, J. F., & Bosk, C. L. (2011). Forbidden knowledge: Public controversy and the production of nonknowledge. *Sociological Forum*, *26*(3), 475–500.

Kinchy, A. J. (2010). Anti-genetic engineering activism and scientized politics in the case of "contaminated" Mexican maize. *Agriculture and Human Values* 27(4), 505–517.

Kinchy, A. J., & Kleinman, D. L. (2003). Organizing credibility: Discursive and organizational orthodoxy on the borders of ecology and politics. *Social Studies of Science*, *33*(6), 869–896.

Kleinman, D., & Suryanarayanan, S. (2012). Dying bees and the social production of ignorance. *Science, Technology & Human Values*, *38*(4), 492–517.

Kriebel, D., Tickner, J., Epstein, P., Lemons, J., Leveins, R., Loechler, E., & Stoto, M. (2001). The precautionary principle in environmental science. *Environmental Health Perspectives*, *109*(9), 871–876.

Kroll-Smith, S. J., & Couch, S. R. (1990). *The real disaster is above ground: A mine fire and social conflict*. Lexington: University of Kentucky Press.

Lau, C. (2015). Perfluorinated compounds: An overview. In J. C. DeWitt (Ed.), *Toxicological effects of perfluoroalkyl and polyfluoroalkyl substances*. Switzerland: Springer International Publishing.

Lyons, C. (2007). *Stain-resistant, nonstick, waterproof, and lethal: The hidden dangers of C8*. Westport, CT: Praeger.

Matz, J., Brown, P., & Brody, J. (2016). Social science-environmental health collaborations: An exciting new direction. *New Solutions*, *26*, 349–358.

Meironyte, D., Noren, K., & Bergman, A. (1999). Analysis of polybrominated diphenyl ethers in Swedish human milk. A time-related trend study, 1972–1997. *Journal of Toxicology and Environmental Health*, *58*(6), 329–341.

Michaels, D., & Monforton, C. (2005). Manufacturing uncertainty: Contested science and the protection of the public's health and environment. *American Journal of Public Health*, *95*(S1), S39–48.

Moore, K. (2008). *Disrupting science: Social movements, American scientists, and the politics of the military, 1945–1975.* Princeton, NJ: Princeton University Press.

Moore, K., Kleinman, D., Hess, D., & Frickel, S. (2011). Science and neoliberal globalization: A political sociological approach. *Theory & Society.* 40(5), 505–532.

Morello-Frosch, R., Zavestoski, S., Brown, P., Altman, R. G., McCormick, S., & Mayer, B. (2006). Embodied health movements: Responses to a 'scientized' world. In S. Frickel & K. Moore (Eds.), *The new political sociology of science: Institutions, networks, and power* (pp. 244–271). Madison: University of Wisconsin Press.

National Institute of Environmental Health Sciences. (2012). *2012–2017 strategic plan: Advancing science, improving health: A plan for environmental health research.* U.S. Department of Health and Human Services. Publication No. 12-7935.

National Research Council, Committee on Improving Risk Analysis Approaches Used by the U.S. (2009). *Science and decisions: Advancing risk assessment.* Washington, DC: National Academies Press.

O'Fallon, L., & Dearry, A. (2002). Community-based participatory research as a tool to advance environmental health sciences. *Environmental Health Perspectives, 110*(S2), 155–159.

Organization for Economic Co-Operation and Development. (2018). Toward a new comprehensive global database of per- and polyfluoroalkyl substances (PFASs). Environmental Directorate. http://www.oecd.org/officialdocuments/publicdisplaydocumentpdf/?cote=ENV-JM-MONO(2018)7&doclanguage=en

Post, G. B., Cohn, P. D., & Cooper, K. R. (2012). Perfluorooctanoic acid (PFOA), an emerging drinking water contaminant: A critical review of recent literature. *Environmental Research, 116,* 93–117.

Raffensperger, C., & Tickner, J. (1999). *Protecting public health & the environment: Implementing the precautionary principle.* Washington, DC: Island Press.

Reynolds, G. M., & Shendruk, A. (2018). *Demographics of the U.S. military.* Retrieved from https://www.cfr.org/article/demographics-us-military

Richter, L., Cordner, A., & Brown, P. (2018). Nonstick science: Sixty years of research and (in)action on fluorinated compounds. *Social Studies of Science, 48*(5), 691–714.

Rosenmai, A. K., Taxvig, C., Svingen, R., Trier, X. A., Vugt-Lussenburg, B. M., Pedersen, M., & Vinggaard, A. M. (2016). Fluorinated alkyl substances and technical mixtures used in food paper-packaging exhibit endocrine-related activity in vitro. *Andrology, 4*(4), 662–672.

Rudel, R. A., Camann, D. E., Spengler, J. D., Korn, L. R., & Brody, J. G. (2003). Phthalates, alkylphenols, pesticides, polybrominated diphenyl ethers, and other endocrine-disrupting compounds in indoor air and dust. *Environmental Science & Technology, 37*(20), 4543–4553.

Senier, L., Hudson, B., Fort, S., Hoover, E., Tillson, R., & Brown, P. (2008). Brown Superfund basic research program: A multistakeholder partnership addresses real-world problems in contaminated communities. *Environmental Science & Technology, 42,* 4655–4662.

Social Science Environmental Health Research Institute. (2017). *PFAS conference presentations.* https://pfasproject.com/pfas-conference-presentations/

Stockholm Convention. (2014). *The new POPs under the Stockholm Convention.* http://chm.pops.int/Convention/ConferenceoftheParties(COP)/Decisions/tabid/208/Default.aspx

Sullivan, M. (2018). *Addressing perfluorooctane sulfonate (PFOS) and perfluorooctanoic acid (PFOA).* https://partner-mco-archive.s3.amazonaws.com/client_files/1524589484.pdf

Sun, M., Arevalo, E., Strynar, M., Lindstrom, A., Richardson, M., Kearns, B., ... U, D. R. (2016). Legacy and emerging perfluoroalkyl substances are important drinking water contaminants in the Cape Fear River watershed of North Carolina. *Environmental Science & Technology Letters, 3*(12), 415–419.

U.S. Environmental Protection Agency. (2006). *Polybrominated diphenyl ethers (PBDEs) project plan*. Washington, DC: U.S. EPA.

U.S. Environmental Protection Agency. (2017a). Polybrominated diphenyl ethers (PBDEs). https://www.epa.gov/assessing-and-managing-chemicals-under-tsca/polybrominated-diphenyl-ethers-pbdes

U.S. Environmental Protection Agency. (2017b). Fourth unregulated contaminant monitoring rule. https://www.epa.gov/dwucmr/fourth-unregulated-contaminant-monitoring-rule

Wallerstein, N. B. D., Oetzel, J., & Minkler, M. (Eds.). (2017). *Community-based participatory research for health: Advancing social and health equity* (3rd ed.). New York: Wiley.

Whyte, W. F. (Ed.). (1991). *Participatory action research*. Thousand Oaks, CA: Sage.

Wikoff, D. S., & Birnbaum, L. (2011). Human health effects of brominated flame retardants. In E. Eljarrat & D. Barceló (Eds.), *Brominated flame retardants. The handbook of environmental chemistry*. Berlin: Springer.

Woodhouse, E. (2006). Nanoscience, green chemistry, and the privilege position of science. In S. Frickel & K. Moore (Eds.), *The new political sociology of science* (pp. 148–181). Madison: University of Wisconsin Press.

Zota, A., Adamkiewicz, G., & Morello-Frosch, R. (2010). Are PBDEs an environmental equity concern? Exposure disparities by socioeconomic status. *Environmental Science & Technology, 44*(15), 5691–5692.

Zota, A., Linderholm, L., Park, J. S., Petreas, M., Guo, T., Privalsky, M. L., & Woodruff, T. J. (2013). Temporal comparison of PBDEs, OH-PBDEs, PCBs, and OH-PCBs in the serum of second trimester pregnant women recruited from San Francisco General Hospital, California. *Environmental Science & Technology, 47*(20), 11776–11784.

Zota, A., Rudel, R. A., Morello-Frosch, R., & Brody, J. (2008). Elevated house dust and serum concentrations of PBDEs in California: Unintended consequences of furniture flammability standards? *Environmental Science & Technology, 42*(21), 8158–8164.

6

HOW COMMUNITY-BASED PARTICIPATORY RESEARCH STRENGTHENS THE RIGOR, RELEVANCE, AND REACH OF SCIENCE

Rachel Morello-Frosch
Carolina L. Balazs

Community-based participatory research (CBPR) is one of multiple names used to describe an array of research methods in the health and social sciences that seek to transform the scientific enterprise by engaging communities in the research process (Brown et al., 2011; Cornwall & Jewkes, 1995; Israel, Schulz, Parker, & Becker, 1998; Minkler & Wallerstein, 2003; Morello-Frosch et al., 2011a). Specifically, CBPR entails academic-community collaboratives in which power is shared among partners in all aspects of the research process—the doing, interpreting, and acting on science. This process elevates community knowledge, challenges traditional power dynamics in the research process, and can directly benefit the communities involved. In particular, scientists and community members who have engaged in CBPR have sought to democratize knowledge production in ways that transform research from a top-down, expert-driven process into one of co-learning and co-production. This has entailed infusing local, community-based knowledge with tools and techniques from disciplinary science, often constructively improvising and shifting the research process to better address community-identified concerns (Corburn, 2005). In the process, CBPR facilitates the translation (that is, application and interpretation) of research findings to community stakeholders and policymakers.

Scholars have taken different approaches to CBPR in terms of the level of community engagement in the research process. This can be thought of as a continuum. On one side, traditional scientific endeavors may collect community information or data but treat community members as passive study participants who have no influence on the research design. As we move along the continuum, community engagement increases, from the community's context influencing the study's questions, to community members participating in the study design, fundraising, data ownership and dissemination of results, and using those results to promote social change. Thus, community members move from being mere study participants to being active research partners. Even on this end of the continuum, however, the extent of community participation varies. (See Figure 6.1.)

The benefits that CBPR generates for community partners have been well documented and include enhanced community empowerment, co-learning between community members and scientists, informing community organizing efforts, and linking research to policy action (Brown et al., 2011; Israel et al., 1998; Minkler, Vasquez, Tajik, & Peterson, 2006; Minkler & Wallerstein, 2003). Less, however, has been written on how CBPR potentially (re)shapes the scientific enterprise itself. This issue has become more salient as federal and private grants supporting CBPR have increased dramatically since 1996, when the National Institute of Environmental Health Science (NIEHS) started funding such research (Wolfson & Parries, 2010).

We argue that communities engaged in environmental health CBPR have helped improve the rigor, relevance, and reach of science, or what we call the "three Rs" (Morello-Frosch et al., 2011a). Rigor refers to the practice and promotion of good science—in the study design, data collection, and

---

PHOTO 1.1: Oil refinery in Richmond, CA.
Photo by Kelly Grindstaff, 2019.
Source: Carolina L. Balazs and Rachel Morello-Frosch. "The Three Rs: How Community-Based Participatory Research Strengthens the Rigor, Relevance, and Reach of Science." *Environmental Justice*, Vol. 6, No. 1, 19 Feb 2013. https://doi.org/10.1089/env.2012.0017

FIGURE 6.1 ■ Schematic of community based participatory research as a continuum of efforts, with varying degrees of community engagement. Levels of engagement increase as community members are transformed from study participants to research partners.

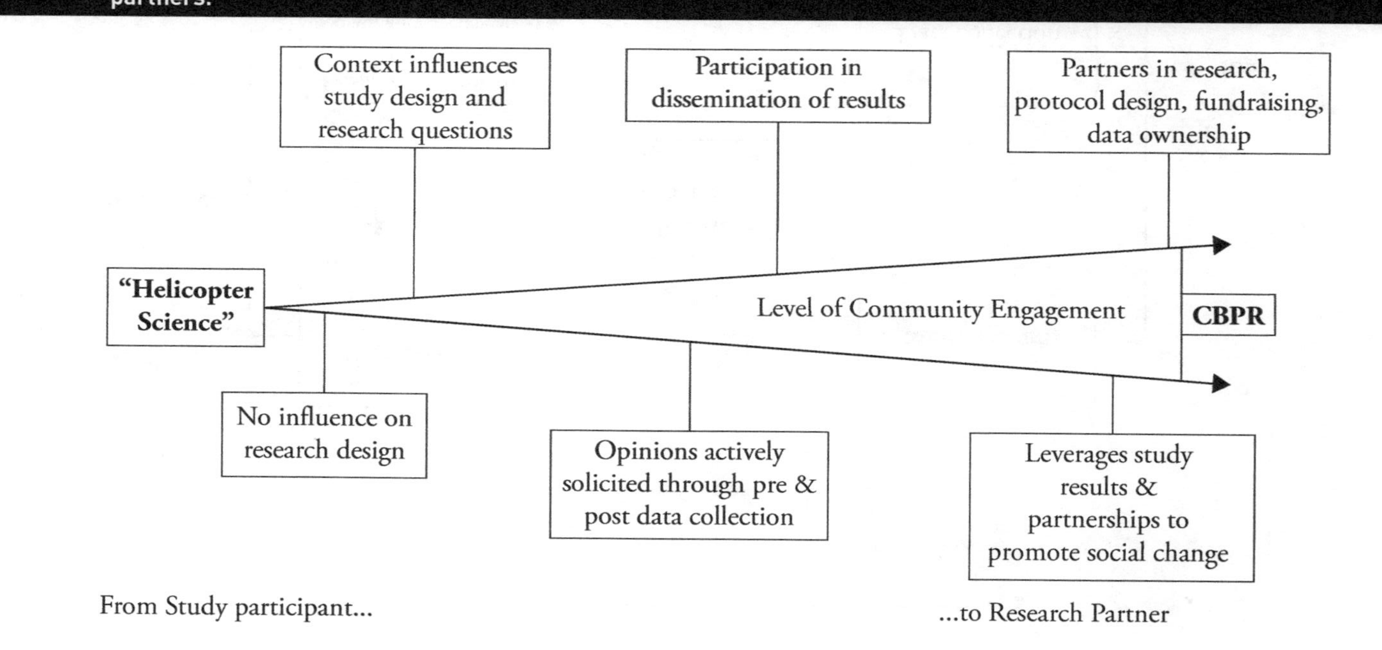

FIGURE 6.2 ■ The 3 Rs (rigor, relevance, and reach) in relation to generalized steps of a community based participatory research approach, where traditional researchers and community members are jointly involved at each step, though participation levels vary.

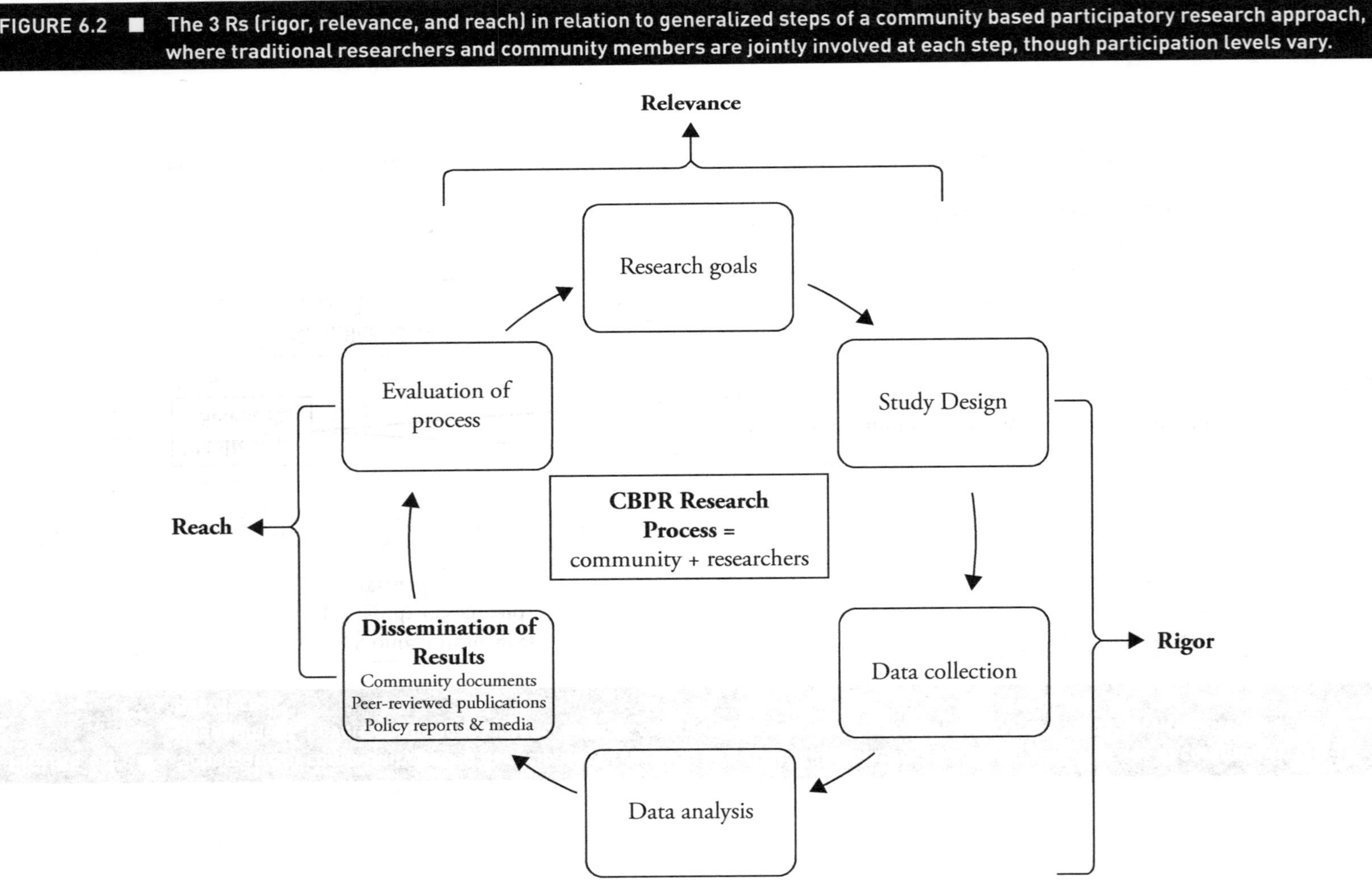

interpretation phases of research. Relevance refers to whether science is asking the right questions. For environmental health—often a significant factor in understanding environmental injustices—relevant research emphasizes appropriate causes of exposure and elucidates opportunities for action or change. Reach encapsulates the degree to which knowledge is disseminated to diverse audiences and translated into useful tools for the scientific, regulatory, policy, and lay arenas. Thus CBPR shapes the relevance, rigor, and reach of the scientific enterprise. (See Figure 6.2.)

CBPR and community engagement in environmental health science has promoted changes in theories of disease causation and new lines of scientific inquiry and helped shape scientific fact-making (Brown et al., 2011; Morello-Frosch et al., 2006; Morello-Frosch et al., 2011a; Morello-Frosch, Zuk, Jerrett, Shamasunder, & Kyle, 2011b). This is exemplified in the cumulative impacts arena. Here, environmental justice advocates have long asserted that chemical-by-chemical and source-specific assessments of the health risks of environmental hazards are scientifically problematic because they do not reflect the cumulative impacts of multiple environmental and social stressors faced by vulnerable communities, which may act additively or synergistically to harm health (CalEPA, 2004; Morello-Frosch et al., 2011b; Sadd, Pastor, Morello-Frosch, Scoggins, & Jesdale, 2011). CBPR has helped advance the science of "cumulative impacts" by elevating the role of structural determinants and their associated social stressors in creating vulnerabilities among certain populations—sometimes called environmental justice communities—to the adverse health effects of environmental hazards (Gee & Payne-Sturges, 2004; Wilson, 2009). Ultimately, this focus on cumulative impacts or the "double jeopardy" of environmental and social stressors is transforming how scientists study environmental health problems (Clougherty & Kubzansky, 2009; Morello-Frosch & Lopez, 2006; National Research Council, 2009). At the same time, advocates have demanded that emerging scientific evidence on cumulative impacts be translated into valid and transparent tools for decision making in environmental regulation and policy even as the science evolves (Morello-Frosch et al., 2011b; Sadd et al., 2011).

This chapter uses two successful cases of CBPR environmental health research to explore CBPR's role in strengthening the three Rs of the scientific process. We examine how community input in CBPR embodied a strategic focus that led to improved science and more direct assessments of policy- and regulatory-relevant questions, in the service of seeking environmental justice.

DESCRIPTION AND RATIONALE OF CASE STUDIES

Our two cases consist of the Northern California Household Exposure Study (HES) and the San Joaquin Valley Drinking Water Study (DWS). The HES involved a research collaboration between an independent research institute (Silent Spring Institute), a regional environmental justice advocacy organization (Communities for a Better Environment, CBE), and two academic institutions (Brown University and the University of California, Berkeley) to characterize indoor and outdoor levels of chemicals in a community bordering a major oil refinery in Richmond, California, and in a rural community in Bolinas, California.[1] The San Joaquin Valley DWS consisted of a collaboration between the Community Water Center (CWC), a community-based organization (CBO) based in the Valley, and researchers from the University of California, Berkeley. Motivated by CWC's desire to understand which residents are most vulnerable to drinking water contamination in the Valley, the DWS examined the association between community-level demographics and water quality in communities across the Valley.

The two cases were selected on the basis of their similarities and differences. Both projects are located in California, which has been an epicenter of environmental justice and environmental health advocacy. Both cases involved CBOs that recognize the links between research, organizing, and advocacy as strategies for improving community environmental health. As such, the CBOs involved motivated the research, and the projects can be placed toward the far right of the CBPR community engagement continuum, though each project varied in its form of community engagement. In both cases, the participatory approaches employed drew from the notion of collaboratively doing, interpreting, and acting on science, a knowledge-production structure that is not linear, but rather cyclical, in that the collective process of acting on science leads to the further doing of science (Brown et al., 2006).

The two cases provide an opportunity for comparison as well. Richmond is located in Northern California's urban San Francisco Bay Area region, whereas the DWS was conducted in a predominantly rural and agricultural region of the state. The HES collected primary data, whereas the DWS relied on secondary data. Finally, although both projects were exposure studies, each looked at a different environmental medium (that is, air versus water).

CASE STUDY 1: NORTHERN CALIFORNIA HOUSEHOLD EXPOSURE STUDY

The Household Exposure Study (HES) entailed a household exposure assessment of air and dust for pollutants from industrial emissions, transportation sources, and consumer products. Recruitment and sampling were conducted in an urban community bordering the Chevron oil refinery in Richmond and a rural community in Bolinas that served as a regional comparison area. The Chevron oil refinery is one of the nation's largest, covering 2,900 acres, employing approximately 1,000 workers, and processing more than 240,000 barrels of crude oil daily into gasoline, jet fuel diesel, and lubricants (http://www.chevron.com). HES partners collected air and dust samples from 50 homes (40 in Richmond and 10 in Bolinas) and from nearby outdoor areas and tested these samples for over 150 analytes, including endocrine-disrupting compounds, as well as particulates, metals, polycyclic aromatic hydrocarbons (PAHs), ammonia, sulfates, and other pollutants originating from nearby industries and which are commonly emitted from refinery activities (Brody et al., 2009).

In addition to the scientific goal of characterizing cumulative pollutant exposures in an environmental justice community and understanding their potential sources, the HES aimed to inform local regulatory decisions regarding oil refinery operations, state chemicals policies, and national decisions about endocrine-disrupting compounds in consumer products (Brown et al., 2011).

All aspects of the HES were designed and implemented collaboratively—from the development of specific study hypotheses to the design of protocols for reporting study results back to community members and other stakeholders. Communities for a Better Environment has a demonstrated history of doing its own scientific work and leveraging the data it collects to push for policy and regulatory change. For example, the organization is well known for pioneering the "Bucket Brigades" for low-cost air sampling, now used widely in California, nationally, and internationally by fence-line communities living near large industrial facilities with hazardous emissions (Lerner, 2005). Additionally, in the San Francisco Bay Area and Los Angeles, CBE has tracked and analyzed flaring activity and emissions from large oil refineries. This scientific work led to the promulgation of a groundbreaking flare-control rule that became a front-page story in the *New York Times* (Marshall, 2005). Thus, for CBE, the HES was as much about producing good science as it was about leveraging the scientific enterprise to conduct community outreach and inform policy.

CBE organizing staff were trained by Silent Spring Institute and university scientists to conduct the indoor and outdoor air monitoring, dust collection, and interviews, thereby enhancing the organization's in-house scientific capacity and ensuring its co-ownership of the research process. Most importantly, CBE's partnership helped the organization demystify science for its constituents by enabling staff to move their data-gathering efforts into the realm of people's homes. For example, as CBE interviewers went through the preliminary exposure questionnaire and set up sampling equipment, the experience encouraged community members to think in new ways about indoor air quality and how contaminants from outdoor pollution sources can penetrate inside the home. These discussions enabled CBE to connect its organizing work with technical and scientific aspects, both of which are central to advancing environmental justice.

CBE helped advance the scientific innovation and rigor of the HES in various ways. The scientific rigor of the study was ensured through collective discussion and negotiation of study design issues such as choosing relevant sampling sites, selecting methods for recruiting study participants, establishing the list of chemicals for analysis, and developing sound protocols for reporting individual study results to study participants and to broader audiences. For example, CBE along with the project's advisory council encouraged the study team to collect a subset of air and dust samples from a community that did not have significant outdoor industrial and transportation emission sources so that these results could be compared to what was found in Richmond. This led to the decision to sample homes in Bolinas. Similarly, CBE encouraged the study team to expand its panel of analytes to include pollutants with sources that are primarily due to oil-combustion activities. This led to the inclusion of target compounds such as vanadium, nickel, and sulfates. As a result of CBE's input, the HES was able to demonstrate indoor penetration of chemicals from heavy oil-combustion activities in Richmond. The study also showed that levels of multiple pollutants tended to be higher inside homes than outdoors (Brody et al., 2009). In particular, the HES found some of the world's highest home dust levels of brominated flame retardants in Richmond and Bolinas (Zota,

Rudel, Morello-Frosch, & Brody, 2008). Flame retardants are commonly used in furniture foam and electronic components and are used in particularly high amounts in California, due to the state's unique and strict flammability standard for furniture foam (State of California, 2000).

The relevance of the HES was bolstered by CBE's critical input in the development of innovative, transparent, and scientifically valid communication materials to report back individual sampling results to all participants who wanted them (Brody et al., 2007; Morello-Frosch et al., 2011a). As environmental justice CBPR projects study the sources and pathways of chemical exposures, they are also faced with the paucity of health effects data for many of the pollutants studied. This situation raises ethical and scientific challenges for whether and how to report results to study participants (Morello-Frosch, Brody, Brown, Altman, & Rudel, 2009). In the context of CBPR, this means ensuring that exposure data are reported in ways that are meaningful and that elucidate potential paths for individual or collective action to protect health. In general, participants tend to want their exposure results, which they often use as a tool for public health advocacy (Morello-Frosch et al., 2009). As a result of CBE's input on this issue, the project created bilingual materials (Spanish/English) including graphic displays for communicating aggregate and individual-level results, scientific uncertainties, and potential strategies for exposure reduction. Ultimately, the project found that the strategy to communicate results contributed to environmental health education and stimulated behavioral change and collective efforts to reduce exposures (Adams et al., 2011).

Finally, CBE's engagement in the HES extended the reach of the HES to broad audiences in order to leverage results to improve regulation and land-use decision making. With the support of scientific partners, CBE, along with some study participants, used data from the HES in their testimony before the Richmond Planning Commission to protest a conditional-use permit application by the nearby Chevron refinery that would have expanded the facility's capacity to refine lower-grade crude oil and significantly increased pollutant emissions. The presentation of the HES results received significant media attention, as well as inquiries from the California Attorney General's Office, both of which compelled the Richmond Planning Commission to allow for more public input on the environmental impact statement of the proposed refinery expansion. Ultimately, the City of Richmond approved the permit and the struggle went into litigation. But a state appeals court decision on the case upheld a lower court's ruling that the environmental impact review for Chevron's conditional-use permit to expand its operations violated state environmental laws for being inadequate and vague about the scope and community health impact of the proposed project.

Ultimately, although CBE's initial focus was on community exposures to pollutants from the Richmond oil refinery, it became clear that demonstrating cumulative impacts of multiple pollutant exposures was also relevant to the organization's mission. CBE, with help from Silent Spring and university partners, received additional funding from the Avon Foundation to conduct a health survey in Richmond with a larger sample (198 respondents provided health data on 722 individuals) than the HES. This project yielded additional data to show disproportionate health challenges in Richmond. These data were disseminated at multiple meetings for community residents, as well as public health and environmental agencies (Cohen, Lopez, Malloy, & Morello-Frosch, 2012). In addition to those presentations and a peer-reviewed publication, a lay report was released from CBE's website (Lopez, Cohen, Zota, & Morello-Frosch, 2009).

CASE STUDY 2: SAN JOAQUIN VALLEY DRINKING WATER STUDY

The San Joaquin Valley Drinking Water Study (DWS) was a CBPR collaboration between two partners: the Community Water Center and the University of California, Berkeley. CWC is one of the only environmental justice organizations based in the San Joaquin Valley that focuses exclusively on addressing drinking water. Comprised of a team of lawyers, policy analysts, and community organizers, CWC works primarily on drinking water advocacy at the local, regional, and state levels. Over the years, CWC has partnered with various research institutions to conduct research projects that address the problems and costs of drinking water contamination.

The DWS sought to answer two main questions: (a) Do community water systems[2] that serve greater percentages of low-income or minority communities have higher levels of nitrate and arsenic in their drinking water, and (b) do these systems also face greater difficulties complying with federal drinking water standards? These questions were motivated by a growing concern regarding drinking water contamination in the Valley and its impact on residents. With its intensive irrigated agriculture, the Valley has two of the most contaminated aquifers in the nation and some of the highest nitrate levels in the country (Dubrovsky et al., 2010). Because nearly 95 percent of the Valley's population relies on groundwater for drinking (California Department of Public Health, 2008), exposure to contaminated groundwater is a particular health risk. This risk is compounded by the fact that, with high costs of mitigation, few systems actually treat for drinking water contaminants.

In 2005, one of the authors (Balazs) began partnering with CWC to study patterns of drinking water contamination and test whether there were social disparities in exposure across the region. In essence, CWC wanted a scientifically rigorous study to assess whether contamination impacts were inequitable and widespread, or limited to just a few communities, or one county alone as policymakers often noted.

The CBPR components of the DWS included vetting and developing the study questions and design with CWC. While community members did not take part in data collection or statistical analysis, at each step of the process, CWC staff gave feedback on study design, definition of key variables, study barriers, and preliminary findings. To answer the study questions, the research team used existing water quality datasets maintained by the California Department of Public Health (CDPH) to estimate exposure and compliance. This represented an alternative way of estimating exposure with existing data, without needing to collect water quality samples at the tap (an effort that would have been beyond the study's budget, given the study area's large geographic scope). In addition, the UC Berkeley and CWC team jointly sought funding from a California foundation to delineate water system boundaries in a geographic information system (GIS) so as to estimate customer demographics of water systems, a critical, yet missing set of data. Using both datasets, the researchers then developed a series of approaches to estimate drinking water quality served by different water systems, and used statistical modeling techniques to examine the association between water quality and customer demographics across the Valley.

The DWS found that communities with higher percentages of Latinos had higher nitrate levels in their drinking water systems, and that those with lower rates of homeownership had higher arsenic levels and greater chances of exceeding federal safety standards (Balazs, Morello-Frosch, Hubbard, & Ray, 2011; Balazs, Morello-Frosch, Hubbard, & Ray, 2012). In sum, the researchers found that water quality was worse in smaller, disadvantaged communities. This was a significant finding as it highlighted a dual burden—not only that small systems face unequal exposure and compliance burdens, but also that the people served by these water systems are socially and economically vulnerable and may be the least able to afford mitigation to reduce exposures.

The CBPR approach enhanced the rigor of the study in several ways. First, it enhanced the study design. Originally, the study was going to examine demographic disparities in safe drinking water access in Tulare County, one of the counties with the highest nitrate levels in the state. But, due to data limitations, the sample size was not large enough to implement a robust statistical analysis. At first, the researchers were uncertain about how best to address this methodological challenge, but community partners viewed this challenge as an opportunity to expand the study to the entire Valley. Not only would this ensure an adequate sample size and wider variability in drinking water quality for assessing potential environmental inequalities, but the implications of this broader scope also would likely be more informative for policymakers and the water regulatory community. By advocating for this enhanced study design, the community partners were, in essence, encouraging researchers to look beyond the known drinking water "hotspots" and analyze drinking water quality issues more broadly for the entire region. This approach, they argued, could elucidate more "upstream" approaches to addressing contamination and remediation issues for a broader and more diverse population of Valley residents.

Second, the CBPR partnership enhanced the rigor of the study by spurring the labor-intensive process of estimating community-level demographics in community water systems. Although the CDPH maintains water quality data, no Valley-wide demographic information on the water system customer base had ever been estimated. By encouraging researchers to secure the resources necessary to fund this extra analytical work, CWC facilitated the development of new analytical methods and data, which have since been shared with and built on by researchers at other universities and research institutes. For example, researchers at the University of California, Davis, integrated some of the results of the water analysis in their most recent report documenting the cumulative impacts of environmental hazards in the San Joaquin Valley (London et al., 2011).

Community involvement ensured the study questions were relevant to pressing policy issues in the region. Valley communities and advocates had lived experiences of high drinking water contaminant levels in unincorporated, highly Latino, farm-working communities. Residents and CWC staff intuited an environmental injustice, hypothesizing that a disproportionate share of this drinking water contamination burden was falling on Latinos and lower-income communities. But their early efforts to convince policymakers and regulatory agencies about the need to address this systemic environmental equity problem were met with skepticism and assertions that these issues were isolated incidents and limited to a small number of places. The CBPR partnership was able to break through this impasse by providing the sophisticated analytical work that demonstrated a regional pattern of systemic environmental inequities in drinking water quality.

With solid scientific results in hand, CWC ensured that the research had a wide reach in two main ways. First, CWC leveraged its connections with decision makers to ensure that the research was presented at key venues. For example, one author (Balazs) was asked to present study results to policy officials, including the United Nations Special Rapporteur for the Human Right to Water and Sanitation and policymakers in Sacramento, and at community-oriented academic conferences. These venues went beyond the traditional academic conferences at which the research team would have otherwise presented (for example, the American Public Health Association, the International Groundwater Conference). In extending the policy reach of the science, findings from the DWS entered policy debates on environmental justice and drinking water via more streamlined paths. In August 2011, for example, the U.N. Rapporteur cited preliminary findings of the drinking water study to the U.S. government and the United Nations (United Nations General Assembly, 2011), well before peer-reviewed findings had been published.

A second, unexpected impact on the reach of the study was that CWC also ensured a broader reach of the research within the research community. Because CWC is a center of expertise and community knowledge on drinking water, throughout the study the center encouraged different research institutes and universities to approach the UC Berkeley team with research questions and collaboration opportunities. This led to formal and informal collaborations on research, data sharing, and methods discussions. It is our belief that, without this facilitating role of our community partner, these efforts may not have developed or would have developed on a much slower basis, only after a peer-reviewed publication had been released. In this sense, CWC served as a catalyst for developing additional research questions and collaborations, helping to break down some of the barriers that exist between research institutes.

CONCLUSIONS: THE IMPACT OF CBPR

This chapter traced the impact of CBPR on improving the 3 Rs—the rigor, relevance, and reach—of research and the scientific enterprise. Academic-community collaboratives are complex endeavors that require significant investment in building relationships to ensure that the goals, objectives, and needs of each partner are clearly addressed. In particular, the willingness of community-based groups to invest significant resources in the scientific enterprise depends on whether this work will advance their short- and long-term interests without straying from their primary organizational missions. The two environmental health cases presented in this chapter highlight the strategic, relevant, and rigorous science that can result from CBPR partnerships. In both cases, it was through the process of democratizing knowledge production that the science was strengthened, its application made relevant, and the reach of its results increased. What's more, the scientific questions and strategic needs of both CBOs were reflected in the design of study protocols, the scope of the analytical work, and the study links to policy and regulation.

Ultimately, both efforts have highlighted new paths for intervention and possibilities for both individual and collective action to reduce exposures to pollutants that are harmful to health. In the HES, CBE's role in the collaborative helped ensure that sampling protocols included chemical analytes that documented for the first time the effects of oil-combustion activities on the indoor air quality of the households of fence-line residents. It also highlighted the cumulative impacts of chemicals from consumer products in homes, including various endocrine-disrupting compounds. The strategy for communicating results to study participants and other diverse audiences helped to elucidate individual and collective strategies for reducing exposures. What began as a focus on disparities in exposure in the DWS has evolved into developing new research directions that address the composite drinking water burden that Valley residents face, including coping costs, compliance burdens, and regulatory failures. Ultimately, these new lines of research will promote multilevel points of intervention to improve drinking water quality at the regional, community, and household levels. In this way, CBPR has helped elucidate innovative lines of scientific inquiry, by linking future research directions with policy interventions. CBPR encourages scientists to specify the implications of their results for regulatory decision making and to communicate results in ways that promote action. It pushes through the gridlock of regulatory paralysis through (over)analysis to elucidate strategies for exposure reduction and precautionary approaches for better protecting community environmental health.

In order to continue exploring the beneficial impacts of CBPR on the scientific enterprise, future research will need to more systematically and precisely document and evaluate the ways in which CBPR improves the rigor, relevance, and reach of science. In addition, this research will need to address how varying degrees of community involvement impact these outcomes.

DEEPENING OUR UNDERSTANDING

The Northern California Household Exposure Study (HES) and the San Joaquin Valley Drinking Water Study (DWS) provide two examples of community-based participatory research.

1. Compare the ways in which the communities participated in both cases.
2. Find another study (preferably local) involving environmental health of a poor and/or minority community. Try to determine the ways in which the community was involved.
3. What recommendations could you make to the researchers to make that study more participatory? Could they have tried to involve the community in similar ways to the HES or DWS? Why or why not?

Acknowledgments

We thank Laurel Firestone, Susana de Anda, and Maria Herrera at CWC. We thank Dr. Christopher Bacon and Tony LoPresti for feedback. We thank Dr. Isha Ray and Dr. Alan Hubbard for collaboration on the DWS.

The HES research was supported by the National Institute of Environmental Health Sciences (R25 ES013258) and the National Science Foundation (SES 0450837).

The DWS research was supported by the NSF Graduate Research Fellowship, the California Endowment (through a collaborative grant between Community Water Center and UC Berkeley), the California Environmental Protection Agency (#07-020), and the Switzer Environmental Fellowship.

References

Adams, C., Brown, P., Morello-Frosch, R., Brody, J., Rudel, R., Zota, A., ... Patton, S. (2011). Disentangling the exposure experience: The roles of community context and report-back of environmental exposure data. *Journal of Health and Social Behavior*, *52*(2), 180–196. doi:10.1177/0022146510395593

Balazs, C., Morello-Frosch, R., Hubbard, A., & Ray, I. (2011). Social disparities in nitrate contaminated drinking water in the San Joaquin Valley. *Environmental Health Perspectives*, *119*(9), 1272–1278. doi:10.1289/ehp.1002878

Balazs, C., Morello-Frosch, R., Hubbard, A., & Ray, I. (2012). Environmental justice implications of arsenic contamination in California's San Joaquin Valley: A cross-sectional, cluster-design examining exposure and compliance in community drinking water systems. *Environmental Health*, *11*, 84. doi:10.1186/1476-069x-11-84

Brody, J., Morello-Frosch, R., Brown, P., Rudel, R., Altman, R., Frye, M. ... Seryak, L. (2007). Improving disclosure and consent: Is it safe? New ethics for reporting personal exposures to environmental chemicals. *American Journal of Public Health*, *97*(9), 1547–1554. doi:10.2105/ajph.2006.094813

Brody, J., Morello-Frosch, R., Zota, A., Brown, P., Pérez, C., & Rudel, R. (2009). Linking exposure assessment science with policy objectives for environmental justice and breast cancer advocacy: The Northern California Household Exposure Study. *American Journal of Public Health*, *99*, S600–S609. doi:10.2105/ajph.2008.149088

Brown, P., Brody, J., Morello-Frosch, R., Tovar, J., Zota, A., & Rudel, R. (2011). Measuring the success of community science: The Northern California Household Exposure Study. *Environmental Health Perspectives*, *120*(3), 326–331. doi:10.1289/ehp.1103734

Brown, P., McCormick, S., Mayer, B., Zavestoski, S., Morello-Frosch, R., Altman, R., ... Senier, L. (2006). A lab of our own: Environmental causation of breast cancer and challenges to the dominant epidemiological paradigm. *Science, Technology and Human Values*, *31*, 499–536. doi:10.1177/0162243906289610

CalEPA. (2004). *Environmental justice action plan*. Sacramento, CA: CalEPA.

California Department of Public Health. (2008). PICME: Permits inspections compliance monitoring and enforcement database. Sacramento: California Department of Public Health.

Clougherty, J., & Kubzansky, L. (2009). A framework for examining social stress and susceptibility in air pollution and respiratory health. *Environmental Health Perspectives*, *117*(9), 1351–1358. doi:10.1289/ehp.0900612

Cohen, A., Lopez, A, Malloy, N., & Morello-Frosch, R. (2012). Our environment, our health: A community-based participatory environmental health survey in Richmond, CA. *Health Education and Behavior, 39*(2), 198–209. doi:10.1177/1090198111412591

Corburn, J. (2005). *Street science: Community knowledge and environmental health justice*. Cambridge, MA: MIT Press.

Cornwall, A., & Jewkes, R. (1995). The use of qualitative methods: What is participatory research? *Social Science and Medicine, 41*(12), 1667–1676. d oi:10.1016/0277-9536(95)00127-s

Dubrovsky, N., Burow, K., Clark, G., Gronberg, J., Hamilton, P., & Hitt, K. (2010). *The quality of our nation's waters: Nutrients in the nation's streams and groundwater, 1992–2004*. Washington, DC: U.S. Geological Survey.

Gee, G., & Payne-Sturges, D. (2004). Environmental health disparities: A framework integrating psychosocial and environmental concepts. *Environmental Health Perspectives, 112*(17), 1645–1653. doi:10.1289/ehp.7074

Israel, B., Schulz, A., Parker, E., & Becker, A. (1998). Review of community-based research: Assessing partnership approaches to improve public health. *Annual Review of Public Health, 19*, 173–202. doi:10.1146/annurev.publhealth.19.1.173

Lerner, S. (2005). *Diamond: A struggle for environmental justice in Louisiana's chemical corridor*. Cambridge, MA: MIT Press.

London, J., Huang, G., & Zagofsky, T. (2011). *Land of risk, land of opportunity*. Davis: Davis: University of California. https://regionalchange.ucdavis.edu/sites/g/files/dgvnsk986/files/inline-files/FINAL-Land%20of%20Risk-Land%20of%20Opportunity%20-2.pdf

Lopez, A., Cohen, A., Zota, A., & Morello-Frosch, R. (2009). . Communites for a Better Environment. https://www.cbecal.org/wp-content/uploads/2012/05/Richmond-Health-Survey.pdf

Marshall, C. (2005, July 21). New emission rule for bay area refineries. *New York Times*.

Minkler, M., Vasquez, V., Tajik, M., & Peterson, D. (2006). Promoting environmental justice through community-based participatory research: The role of community and partnership capacity. *Health Education and Behavior, 35*(1), 119–137. doi:10.1177/1090198106287692

Minkler, M., & Wallerstein, N. (2003). *Community-based participatory research for health*. San Francisco, CA: Jossey-Bass.

Morello-Frosch, R., Brody, J., Brown, P., Altman, R., & Rudel, R. (2009). Toxic ignorance and right-to-know in biomonitoring results communication: A survey of scientists and study participants. *Environmental Health*. 8(6). doi:10.1186/1476-069x-8-6

Morello-Frosch, R., Brown, P., Brody, J., Altman, R., Rudel, R., & Zota, A. (2011a). Experts, ethics, and environmental justice: Communicating and contesting results from personal exposure science. In G. Ottinger (Ed.), *Environmental justice and the transformation of science and engineering* (pp. 93–119). Boston, MA: MIT Press.

Morello-Frosch, R., & Lopez, R. (2006). The riskscape and the color line: Examining the role of segregation in environmental health disparities. *Environmental Research,102*(2),181–196. doi:10.1016/j.envres.2006.05.007

Morello-Frosch, R., Zavestoski, S., Brown, P., McCormick, S., Mayer, B., & Altman, R. (2006). Social movements in health: Responses to and shapers of a changed medical world. In S. Frickel & K. Moore (Eds.), *The new political sociology of science: Institutions, networks, and power*. Madison: University of Wisconsin Press.

Morello-Frosch, R., Zuk, M., Jerrett, M., Shamasunder, B., & Kyle, A. (2011b). Synthesizing the science on cumulative impacts and environmental health inequalities: Implications for research and policymaking. *Health Affairs, 30*(5), 879–887. doi:10.1377/hlthaff.2011.0153

National Research Council. (2009). Implementing cumulative risk assessment. In *Science and decisions: Advancing risk assessment* (pp. 213–239). Atlanta, GA: National Academies Press.

Sadd, J., Pastor, M., Morello-Frosch, R., Scoggins, J., & Jesdale, B. (2011). Playing it safe: Assessing cumulative impact and vulnerability through an environmental justice screen method in the South Coast Air Basin, California. *International Journal of Environmental Research and Public Health, 8*, 1441–1459. doi:10.3390/ijerph8051441

State of California. (2000). *Requirements, test procedure, and apparatus for testing the flame retardance of resilient filling materials used in upholstered furniture.* Sacramento, CA: Department of Consumer Affairs.

United Nations General Assembly. (2011). *Report of the Special Rapporteur on the Human Right to Safe Drinking Water and Sanitation. Doc. no. A/HRC/18/33/Add.4. UN Human Rights Council.*

U.S. Environmental Protection Agency. (2010).*Public drinking water systems: Facts and figures.* http://water.epa.gov/infrastructure/drinkingwater/pws/factoids.cfm

Wilson, S. (2009). An ecological framework to study and address environmental justice and community health issues. *Environmental Justice, 2*(1), 15–23. doi:10.1089/env.2008.0515

Wolfson, M., & Parries, M. (2010). The institutionalization of community action in public health. In J. Banaszak-Holl, M. Zald, & S. Levitsky (Eds.), *Social movements and the development of health institutions* (pp 117–127). New York: Oxford University Press.

Zota, A., Rudel, R., Morello-Frosch, R., & Brody, J. (2008). Elevated house dust and serum concentrations of PBDEs in California: Unintended consequences of furniture flammability standards? *Environmental Science and Technology, 42*, 8158–8164. doi:10.1021/es801792z

Notes

1. The Northern California HES was primarily funded through the Environmental Justice Program of the National Institute of Environmental Health Sciences, and by a grant from the National Science Foundation. One academic collaborator had a long-standing relationship with the Silent Spring Institute and another had a long-standing relationship with Communities for a Better Environment.
2. Community water systems are public water systems that serve at least 25 customers or 15 service connections year-round (U.S. Environmental Protection Agency, 2010).

7

EMOTIONS OF ENVIRONMENTAL JUSTICE

J. M. Bacon

Kari Marie Norgaard

Collaborations between academics and activists in the 1980s led to the formulation of theoretical frames of *environmental racism* and *environmental justice* that are now at the center of environmental sociology, environmental studies, eco-criticism, food studies, and many related fields. Over the past decades, environmental justice scholars and activists have expanded public understanding of what gets called "the environment." At the same time, the types of claims for justice expanded from an initial emphasis on proximity to toxic sites, to consideration of more and more dimensions of environmental inequalities (Agyeman, Schlosberg, Craven, & Matthews, 2016; Brulle & Pellow, 2006; Pellow, 2017; Schlosberg, 2013). Yet still today environmental justice work continues almost exclusively to address unequal *physical health* impacts from environmental degradation (see Brulle & Pellow, 2006; Crowder & Downey, 2010). There is little development within this literature (or the movement) of the notion of unequal *mental or emotional harms*, how the psychological impacts of racism might be part of environmental justice, or symbolic dimensions of how power operates through environmental degradation more generally. Most important, there is little interrogation of what the concept of emotional harm might mean, or about the relationships between emotions, environmental change, other features of social structure, and the production of environmental injustices. There is little engagement with theories of emotions.

As sociologists, we contend that emotions lie at the heart of social organization. Emotions are a key link between personal and political. Emotions are part of culture, cognition, social interactions, identity formation, and more. Because emotions link micro-level agency to macro-level social structure, they operate in the embodiment of power and the meaning of environmental justice in a variety of ways. Shared emotional experiences are integral to the construction of social order itself. When it comes to social movements, emotions matter for identity formation, social suffering, and social movement mobilization. What role do emotions play in the embodiment of power, oppression, and resistance concerning environmental justice? In what sense can emotions be part of a racialized experience of the environment, or might emotional experiences themselves constitute an occurrence of environmental injustice? What role do emotions play as environmental degradation inscribes racialized power relations, advances assimilation and genocide, or does the work of colonial violence? How do emotions animate environmental justice activism, solidarity work, or quiescence?

This chapter will highlight the importance of emotions in environmental justice through a series of case studies from the authors' work. We will not attempt an exhaustive survey of the (likely infinite) ways that emotions matter for the field of environmental justice. Rather, we provide a range of examples from our own work to illustrate how emotions shape *the production of environmental injustice*, how emotions matter for the ways communities and individuals *experience environmental injustice,* and how emotions are part of people's *responses to environmental injustices.*

PHOTO 7.1
Education Images / Contributor / Getty Images

EMOTIONS AND THE PRODUCTION OF ENVIRONMENTAL JUSTICE

In the same way that racism is structural or institutional, environmental justice scholars emphasize that the production of environmental inequalities is part of the fabric of modern society. At the broadest macro level, scholars including Laura Pulido (2017), Hilda Kurtz (2009), David Pellow (2017), and Kyle Whyte (2018) describe racial capitalism and the notion of a racial and settler state in the production of environmental injustice. Much attention has also been paid to the meso-level operation of power in the production of inequality. City and county zoning practices, the particulars of policy implementation, the policies and practices of agencies such as air quality districts, and historical practices of redlining, among others, structure the distribution of environmental harms and the terrain within which communities seek to mobilize. Lastly, cultural practices of the general public, including avoidance and denial, legitimate environmental inequality through its normalization (Pulido, 2000). Emotions play a role in the structuring of these and other aspects of the social production of environmental racism and environmental inequality, from the operation of the racial state and racial capitalism, to the cultural norms and practices that legitimate inequalities. While emotions matter across each of these dimensions, we will use Norgaard's (2011, 2012) work on emotions in the social organization of climate denial and the construction of innocence, and work by Sutton and Norgaard (2013) on the cultural practices at play in normalization of human rights atrocities, to illustrate examples of emotions in the production of environmental injustice at the meso and micro levels.

Why and how do individuals distance themselves from information about their government's participation in torture and other human rights violations? When citizens fail to speak out, this (non)response implicitly legitimates and thus facilitates the continuation of abusive state actions. Work on this process from Sutton and Norgaard (2013) can be applied to illustrate the ways that emotions are part of the normalization of environmental injustice by more privileged social actors who fail to intervene. Sutton and Norgaard explore how sociocultural contexts and practices mediate individuals' avoidance, justification, normalization, silencing, and outright denial of human rights abuses in Argentina during the last military dictatorship (1976–1983) and in the United States during the "war on terror" after September 11, 2001. Their work shows the roles of patriotic and national security ideologies and practices of silence and talk as organizers of cultures of denial. In both Argentina and the United States, interviewees offered rich descriptions of how they grapple with disturbing information about human rights abuses. Sutton and Norgaard describe how emotions of fear, guilt, and shame structure the reactions of bystanders and the discourses of both resistance and normalization of atrocities. These emotions shape not only what citizens talk about or fail to talk about, but *how* they do so. Sutton and Norgaard detail the dynamics in which subtle forms of denial take place through idiomatic expressions and language that foster normalization and use rhetorical tricks (such as euphemisms) to invite avoidance by trivializing or preventing full recognition of atrocious events.

EXPERIENCING ENVIRONMENTAL INJUSTICE: EMOTIONS AND THE MEANING OF ENVIRONMENTAL HARMS

> "The most powerful weapon in the hands of the oppressor is the mind of the oppressed."
>
> —Stephen Biko

The most explicit way that environmental justice scholars have grappled with emotions is through the notion that emotional experiences are a form of harm. Robert Bullard's (1990) landmark text *Dumping in Dixie* included discussion of psychological impacts of toxic exposure, and this theme continues in work by Erikson, Eddelstein, and Picou and in other key work in the development of the field. Literature on mental health and environmental decline has grown with recent studies on the psychological consequences of Hurricane Katrina and the Deepwater Horizon oil spill in the Gulf of Mexico. For example, Picou and Kenneth (2010) describe posttraumatic stress disorder (PTSD) as an outcome for survivors of Hurricane Katrina and detail how race, gender, and class intersect in this experience. Disasters represent very specific social circumstances, however, and the role of the natural environment has yet to be integrated into generic theory on social processes. Although literature in the field of sociology of disasters details the negative psychological consequences of environmental degradation and their unequal distribution along the lines of race, class, and gender, mental health impacts are covered only sparsely within the environmental justice framework.

Work in the area of disasters details important building blocks for any understanding of mental health and environmental decline. In their work on the Exxon Valdez oil spill, for example, Steven Picou and co-authors describe how, in addition to economic and subsistence impacts, the spill "shook the cultural foundation of Native life." Their work places the psychological impacts of the spill front and center, noting the prevalence of chronic levels of psychological stress, instances of PTSD, and depression. Still, there are key differences between disaster events and protracted environmental decline, and even bigger differences between the diagnostic and positivist assumptions within the mental health literature, and the interpretive theorizing of how emotional experiences reproduce social structure within sociology of emotions. Though extremely important, existing studies on the impacts of disaster events and contamination emphasize emotions related to personal contamination or economic losses, rather than the grief or other emotions related to damaged relationships with land, or connections to impacted species.

Furthermore, there is no interrogation within this literature of the concept of emotional harm, or the relationships between emotions, environmental change, other features of social structure, and the process of inequality formation. In other words, a significant gap exists between literature that considers emotions as socially constructed embodiments of power linking micro-level agency to macro-level social structure, clinical scholarship on emotions and mental health as mentioned above, or work on how emotions function to reproduce colonialism or racism (see Thoits, 2012). On the one hand, racism, classism, and other forms of oppression are understood to manifest as negative mental health outcomes (Brown, 2003; Thoits, 2010). On the other hand, theoretical literature within sociology of emotions conceptualizes emotional harm and points to relationships between emotions and mental health (Scheff, 2014). If negative emotional states can embody oppression, what exactly is the "harm" of these experiences? What might we understand about the production of inequality by examining disruptions to relationships between nature, emotions, and society?

In this next section, we draw upon work by Norgaard and Reed (2017) in the context of the Klamath River dam license application. Using interviews, public testimonies, and survey data, the authors describe how the natural environment is a central influence on emotional experiences, including individuals' internalization of identity, social roles and power structures, and resistance to racism and ongoing colonialism, for Karuk people. Norgaard and Reed articulate how the production of inequality occurs through disruptions to relationships between nature, emotions, and society. Grief, anger, shame, and powerlessness and hopelessness associated with environmental decline serve as signal functions confirming structures of power.

Grief: "Just Like Tearing My Heart Out"

The most frequently expressed emotion in the face of the degraded water quality and diminished quantity of fish in the river was grief. Nearly everyone we spoke with conveyed this emotion with intensity. For Rabbit, a father and traditional fisherman in his early 30s, "It gets pretty emotional for me, you know when I see salmon *dying* because of the algae, or the river is too low." He went on to elaborate:

> It saddens me...to see all the algae and all the toxins, it just saddens me that they continue to allow this to happen. They know the long term effects it's going to have...it's going to be devastation...you know for everything in that river, not only the salmon, for everything in that river.

One younger fisherman, Ron, relates the intense grief of not being able to provide fish to Elders: "If I couldn't be able to dip for my Elders, it would just break my heart, you know, if I couldn't go down there and gather up some eels for them, it would just be like tearing my heart out." On the other hand, experiences of a degraded river are also associated with a long-felt awareness of Karuk culture and life under attack, and a long-standing sense of their imminent destruction. Emotional responses to the destruction of the environment reanimate these histories of physical genocide. That Karuk life and culture could come to an end is the grim background cadence of people's everyday sensibilities. In the course of our interviews, many people shared stories such as the following from Ron's cousin:

> The Karuk people actually believe that if the Salmon quit running, the world will quit spinning, you know. Maybe the human race as we know it may be non-existent.... If the river quits flowing, it's over. If Salmon quit running, it's like the sign of the end.

TABLE 7.1 ■ Grief in Relation to Identity, Social Interactions, and Social Structure

"Identity"	"Social Interactions"	"Social Structure"
"like tearing my heart out"	"sadness because quiet own at the Falls"	"sadness that Karuk people may disappear"

Source: Kari Marie Norgaard (2019). *Salmon and Acorns Feed Our People: Colonialism, Nature and Social Action.* Rutgers University Press.

Or this reflection from Earl Aubry:

> My Grandma, she said the deer would probably go first, according to what the medicine people talked about when she was real small. She said from what she could understand the animals were going to let us know when the end is here. Because they'll disappear.

The fear and dread interwoven in these stories form a foreboding backdrop to the many layers of struggle associated with the degraded river, as changing environmental conditions become the leading edge of genocide (Table 7.1).

Anger: "I Get Pissed Off"

> ***"***
> ***now when I get pissed off, now what I do? I go out and drinking. But what if I had on?... If I had a sweat lodge back of my house...or d a fishery that we had gh fish and if I could go o my mother and my kids ne way we need to live... what we're looking for.***

Although grief was the emotion people most frequently conveyed, many also expressed anger over the degradation of the Klamath River. One man in his early 50s noted, "You might be pissed off. You might be really super angry. You'd be super angry, like I was, and not really know what the hell you were angry about." Another man from the same community added, "When you don't have something that you feel like you have a right to have [fish, access to a healthy life] you're disenfranchised. You're angry." The anger voiced here is not just about what is happening to the river, but how both are intertwined with a sense of "denied access" to a variety of important activities and responsibilities that lie at the very heart of being Karuk.

Shame: "That Puts You in This Little Down Feeling"

Shame has been considered one of the most important emotions in the formation of social structure and stability (Scheff, 1994, 2000). Here, too, shame operates to inscribe ongoing racism and colonialism across spheres of social action, from relations with individual identity to social interactions and structure. According to Scheff (2000), shame arises when subjects fail to achieve social ideals, or diverge from certain social standards. People described shame in relation to personal identity in several ways. First was a sense of direct identification with the river as a contaminated entity, as Frank Lake poignantly articulates:

> I think particularly for indigenous people, social, cultural, and community wellness reflects the ecological quality of their environment. So when the river's degraded, and it's liquid poison in some ways, and

> you're supposed to draw all your sustenance and your identity as a river Indian or a river person, then of course that weighs on you. It's like, you want to be a proudperson and if you draw your identity from the river and theriver is degraded, that reflects on you.

Conceptualizing the nature of physical harms from environmental decline leading to cancer or lead poisoning appears more straightforward, but what exactly is the nature of emotional harm? The ability to maintain a coherent meaning system is considered a vital component of mental health and psychological well-being (Mirowsky & Ross, 1989; Thoits, 2010). By contrast, Ron Reed described a sense of shame and emptiness:

> When we don't have a way of life, you're left with emptiness...if you don't know the creation stories, if you don't know tribal philosophy—there is a big void in your life. I think there is a level of embarrassment with the lack of knowledge, with the lack of presence in the culture. With all that, you end up with low self-esteem. You can't fish. You can't hunt. You don't know how to pray. People aren't really eager to talk about something that embarrasses them. And it must be an embarrassing moment not to be coming down to the Falls, or going to the ceremonies.

As Ron's words reveal, it is the emotions of shame related to one's inability to carry out valued social roles in the face of the degraded environment that inscribe ongoing racism and colonialism. The emotional "harm" is a function of their cognitive dimension in the inscription of social power (Table 7.2).

Powerlessness and Hopelessness: "The Natural Thing Is to Feel Hopeless"

Grief, anger, and shame were also mixed with feelings of powerlessness in the face of institutional forces working against the health of the Klamath River. Leaf Hillman described:

> People say, "Do you really think they are going to take out the dams on the Klamath River? You'd be out of your mind to think that." Well I don't know. Do I really think there is justice in the world? No. That's an easy one. Do I ever think that they'll be justice? No. Do I think there is any hope? I don't know. People say, "How can you be even the slightest bit optimistic?" It's not easy to be optimistic about any of

TABLE 7.2 ■ Shame in Relation to Identity, Social Interactions, and Social Structure

Identity	Social Interactions	Social Structure
if you draw your identity as a river Indian but the river is contaminated that reflects on you	shame that can't provide for elders or family, cannot perform responsibilities to other species	shame that cannot find way in "modern society"

Source: Kari Marie Norgaard (2019). *Salmon and Acorns Feed Our People: Colonialism, Nature and Social Action*. Rutgers University Press. Copyright © 2019 by Kari Marie Norgaard and the Karuk Tribe.

> these things that I'm talking about. The easy, and I think the natural thing, is to feel hopeless.

Frank Lake vividly described the experience of living with the degraded river as "enduring an assault on one's relations" yet being powerless to fully stop it:

> You know, that spiritual tie, kind of more like kinship or family type of relationship. That's where I think the grief comes in. It's like, a sense of powerlessness. You know, and yet what can you do?... You basically see this assault or this attack on your family, either directly as humans, but also the extension of your family relationship and the tribal perspective of seeing that with salmon, you see this attack. You see this, you know, and there is this constant—I guess the only word I can think of is assault—on them. And there are certain things you can do within your capacity, and then some things are so broad outside of the influence, that it's hard to comprehend what's going on.

The experiences Karuk people articulate are similar to those Downey and Van Willigen (2005) found in their work on proximity to environmental contamination as generating personal powerlessness, and the work of Shriver and Webb (2009), who describe how an "endless battle to validate health and environmental concerns, along with the constant assault on Native American values, has fostered a sense of apathy and hopelessness among some tribal members" (p. 282). These themes around emotional harms are being further developed in recent literature. For example, Debra Davidson (2017) illustrates the critical role of emotions in exposure trauma and environmental reflexivity. Auyero and Swistun's (2009) work on "environmental suffering" builds on Bourdieu's (1999) concept of social suffering to articulate how long-term environmental decline interacts with cultural practices and state and corporate power in an Argentine shantytown. Similarly, Shriver and Webb (2009) use the "ecological-symbolic" perspective (Kroll-Smith, Couch, & Stephen, 1993) to describe how Ponca tribal members interpret and develop diagnostic frames concerning air contamination. These studies move closer to engaging the complexity of how the social meanings of emotional experiences can be part of the operation of power.

EMOTIONS AND ENVIRONMENTAL JUSTICE MOBILIZATION

Environmental justice activism, like all social movements, is fueled by strong emotions. Emotions matter for identity formation and social movement mobilization. Rage, fear, hope, and betrayal feature commonly in the comments of those participating in environmental justice activism. Grief and anger, for example, operate in complex ways in relation to cognition identity and moral understandings. These emotions and others (for example, hope) are critical catalysts for social movement activity (Jasper, 2011). Emotions are also key components of the success and failure of environmental justice movements. For example, emotions organize and shape the motivations and experiences for allies in solidarity work (Bacon, 2017), thereby influencing the effectiveness of solidarity efforts (Grossman, 2017). Emotions such as guilt, fear, and helplessness also structure movement nonmobilization in cases of environmental privilege (Norgaard, 2011) and also as a component of environmental suffering (Auyero & Débora, 2009). Here we provide examples from our data on how emotions structure settler-Indigenous solidary work, how emotional experiences become a form of environmental harm for Karuk people on the Klamath River, and how emotions may prevent movement mobilization.

Emotions and Indigenous Solidarity

We excerpt first from Bacon's (2017) work with non-Indigenous activists who engage in solidarity with Indigenous peoples on environmental issues, where emotions powerfully inform practices of solidarity. In the case of intergroup solidarity for environmental justice, emotions can take on particular forms. While diverse, sites of Indigenous environmental justice struggle are different from other environmental justice movements in that they are structured by conflicts with the settler state, and with colonial occupation itself. Furthermore, because of the way settler colonialism is erased from public consciousness through miseducation, and a general lack of media attention, activists who wish to engage in solidarity often experience profound shocks related to their own knowledge gaps about the issues they are contending with and the people they are attempting to support. Prior participation in either environmentalism or social justice do not adequately prepare activists for solidarity with Indigenous-led environmental justice movements. The knowledge gaps and cultural differences involved in solidarity with Indigenous-led environmental justice movements generate strong and often challenging emotions for non-Indigenous activists. Emotions permeate the reflections of these "solidarity activists," and some of the main tropes we will explore here are the emotions related to ignorance, stigmatized identities, and the environment itself.

One of the main emotional tropes evinced by activists is their response to learning about the injustices that have produced the environmental benefits they have come to enjoy in their day-to-day lives. This is especially pronounced in the case of white, economically privileged solidarity activists, but it is also part of a widespread trend. This early stage of coming to know about settler colonialism and colonial ecological violence is often marked by periods of guilt and shame that are combined with a deep sense of shock and betrayal. Although many interviewees enter into the solidarity work with a preexisting distrust of authorities, the academy included, the realization of the profound knowledge gaps generated by settler colonialism still creates emotional upheaval.

This shock is conveyed in a variety of ways but is almost always comingled with guilt or shame. Interviewees describe the "personal turmoil" they felt while first learning about their own privilege relative to Indigenous peoples. This turmoil was expressed not only by those with majority identities but also by those with marginalized identities. Although some saw their own experiences of marginalization as a linkage between themselves and other movements such as Indigenous movements, some found it difficult to cope with their newfound identity as a settler or colonizer. For example, Eiko, a Japanese American interviewee, said,

> When [my friend] told me I'm a colonizer, it undermined that [sense of belonging] and it put me back in a space of "I don't belong anywhere" and so I was uncomfortable.

Despite this discomfort, Eiko committed to participation in solidarity. In many cases, engagement with solidarity actions is described as a way to "take it to the next level" or to move beyond "just sulking in your own…self-doubt and guilt but actually moving on to make a difference."

While solidarity may be seen as a way to "move beyond wallowing," moments of emotional turmoil and discomfort in the course of solidarity are virtually inevitable given the enormous knowledge gaps and cultural differences. This can occur during what participants often refer to as "call outs," which are direct and open disapprovals of specific behaviors on the part of solidarity activists. Turmoil and discomfort can also occur more diffusely through a lingering sense of confusion or disorientation, which permeated many accounts.

Solidarity activists' desire to "be useful" often contributes to fears about how to best perform solidarity. These fears may result from call outs but more often are just part of an awareness of cultural difference and internalized shame about belonging to the settler class. Reflections on confusion and anxiety differ. Some, like Iris, are light-hearted even while expressing concern. She says:

> I was definitely worried I was going to upset some of the elders just because of cultural things I wasn't paying enough attention to. In a sense I had no idea of what was going on or cultural norms…. [S]o you know, I was just like "ok, I'm just gonna do this and smile." Because I have no idea what I'm supposed to do, but here I am.

Iris takes a can-do approach, doing what she thinks is right and hoping for the best, but for others, the anxiety provoked by "not knowing" seemed much less manageable, and some participants even indicated their desire to "leave camp early" in an effort to avoid making mistakes or accumulating further stigma or shame.

Direct call outs tended to exacerbate the levels of anxiety about "not knowing" among interview participants, but even those on the receiving end of such call outs tended to acknowledge the need for them. Ideally, call outs lead to self-reflection and transformed behaviors, but they also tend to generate feelings of guilt, shame, and confusion, and in some cases hostility or disaffection. Yet, even in best-case scenarios where behavior is positively adjusted to meet the expectations of the group, emotions like pride or shame continue to be an important driving force in maintaining these adjustments. One interviewee who became a repeat participant in Indigenous solidarity events notes:

> That first year…there were people who were with our group whose conduct I was not super proud of. Not that I'm a perfect example, but there are some pretty glaring things that you should not be doing when you're there for solidarity. So there's a huge difference between that year and this year. I think it's actually been a number of the same people but we all learned a lot and had specific ideas of how we wanted to be. Either how we wanted to present ourselves or like watch each other's backs.

In addition to call outs, the situationally relevant stigma associated with settler identity and/or whiteness can be its own source of emotional challenge for solidarity participants. In his work on emotions and social movements, James

Jasper (1997) writes that "[d]emonization fuels powerful emotions for social movements, such as hatred, fear, anger, suspicion, and indignation." Although this is undoubtedly true, the capacity for this affective propagation is limited in work where settlers join a movement to disrupt settler colonialism.

In the case of settler solidarity, participants are profoundly implicated in the system they resist; this fact does not go unnoticed by those with whom they are in solidarity nor by the participants themselves. This position circumscribes the limits of collective identity, and expressing this boundary between themselves and the Indigenous people they work with becomes a common component of solidarity participants' narratives. It is also a narrative charged with emotion. A 21-year-old solidarity participant from Marin, California, recounted a few moments where social difference was made explicit during her solidarity experiences. While she makes light of the experiences, often relating the stories with a laugh, it is clear that these experiences were emotionally challenging. For example, she recounts a moment of talking with her friend Jay and a Native man they met at the event:

> He was like I don't mean to offend you but I hate the U.S. government and I hate white people, and we were like "woah." He was like I don't mean to offend you, and it's like I'm not offended. People are mean, particularly white people. So, um it was pretty much just out of the blue, it was just pretty random.

While she logically reasons that hating white people makes sense because white people are "mean," it seems unlikely that this interaction required no emotion management. In fact, even the retelling of it in our interview felt highly managed to demonstrate uncritical acceptance of the Indigenous perspective that identified her group, and by extension herself, as hate-able. Later in our conversation, this same participant reflected more on the emotional challenges of solidarity. She said,

> I would say...you feel things not because people make you feel things but because you feel them. So if I felt uncomfortable because of my whiteness it's because that was how I was feeling because I'm in a native space and I'm there because of shitty things that white people have done.

In this reflection, she takes responsibility for her own emotions but also seems to conveniently forget that there were several instances where whiteness was actively stigmatized in conversations with her. Although the internalized shame of belonging to a group that is responsible for the very issues one is trying to confront can be mobilizing, over the long term, participation in a movement in which one's identity is stigmatized may be difficult to sustain.

Some participant reflections suggest that this shame can be approached not through a focus on the stigmatized self but on the generosity of those with whom you are working. For example, Samuel, who reflected on his family's participation in the Gold Rush, contended that the tribe he works in solidarity with is giving him a chance to have a "better relationship" with Indigenous people and the land. Similarly, Meg, in her reflections on a celebratory salmon dinner, contended that the invitation to participate was stunning in its generosity. She says she was thinking,

> Wow, you want to share your salmon with me? Even though my ancestors came here and probably fucked up a lot of shit…even though I don't really know those people and I probably don't think as they do that's still…like I'm fucking white and...that's real and so I think…it feels like a very special thing to be able to share…. It felt very special.

For these participants, shame of their ancestors is a backdrop for appreciation of the opportunity to create a different and better pattern of relations with Indigenous peoples.

However, other participants suggested that any sense of positive self-worth gained by participation in solidarity was itself colonial and inappropriate. For this group, not only did they police this tendency in themselves, but they were also vocally critical of it in others. Anyone who expressed pride in their commitment to Indigenous solidarity was seen as "problematic." Furthermore, anyone who spoke of learning from their experiences was also considered to be "exploiting" or "extracting" from Indigenous peoples. In this context, there is no positive way to regard oneself or one's actions. This pattern raises serious concerns about the sustainability of engagement and the emotional health of those committed to this perspective.

Given the presence of this perspective, it is not surprising that some participants expressed nihilistic sentiments, such as a need to justify their own existence. One participant, a university senior from southern Oregon, in reflecting on his knowledge gaps and his own place in the world after learning about settler colonialism, said,

> I didn't think Indigenous people were still struggling to have religious freedoms or practice cultural traditions on their native lands…. It's just so heartbreaking.... Just thinking about that is heartbreaking because I don't want anyone to feel like that…. And I'm always trying to justify my life.

Emotions and Indigenous Resistance to Environmental Injustice

A second example of emotions in environmental justice activism comes from Norgaard's work with the Karuk tribe on the Klamath River dam removal. While the original intent of the work was to document social impacts from the Klamath River dams as part of a legal and policy process to redress harms, emotions were clearly a part of the widespread community mobilization that was taking place.

Indigenous resistance to a wide range of ways that colonialism and assimilation occur has been continuous, pervasive, and diverse. People resist by engaging in direct action via protest events, taking legal action against federal and state agencies, participating in natural resource policymaking, testifying in public hearings, and participating in dances and ceremonies. People use personal prayer, continue to hunt and fish according to tribal law despite personal risks, learn and teach the Karuk language to their children, and develop educational curriculum.

The relationship between emotions and the environment are fundamental to any valid understanding of resistance. Anger is an emotion that is often associated with political agency, but the dynamic between fear and hope, and other emotions associated with group solidarity, are part of resistance as well. One man alludes specifically to the dynamics between fear and hope, as he moves from reflecting on things that are lost to referencing hope for the future. Hope is his motivation for traveling outside the area to engage in protest events such as when the tribe went to the shareholder's meetings of the companies that owned the dams, first to Edinburgh, Scotland, and later to Omaha, Nebraska:

> But we never give up hope. That's why I went to Scotland. That's why I went to Omaha, twice. Because we have hope to fix this river. Which is one step to getting the Karuk people back to where they once were.

Another man in his 40s describes how work on the fisheries crew keeps him going and serves as an avenue to fulfill cultural responsibilities:

> My job, for one...bringing the salmon back and restoring the health of our river, if everybody reaches out and does their own little part...that's kind of how I'm feeling right now with my job, you know. I'm trying to restore the river, doing these fish surveys, and create a positive effect for the fish.

A mother in the community notes the importance of activism against the dams for the identity of young people:

> Now there is such a big push because we see an opportunity. I think in a lot of ways that this provided people with a role. These younger folks have a purpose, knowing that they can make a difference in the Klamath Campaign. I think that that makes a difference.

Although the emotions implicit in these passages are more subtle, the natural environment clearly is a motivator for social action, and emotions in response to environmental change operate here in the form of what Jasper (2011) calls "moral batteries," whereby "fear, anxiety and other suffering in the present" is combined with "hope for future change" to motivate action (p. 14.7).

Emotions, Environmental Privilege, and Nonmobilization

Global capitalism is currently producing wider divides between the material conditions of the lives of haves and have-nots. At the same time, cheap airfares and quality digital Internet images close gaps in space and time, bringing privileged people ever closer to the worlds of those we exploit. Through tools of order and innocence, a safe, secure mental world is actively produced for the privileged, and environmental problems are kept invisible to those with the time, energy, cultural capital, and political clout to generate moral outrage. Wastes and hazards are exported to other nations, other places, other populations. Privileged people are protected from full knowledge of environmental (and many other social) problems by national borders, gated communities, segregated neighborhoods, and their own fine-tuned yet unconscious practices of not noticing, looking the other way, and normalizing the disturbing information they constantly come across. The intersection of environmental justice and privilege is an extremely important concept that is only now beginning to receive attention.

If we are to really grapple with the how and why of the reproduction of environmental privilege, we must take seriously the experiences of privileged social actors themselves, acknowledging the complexity and difficulty of occupying this social location. Privilege may act like a breeze at one's back, making life easier. Yet people in privileged situations also experience the complex and ambivalent mental and emotional landscape of denial, with all of its feelings of guilt, responsibility, and hopelessness. Despite being in positions of power, privileged people may feel powerless. People occupying privileged social positions encounter "invisible paradoxes," awkward, troubling moments that they seek to avoid, collectively pretend not to have experienced (often as a matter of social tact), and forget as quickly as possible once they have passed.

Raoul Lievanos (2010) defines environmental privilege as "the taken-for-granted structures, practices, and ideologies that give a social group disproportionately high level of access to environmental benefits." Park and Pellow (2011) write that

> environmental privilege results from the exercise of economic, political and cultural power that some groups enjoy, which enables them exclusive access to coveted environmental amenities such as forests, parks, mountains, rivers, coastal property, open lands and elite neighborhoods. Environmental privilege is embodied in the fact that some groups can access spaces and resources which are protected from the kinds of ecological harm that other groups are forced to contend with everyday. (p. 4)

Park and Pellow (2011) describe environmental privilege emerging from "the exercise of economic, political and *cultural* power" (p. 4, emphasis added). Culture is implicated in the reproduction of social relations, but exactly how does this work? How do privileged people create a positive sense of self, normalcy, and even sense of their innocence, in the face of knowledge of their high carbon footprints? In fact, societies develop and reinforce a whole repertoire of techniques or "tools" for ignoring disturbing problems. Norgaard's work in Norway describes how cultural avoidance strategies were directed toward two basic fears: those associated with security and concerns for the future, and those connected to guilt and feelings of responsibility for the problem:

> Yes, if you take for example this with cars, we drive a lot of cars, in my family that is. We go on vacation and we go shopping, and my partner drives to work every day. And I drive often up here (to his office) myself. It gives us flexibility and so forth. And then we experience...we don't like it. We feel that we must do it to make things work in a good way, on a practical level, but we have a guilty conscience, a bit of a guilty conscience.

Normalization of climate change occurred by using "tools of order" to recreate order and security and "tools of innocence" for the "construction of innocence." In the course of normalizing a troubling situation, residents simultaneously reproduce transnational environmental privilege. The construction of denial and innocence work to silence the needs and voices of women and people of color in the Global South and thus reproduce global inequality along the lines of gender, race, and class.

IN CLOSING

Environmental justice work has focused on physical health impacts of environmental decline, but emotional or mental health impacts represent a crucial, if less analyzed, dimension of power that is highly relevant to many aspects of environmental justice. We hope that the examples we have highlighted in this chapter will prompt other scholars and activists to pay explicit attention to the diverse ways that emotions matter for environmental justice movements and scholarship.

DEEPENING OUR UNDERSTANDING

Choose a case of environmental injustice of interest you, for example, another case from this text, or something close to your home.

1. Are emotions addressed in this case?
2. How might emotions matter for the ways that community and/or individuals *experience* environmental injustice?
3. How might emotions be a part of people's *responses* to that environmental injustice?

References

Agyeman, J., Schlosberg, D., Craven, L., & Matthews, C. (2016). Trends and directions in environmental justice: From inequity to everyday life, community, and just sustainabilities. *Annual Review of Environment and Resources*, *41*, 321–340. doi:10.1146/annurev-environ-110615-090052

Auyero, J., & Débora, A. S. (2009). *Flammable: Environmental suffering in an Argentine shantytown*. New York: Oxford University Press.

Bacon, J. M. (2017). "A lot of catching up", knowledge gaps and emotions in the development of a tactical collective identity among students participating in solidarity with the Winnemem Wintu. *Settler Colonial Studies*, *7*(4), 441–455. doi:10.1080/2201473x.2016.1244030

Bourdieu, P. (1999). *The weight of the world: Social suffering in contemporary society*. Palo Alto, CA: Stanford University Press.

Brown, Phil. (2003). Qualitative methods in environmental health research. *Environmental Health Perspectives*, 111, 1789–1798. doi:10.1289/ehp.6196

Brulle, R. J., & Pellow, D. N. (2006). Environmental justice: Human health and environmental inequalities. *Annual Review of Public Health, 27*, 103–124. doi:10.1146/annurev.publhealth.27.021405.102124

Bullard, R. D. (1990). *Dumping in Dixie: Race, class, and environmental quality*. Boulder, CO: Westview Press.

Crowder, K., & Downey, L. (2010). Interneighborhood migration, race, and environmental hazards: Modeling microlevel processes of environmental inequality. *American Journal of Sociology, 115*(4), 1110–1149. doi:10.1086/649576

Davidson, D. J. (2019). Emotion, reflexivity and social change in the era of extreme fossil fuels. *The British Journal of Sociology, 70*(2), pp.442-462.

Downey, L., & Marieke, V. W. (2005). Environmental stressors: The mental health impacts of living near industrial activity. *Journal of Health and Social Behavior, 46*(3), 289–305. doi:10.1177/002214650504600306

Grossman, Z. (2017). *Unlikely alliances: Native nations and white communities join to defend rural lands*. Seattle: University of Washington Press.

Jasper, J. M. (1997). The art of moral protest: Culture, biography, and creativity in social movements. *American Journal of Sociology, 104*(6), 1835–1837. doi:10.1086/210235

Jasper, J. M. (2011). Emotions and social movements: Twenty years of theory and research. *Annual Review of Sociology, 37*, 14.1–14.19. doi:10.1146/annurev-soc-081309-150015

Kroll-Smith, J., Couch, J., & Stephen, C. (1991). What is a disaster? An ecological-symbolic approach to resolving the definitional debate. *International Journal of Mass Emergencies and Disasters, 9*(3), 355–366.

Kurtz, H. E. (2009). Acknowledging the racial state: An agenda for environmental justice research. *Antipode, 41*(4), 684–704. doi:10.1111/j.1467-8330.2009.00694.x

Mirowsky, J., & Ross, C. (1989). *Social causes of psychological distress*. New York: Aldine de Gruyter.

Norgaard, K. M. (2011). *Living in denial: Climate change, emotions, and everyday life*. Cambridge, MA: MIT Press.

Norgaard, K. M. (2012). Climate denial and the construction of innocence: Reproducing transnational environmental privilege in the face of climate change. *Race, Gender & Class, 19*, 80–103.

Norgaard, K. M., & Reed, R. (2017). Emotional impacts of environmental decline: What can native cosmologies teach sociology about emotions and environmental justice? *Theory and Society, 46*(6), 463–495. doi:10.1007/s11186-017-9302-6

Park, L., & Pellow, D. (2011). *The slums of Aspen*. New York: New York University Press.

Pellow, D. N. (2017). *What is critical environmental justice?* New York: Wiley.

Picou, S. J., & Hudson, K. (2010). Hurricane Katrina and mental health: A research note on Mississippi Gulf Coast residents. *Sociological Inquiry, 80*(3), 513–524. doi:10.1111/j.1475-682x.2010.00345.x

Pulido, L. (2000). Rethinking environmental racism: White privilege and urban development in Southern California. *Annals of the Association of American Geographers, 90*(1), 12–40. doi:10.1111/0004-5608.00182

Pulido, L. (2017). Geographies of race and ethnicity II: Environmental racism, racial capitalism and state-sanctioned violence. *Progress in Human Geography*. 41(4), 524–533. doi:10.1177/0309132516646495

Scheff, T. (1994). *Microsociology: Discourse, emotion, and social structure*. Chicago: University of Chicago Press.

Scheff, T. (2000). Shame and the social bond: A sociological theory. *Sociological Theory, 18*(1), 84–99. doi:10.1111/0735-2751.00089

Scheff, T. (2014). The ubiquity of hidden shame in modernity. *Cultural Sociology, 8*(2), 129–141. doi:10.1177/1749975513507244

Schlosberg, D. (2013). Theorising environmental justice: The expanding sphere of a discourse. *Environmental Politics, 22*(1), 37–55.

Shriver, T., & Gary, W. (2009). Rethinking the scope of environmental injustice: Perceptions of health hazards in a rural native American community exposed to carbon black. *Rural Sociology, 74*(2), 270–292. doi:10.1111/j.1549-0831.2009.tb00392.x

Sutton, B., & Norgaard, K. M. (2013). Cultures of denial: Avoiding knowledge of state violations of human rights in Argentina and the United States. *Sociological Forum, 28*(3), 495–524.

Sze, J. (2006). *Noxious New York: The racial politics of urban health and environmental justice*. Cambridge, MA: MIT Press.

Sze, J., London, J., Shilling, F., Gambirazzio, G., Filan, T., & Cadenasso, M. (2009). Defining and contesting environmental justice: Socio-natures and the politics of scale in the Delta. *Antipode, 41*(4), 807–843. doi:10.1111/j.1467-8330.2009.00698.x

Thoits, P. (2010). Stress and health major findings and policy implications. *Journal of Health and Social Behavior, 51*(1_suppl), S41–S53. doi:10.1177/0022146510383499

Thoits, P. (2012). Emotional deviance and mental disorder. *Emotions matter: A relational approach to emotions*. Toronto: University of Toronto Press.

Whyte, K. (2018). Critical investigations of resilience: A brief introduction to indigenous environmental studies & sciences. *Daedalus, 147*(2), 136–147. doi:10.1162/daed_a_00497

PART III

POLICY AND ENVIRONMENTAL PROTECTION

8

REGULATORY CULTURE

Racial Ideologies and the Fight for Environmental Justice within Government Agencies

Jill Lindsey Harrison

Environmental regulatory agencies like the U.S. Environmental Protection Agency (EPA) play a crucial role in managing environmental hazards, providing protections that industry and individuals will not or cannot provide. Regulatory agencies establish acceptable industrial practices and pollution levels, assign permits to restrict the pollution of industry and other actors, enforce environmental laws and regulations, monitor air and water quality, clean up contaminated sites, and help fund state and tribal agencies' environmental programs. They do so in an increasingly hostile context. Antiregulatory elites have eroded environmental regulatory agencies' operating budgets and legal authorities to restrict environmental hazards for several decades (Centeno & Cohen, 2012; Faber, 2008; Harrison, 2014; Harvey, 2005; Mascarenhas, 2012). The Trump administration has accelerated this in precedented ways, much to the dismay of EPA staff and others who fight to defend environmental regulation (Davenport, 2017; Dillon et al., 2018; Fredrickson et al., 2018; Gardner, 2017; Save EPA, 2017).

Under President Obama's first EPA administrator, Lisa Jackson, the EPA developed environmental justice (EJ) programs and provided EJ guidance to federal, state, and tribal agencies, many of which have started to integrate EJ principles into their own policies, programs, and practices. Agencies undertaking EJ efforts have created formal EJ policies, convened EJ advisory committees to advise the agency on integrating EJ principles into agency practice, developed EJ trainings for staff, developed public participation guidelines to disseminate information more widely and solicit community input on regulatory decisions, and tasked "EJ staff" with proposing EJ reforms to regulatory practice—including into aspects of regulatory work that directly limit what polluters can and cannot do. Additionally, a few agencies have developed EJ screening tools to identify communities that are environmentally overburdened and socially vulnerable; funded cross-agency and multiagency projects to reduce cumulative risks in certain overburdened communities; created EJ grant programs allocating funds to community-based and tribal organizations aiming to reduce environmental inequalities; and started to develop EJ protocols for the core regulatory functions of permitting, enforcement, and rulemaking. EJ advocates inform these processes through agencies' EJ advisory committees and public meetings, and occasionally by being hired to administer EJ programs.

Although EJ advocates and agencies' EJ staff have proposed and fought hard for many important regulatory reforms that could protect overburdened and vulnerable communities from dangerous environmental hazards, agencies have institutionalized few of them. Scholars evaluating agencies' EJ efforts have concluded that they fail to change core regulatory practices (rulemaking, permitting, and enforcement) in ways that would reduce cumulative risks in overburdened and vulnerable communities or create greater environmental equity across communities, lack adequate accountability measures, and, in some cases, may undermine critical EJ advocacy (Banzhaf, 2011; Bruno & Jepson, 2018; Bullard, Mohai, Saha, & Wright, 2007; Eady, 2003; Engelman

---

Photo 8.1
bakdc/Shutterstock

Lado, 2017; Foster, 2000; U.S. Government Accountability Office, 2005; Gerber, 2002; Harrison, 2015, 2016, 2017, 2019; Holifield, 2004, 2012, 2014; Kohl, 2015, 2016; Konsiky, 2015; Lashley, 2016; Liévanos, 2012; Liévanos, London, & Sze, 2011; London, Sze, & Liévanos, 2008; Payne-Sturges et al., 2012; Shilling, London, & Liévanos, 2009; Targ, 2005; Vajjhala, 2010; Vajjhala et al., 2008). Agencies' EJ efforts have required few substantive concessions from industries that profit off of hazards and little systematic, meaningful change to the regulatory practices that authorize them to do so.

This is not to say that EJ-supportive staff within these agencies have not tried to do more. Informed by EJ advocates, agencies' EJ staff and EJ-supportive coworkers have proposed many regulatory reforms that could help protect working-class, racially marginalized, and Native American communities from dangerous environmental hazards. They express strong commitments to and respect for the EJ movement, care deeply about getting government agencies to address environmental injustices, and in many cases bring into their jobs personal experiences of oppression and living in contaminated areas, which their coworkers, on average, do not share. They are, in many ways, activist insiders—people working to change regulatory agencies from the inside. They are also deeply frustrated with their agencies' progress on implementing EJ reforms to regulatory practice. I wanted to know what it is like to try to fight for EJ from the inside of regulatory agencies. What constraints do EJ-supportive staff face?

Other EJ scholars have contributed important insights into why government agencies' EJ efforts fall so short of the principles of the EJ movement that fought for them. Some scholars take a political economic approach, emphasizing that elected officials have neutered EJ programs and restricted their funding due to pressure from industry and hostility to the new regulatory restrictions those programs might bring (Eady, 2003; Faber, 2008; Gauna, 2015; Gerber, 2002; Liévanos, 2012; Liévanos et al., 2011; London et al., 2008; Shilling et al., 2009). Others, taking a legal studies approach, highlight how weak EJ policy undermines agencies' EJ efforts by underspecifying their authority to implement more robust EJ reforms to regulatory practice and directing staff to focus on practices that reflect narrow interpretations of EJ (Gauna, 2015; Holifield, 2004, 2012; Konisky, 2015b; Noonan, 2015). Scholars have also attributed agencies' disappointing EJ efforts to technical limitations, such as the inadequacy of EJ screening tools and cumulative impact analysis techniques that would more accurately identify the most environmentally burdened communities and the full scope of harms they face (Eady, 2003; Gauna, 2015; Holifield, 2012, 2014; Payne-Sturges et al., 2012; Shadbegian & Wolverton, 2015).

My research shows that the slow pace of environmental regulatory agencies' EJ efforts *also* stems from elements of regulatory workplace culture that transcend institutions and changes in leadership: everyday practices through which staff reject EJ reforms as violating what they think the goals and priorities should be of the regulatory organizations they work for and to which they feel very committed. Elsewhere, I detail a wide range of internal cultural constraints facing agencies' EJ efforts (Harrison, 2015, 2016, 2017, 2019). In this chapter, I focus specifically on how government agency staff engage in racialized claims and actions that defend existing regulatory practice and undermine proposed EJ reforms. In so doing, they bolster a regulatory system that has long disproportionately protected white, middle-class, and wealthy communities. Throughout, I suggest various factors that may motivate such practices.

The narratives I showcase here reflect dominant forms of racial ideology in the United States today. Bonilla-Silva (2014), Lipsitz (1995), Omi and Winant (2015), and other contemporary race scholars show that popular "colorblind" narratives like "I don't see race," and "post-racial" claims that success comes from hard work and hardships from one's own failings rather than racism or government policies, frame racism as limited to conscious prejudice and located in the past rather than as an ongoing systemic phenomenon. Such "colorblind" narratives assert a commitment to abstract liberalist notions of racial equality while ignoring the "racialized social system" in which industry practices, social democratic government reforms, and neoliberal regulatory rollback have systematically, disproportionately afforded material resources to whites (Bonilla-Silva, 2014). These narratives recast race-conscious civil rights policies (such as affirmative action programs and agencies' EJ reforms) as "reverse racism"—unfairly discriminating against and taking resources away from whites. Importantly, scholars show that people wield these narratives "even when they are well-meaning and intend to be non-racist," thereby "debunking the idea that racial reproduction rests on intentional, malevolent, or politically conservative white actors" (Mueller, 2017, p. 221).

Building on scholarship showing that government institutions, including their staff, play key roles in producing and policing racial categories and racial meanings (Bonilla-Silva, 2014; Goldberg, 2002; Kurtz, 2009; Lipsitz, 1995; Mascarenhas, 2016; Moore & Bell, 2017; Omi & Winant, 2015; Park & Pellow, 2004; Pellow, 2018; Pulido, 2000, 2015; Richter, 2018; Taylor, 2014), my findings demonstrate how some environmental regulatory agency staff use colorblind and post-racial narratives to reject EJ reforms, in turn bolstering a regulatory system that has disproportionately protected whites.

METHODS

To understand the challenges EJ-supportive staff face in trying to promote EJ reforms within their organizations, I conducted confidential, semistructured interviews with staff from environmental regulatory agencies and ethnographic observation at agency meetings in the United States from 2011 to 2018. Ethnographic observation and interviews enabled me to identify how different staff understand their responsibilities and their agency's identity, how they interpret what "EJ" means, and how staff react to proposed EJ reforms. The semistructured nature of the interviews allowed me to pursue certain themes of interest while also allowing the participants to narrate and interpret their experiences. It also allowed me to develop the rapport necessary to discuss politically controversial issues. Confidential interviews and internal (not public) meetings give staff the space to express beliefs they do not (and cannot) express in formal agency documents or public events.

I interviewed 89 current and former agency representatives. About half of these are or were "EJ staff," formally tasked with administering agency EJ programs. The majority of these individuals are persons of color, which appears to reflect the broader population of agencies' EJ staff. The other half were not "EJ staff." Their sentiments about EJ reforms varied from enthusiasm to ambivalence to contempt, some participated in their agency's EJ efforts as their time and interests allowed, and most are white, as are most staff in environmental regulatory agencies. I asked my interview participants to describe their involvement with agency EJ efforts, which EJ reforms (if any) they view as important or disagree with, what challenges agency EJ efforts face, how staff react to proposed EJ reforms, and how those reactions affect agencies' progress on EJ reforms. Interview participants included representatives from the U.S. EPA (including EPA headquarters, eight of its regional offices, and one other satellite office); the Department of Justice; the Department of the Interior; seven state environmental regulatory agencies (California Environmental Protection Agency, Colorado Department of Public Health and Environment, Connecticut Department of Energy and Environmental Protection, Minnesota Pollution Control Agency, New York State Department of Environmental Conservation, Oregon Department of Environmental Quality, and Rhode Island Department of Environmental Management); and city or county agencies in California, Colorado, and Missouri. These agencies vary in how long-standing, well developed, and publicized their EJ efforts are; the number of designated EJ staff; and whether they have a formal EJ policy endorsed by agency leadership or the legislature. Given that all agencies' EJ efforts fall short of EJ advocates' expectations, the multisited study design enabled me to identify conditions that transcend this diverse array of institutions and to identify more universal factors shaping their EJ efforts.

I also observed numerous agency meetings related to their EJ efforts. These included public meetings; informational sessions about EJ grant programs; EJ advisory committee meetings; a public participation event about an EJ controversy; and various other agency-convened EJ webinars, workshops, and conference calls. These varied from 90 minutes to two full days. These helped me meet and recruit interview participants, observe how staff talk to EJ advocates, and observe how EJ advocates advise agencies on what their EJ efforts should entail. My observations also included internal (not public) agency EJ-related meetings. These included four internal EJ planning meetings and one internal EJ training session for staff. These lasted from one to two hours and included about 20 to 30 agency staff. These internal meetings were especially valuable, as they allowed me to observe how bureaucrats react to and characterize proposed EJ reforms. Whereas in public meetings bureaucrats tend to present a united front and be very circumspect, in internal meetings they debated the merits of proposals and defended competing perspectives.

I analyzed my interview transcripts, fieldnotes, published scholarship, and other documents for what they revealed about constraints to regulatory agencies' EJ efforts and techniques through which EJ-supportive staff strive to circumvent or confront those constraints. I endeavor to be clear about how much I can generalize from my data. In some cases, I have sufficient evidence that an element of regulatory culture is fairly widespread. Notably, I interviewed such a large portion of all EJ staff in the United States, and their accounts have so much in common, that I feel confident making claims about the experiences of EJ staff in U.S. environmental regulatory agencies. Additionally, my own findings and other scholarship indicate that environmental regulatory agency staff tend to identify as public servants and environmental stewards, take great pride in their work, and often make sacrifices for their jobs. In contrast, some of the practices and beliefs I describe in this chapter have not been studied by other scholars in the context of environmental regulatory agencies. Thus, while extant scholarship shows that these racialized narratives are widespread in U.S. society, and my own data allow me to confidently state that these practices widely affect EJ staff, I am unable to specify how widely held these ideas are by staff in general.

I use the term "EJ staff" to refer to those tasked with leading their agencies' EJ reform efforts; they are very few in number and usually have little authority. I refer to other agency representatives as "other staff," "colleagues," "coworkers," or "bureaucrats" (used with no negative connotation); these catch-all terms refer to staff of all ranks, both career and politically appointed individuals, and any degree of support for EJ reforms (some are strongly supportive, others ambivalent, and others staunchly opposed). All names are pseudonyms. All uncited quotations are from my own interviews or observations.

STAFF MEMBERS' RACIALIZED RESISTANCE TO AGENCY ENVIRONMENTAL JUSTICE REFORM EFFORTS

EJ scholars rightly explain that the slow pace of agencies' EJ efforts stems in part from limited resources, regulatory authorities, and analytical tools—factors beyond the control of agency staff. All of the staff I interviewed described ways in which these factors limit agencies' progress on EJ reforms. But I found that other dynamics are also at play—additional factors internal to agencies and within the control of staff that stymie agencies' EJ programs and practices.

EJ staff are few in number and possess minimal authority within their organizations. Thus, they are unable to *require* coworkers to implement EJ reforms to regulatory practice. Instead, they must educate their colleagues about EJ, try to convince them that EJ reforms are worth their support and cooperation, and negotiate EJ reforms their colleagues will accept. While some bureaucrats become supportive of EJ reforms, others actively denigrate them. Nearly all of the staff I interviewed who lead and/or help implement their agency's EJ efforts described facing pushback to EJ reforms from coworkers. They described their EJ work as "a battle" or a "struggle" and that they face considerable resistance from other staff. For example, Barbara stated, "Lots of folks at [this agency] think that we should not be doing EJ." Nicky summed up her years of experience as an EJ staff person at multiple agencies as, "It's just this total resistance."

At all agencies in my study, much of the discursive resistance to EJ that I witnessed and was told about focused on the fact that proposed EJ reforms identify *racial* environmental inequalities and seek to reduce them (among other goals). For example, when polluting facilities apply for permits to expand their operations, EJ advocates argue that government agencies should conduct extra community outreach and more thorough analysis of potential adverse impacts when the permits affect racially marginalized communities (as well as working-class, Native American, and overburdened communities). EJ advocates also argue that agencies should create "EJ grant" programs that allocate funds to organizations working to improve environmental conditions in racially marginalized, overburdened communities (and others that are disproportionately susceptible to those harms because of poverty or cultural oppression).

Below, I discuss *racialized* forms of staff pushback against proposed EJ reforms: colorblindness, post-racial claims, and racial prejudice. To be clear, EJ staff emphasize that not all of their colleagues engage in these practices, that many coworkers express considerable concern about EJ issues, and that some are very supportive of EJ reforms. However, EJ staff emphasized that *some* colleagues cast EJ reforms as contrary to regulatory practice and derail the efforts of EJ staff. The point is that this pushback is frequent enough to stymie EJ-supportive staff members' efforts to integrate EJ principles into regulatory practice. Additionally, because these narratives are widespread in U.S. society, *anyone can* deploy them in these ways.

Colorblindness and Bureaucratic Neutrality

Colorblind racial ideology is the belief that racism is limited to conscious, intentional discrimination. According to this perspective, bureaucrats must ensure that their decisions (that is, in relation to a permit, rule, or enforcement action) are not influenced by the demographic characteristics of the communities affected by that decision. In other words, ensuring one has not been "racist," under this perspective, requires *not* taking race into consideration—to be "colorblind." Accordingly, the race-conscious nature of EJ reforms violates this interpretation of *neutrality*.

Some staff use colorblind narratives to cast race-conscious EJ reforms as unwarranted and unjust. For example, Elizabeth stated that "some staff and managers" disparage proposed EJ reforms by characterizing them as "reverse racism. [They say,] 'Why should those communities of color get this extra treatment? We need to protect that middle-class white community, too."' From this viewpoint, neutrality, which is important in government agencies, means *ignoring* racial inequality.

One EJ staff person who has led video-guided EJ trainings for staff showed me the following anonymous feedback submitted by a coworker, who used colorblind arguments to reject as unfair the EJ training and other proposed EJ reforms designed to reduce racial environmental inequalities:

> The whole idea of this thing is based on a lie. There are many people that I have spoken to about this training that fundamentally disagree with what EJ purports to do. In the minds of many common sense folk this is nothing but propaganda. The lady in the video basically alludes that everything we do is racist, whether that's in the work environment, during our leisure time, or just as individuals. Not true at all. ... The training also concludes that when in doubt just blame a white person for your life circumstances if they are bad. I think that is the most racist thing I've seen in a long time. What I would change would [be] to have this training be taken out of the department. Having it be mandatory is the type of social engineering BS that the department and country does not want. It's Hitler-eske. ... Working hard and getting ahead in life is color-blind and not racist, just like the vast vast majority of people in this country.

By stating unequivocally that "the department and country" are "not racist," this staff member frames the EJ training and proposed EJ reforms that acknowledge racial inequalities as themselves racist and thus unjust. Rather, justice requires being "color-blind."

I observed other staff—all white—use such claims to reject EJ reforms. Tim, a manager, told me that he and other staff oppose EJ reforms because they are committed to being colorblind. When I asked him how receptive staff are to EJ reforms, he responded: "People are not that receptive here. Not because they don't know what it is, but because they *do* know what it is.... A lot of what we do here is remediation of contaminated sites. We are not worried about the color of the skin, or the thickness of the wallet in the house that's next door."

Other scholars have observed environmental regulatory agency staff use colorblind arguments to explain their failure to enact EJ reforms. Notably, Ryan Holifield (2004, pp. 293, 295) notes that EPA staff "took pains to emphasize that they never based their remedial decisions on the racial, ethnic, or economic composition of the community." This is in contrast to EJ advocates' assertions that agencies should do more thorough analysis and outreach in communities that have historically been marginalized from regulatory decision-making processes. For instance, he quotes one staff member as saying, "We select a remedy based on *science*, regardless of who lives around the site. Race does not play a part in my decision" (p. 293). Holifield concludes that, in part because of staff members' insistence on taking a colorblind approach to regulatory decision-making, "the Clinton EPA's approach to environmental justice did not include an effort to engineer a 'fairer' distribution of Superfund risk along lines of race or class" (p. 293).

Key here is the fact that bureaucracies and their staff are expected to be neutral. Thus, even though these organizations' procedures have disproportionately benefited whites, organization members reject EJ reforms as violating the guiding principle of bureaucracy itself: impartiality. Race-conscious policies like EJ reforms are confusing to many staff and something to defend their organization against. Wendy Moore and Joyce Bell (2017) explain that people widely appeal to impartiality and neutrality to preclude acknowledgment that government institutions have ignored and reinforced racial inequality and to reject calls for institutional reforms that would reduce those inequalities: "Colorblind racist discourse...often takes place through an espoused commitment to 'abstract liberalist' discursive tenets, which profess rhetorical commitment to equality of opportunity and, at the same time, minimize the contemporary relevance of the history of explicit racial oppression as well as contemporary institutional and structural mechanisms that perpetuate racial inequality" (p. 101; see also Evans & Moore, 2015).

Accordingly, staff can critically portray current regulatory practice as unbiased and those promoting EJ reforms as biased and/or selfish. Numerous EJ staff told me that colleagues vaguely disparage EJ reforms as unjustifiably trying to hoard resources for a particular group of racial minorities. For instance, Peter recounted that coworkers would deride elected officials advocating EJ policies as doing so "for their own political purposes...for getting and maintaining the support of their constituents." Similarly, they would accuse Peter of using EJ as a way to "self-promote" or "opportunistically look after my own self interests." Such narratives delegitimize EJ reforms as unwarranted and those who promote them as self-serving.

Moreover, some white staff have defended this colorblind interpretation of bureaucratic neutrality so stridently that they have filed formal complaints against EJ staff and their EJ reform efforts. Peter described how his coworkers halted some of his EJ efforts by filing a formal civil rights complaint, alleging that an EJ training he and fellow EJ staff administered violated staff members' rights, which prompted the agency's civil rights office to conduct a formal investigation. Notably, although this allegation was ultimately disproven, the investigation halted their EJ trainings indefinitely: The investigation took a long time to complete, and the new management that came in during that time were not ready to restart the trainings. Peter lamented, "the formal investigation shut down the training, which was what the complainant wanted to achieve: sabotage the training.... The investigation unfortunately severely limited our ability to provide foundational understanding and valuable training to staff, which would have furthered their understanding of EJ and underserved communities." EJ staff must strike while the iron is hot—they get permission to do things like in-person, group EJ trainings only when management is supportive and thus must take advantage of those opportunities when they are available. In this case, by the time the investigation was finally concluded, the new managers did not support the EJ trainings.

Post-racial Ideology

Many EJ-supportive staff also attributed colleagues' resistance to EJ to *post-racial ideology*, the belief that racism is a thing of the past and thus that government efforts to reduce racial inequalities are unnecessary and unjustified. Despite extensive evidence of racial discrimination, bias, and associated material racial inequality in every sector of American life—employment, housing, crime and punishment, the marketplace, education, health care, environmental regulation, and beyond—wide swaths of the American public assert that we live in a post-racial society. EJ-supportive staff—including white staff as well as staff of color—explain that coworkers invoke such arguments to denounce EJ reforms instructing staff to take the racial characteristics of communities into account.

For example, Anne told me that, in the EJ meetings attended by representatives from the various departments, "We spent *months* just trying to convince some folks that inequity existed and was real." Scott, an EJ staff person at a different agency, said that his colleagues express a lot of "skepticism that this is an issue at all." Brian told me that some of his colleagues disparaged his EJ reform efforts by asserting that racism is no longer a problem in U.S. society:

> The bureaucrats say, "Oh, and EJ's important now again.... Here we go again."... We have so many of these people who've been around here for years.... Layer after layer of people who are just gatekeepers and do not think that racism exists [or that the] color of your skin has anything to do with anything, that that's all back in the '60s, [that] we shouldn't be worried about that stuff, and you should speak English. Not in leadership, but in the ranks, and the ranks control a lot of stuff. They call themselves the "we-bes": "We be here before you; we be here after you."...They say, "We'll wait out this administration. We'll wait out the legislators. Let's just wait them out. They're going to be turned out in a year. We'll just punt it."

Such assertions trivialize EJ reforms as a recurring fad, and also as obsolete because racism is no longer salient. All of these accounts convey that EJ-supportive staff have to spend precious time debating the existence of racial inequality, which stymies their progress on EJ reforms. Additionally, Brian's account depicts staff wielding these narratives to proudly declare they will refuse to implement EJ reforms to regulatory practice and implies that such behavior is acceptable and long-standing.

When reflecting on coworkers' post-racial narratives, most EJ-supportive staff noted that most agency staff are white and from predominantly white, middle- or upper-middle class neighborhoods. They asserted that these types of privilege make staff perceive environmental and other forms of inequality as not serious and thus EJ reforms as unnecessary. For example, Michael asserted that "many offices at [this agency] are homogeneous in terms of race and the orientation of the staff regarding the realities that real people face. This homogeneous complexion of the [agency's] staff contributes to a problem of groupthink—namely, the groupthink that EJ is irrelevant or not worth people's time." Similarly, here is how Anne explained why bureaucrats in her agency repeatedly questioned the existence and scope of racial environmental inequality:

> They haven't really understood, experienced, seen, lived the life that people who are disadvantaged live. So, they don't always get that justice is an important issue, because they have lived a life of privilege.... [They hold the mindset of] "pull yourself up by your bootstraps;"... "everybody has equal opportunity;" [and] "it's a level playing field."... If you come into the world with that kind of mindset and you have been blessed by those opportunities, then you don't really see it.

Many EJ-supportive staff revealed their own personal experience of racial oppression, having a diverse social network, and/or living in environmentally hazardous communities. They speculated that staff without such life experience are less likely to appreciate the severity of environmental inequalities and thus less likely to view EJ reforms as urgent or even necessary. For example, Jamie specified that one of the major "challenges" facing EJ policy implementation at his agency is "an issue around the diversity of thought and a greater context of diversity of experiences" among staff. He noted that he was raised in an industrially polluted community with high rates of cancer and other diseases linked to environmental toxics, and that many members of his own family died from cancer. He explained how this experience fuels his commitment to EJ reforms, and how his coworkers' antipathy for EJ stems in part from their lack of such experience:

> I have also watched a lot of people get sick and die over the years who are waiting for those incremental steps [of EJ reforms] to become more than baby steps, to come at a quicker pace. And especially those of us that come from these communities, we understand that this is a marathon, but at the same time we also want to see progress happen much quicker.... We bring really smart people in who have no connection to communities. So, therefore, the things they create many times miss what is really going on, on the ground, in people's lives.

Jamie points to his experience growing up in an overburdened community to explain why he views EJ reforms as *crucial and imperative*. In contrast, the fact that most other staff lack that experience means they lack that sense of urgency and make decisions that fail to adequately address the conditions people face in overburdened communities. Other scholars have similarly argued that bureaucrats' own experiences living in overburdened and vulnerable communities have motivated their commitment to EJ-type reforms (Engelman Lado, 2017, p. 69 fn. 95; Kohl, 2015).

Racial Prejudice

Some staff delegitimize EJ reforms by using racially prejudiced arguments that working-class communities, those of color in particular, are irresponsible. For example, Rose told me that, when she first proposed an EJ grant program to fund community-based organizations working to reduce environmental hazards, "Inside the agency, they started screaming and hollering: 'Are you crazy?.... They don't know how to do paperwork.... They'll be out buying cars and televisions."' This stereotype-fueled pushback was so severe that Rose had to integrate unprecedented restrictions on how the grants could be spent. Additionally, this motivated her to keep the grant small in size: "To keep people off my back, I'm going to make sure that that the money isn't a lot. Maybe about $20,000, $25,000—no more than that. I didn't want to go beyond that. Because I knew the oversight, it was going to be terrible."

I witnessed staff discursively wield racial prejudice to disparage proposed EJ reforms and other government assistance for working-class communities. I was invited to a quarterly meeting of about 25 county-level public health and environment officials to tell them about my research project and ask for their comments. Although the group does not regularly discuss "environmental justice" per se, the person who invited me felt that the group members do work that is highly relevant to environmental inequalities and thus that they would be interested in learning about my research. After introducing them to environmental inequalities and basic EJ principles, I asked what they thought agencies' EJ efforts should entail. As in my one-on-one interviews with staff, this group's responses were varied, and they insisted that agencies' abilities to address environmental inequalities are constrained by regulatory authority and dwindling resources. Yet some also scoffed at the idea of providing extra state resources to low-income communities—and low-income Native American and immigrant communities, in particular—by asserting that community members would squander those investments. To illustrate, here is an excerpt of the discussion as I recorded it in my fieldnotes:

> Person A [white man]: "Low-income populations seem to make the worst choices." They don't make much money, but then they buy "cigarettes and booze. Inappropriate choices!"
>
> Person B [white man]: We should get "the WIC ladies" to give you a presentation. "It is a ridiculous waste of money we are throwing at food stamps. They're buying *frozen dinners*!"
>
> Person C [white woman]: "*Ice cream!!*"

> Person B: "They can buy all that stuff now. To me, *that's* an injustice. Our government programs are creating the problem."
>
> Person A: He ranted about "program abuse"—"alcohol abuse" among Native Americans, "who learn that the more kids you have, the more money they can get from the government and that it's easier to get a check from the government than get a job." The system is "easy to abuse."

All of these individuals have leadership roles in county public health agencies and thus decide whether their subordinates take inequalities seriously. Because their organizations have not been formally directed to implement EJ reforms and programs that provide extra assistance to working-class and racially marginalized communities, the extent to which they do so will be limited by these officials' discursive work that casts those communities' residents as irresponsible.

Numerous EJ staff and coworkers who support them feel that some coworkers' pushback against EJ reforms also stems from racist prejudice against staff of color—particularly against those who are black. Several black EJ staff elaborated that they and other staff of color have experienced racial discrimination in the workplace in ways that have undermined their work. One black EJ staff person spent most of our interview describing coworkers' racist hostility toward him throughout multiple decades at the U.S. EPA. He described instances of other staff and managers bullying him, physically threatening him, and actively limiting his career opportunities. Other staff of color have told him about similar experiences. For example, when he started working at the agency, a black woman who works there and had known him through his previous job called him and said, "Didn't anyone tell you that EPA is a hard place for a black person to work?" He noted that former EPA staff member Marsha Coleman-Adebayo won a discrimination lawsuit against the agency in 2000 on the grounds of racial and sex discrimination (Coleman-Adebayo, 2011), that Congress eventually passed the Notification and Federal Employee Antidiscrimination and Retaliation (No FEAR) Act in 2002 to discourage unlawful discrimination and retaliation against employees, and that, subsequently, many EPA staff of color shared their experiences of discrimination and formed support groups for staff of color. He emphasized that his white colleagues do not share these experiences. When he tells white colleagues about the discriminatory experiences he and other staff of color have endured at EPA, they express utter surprise, saying, "Here?! Really?! I have such a great experience working here." His account conveys that many staff of color feel beleaguered by racial prejudice at work. Enduring such hostility is painful and pulls them away from their own job responsibilities. He emphasized that this is compounded for EJ staff of color, who sustain hostility not only for being persons of color but also for challenging the agency's role in racial environmental inequalities.

In a pained conversation with me, Michael explained that he and other EJ staff are informally bullied for their EJ work, that such treatment is increasingly difficult to prove, and that it is tied to the fact that they are persons of color:

> There are plenty of examples, ... instances in which I know persons have said that a manager has told them, "Don't work on EJ, because that may affect your career." ... People have been called "trouble-makers."... When you have a manager that just says it when you

> are walking down the hall and you haven't talked to that manager in weeks, you don't follow up with them to say, "Why would you say something like that?" They don't say, "Still causing trouble in that particular project?" No! He just says, "Still causing trouble?"... They weren't precise when they said it. That's how most discrimination occurs. That's why you can't prove it. It's because when someone does something that is discriminatory, they are not trying to be obvious about it.... They don't want to give you an obvious red flag by where you would record and document what they did. It could be written off as, "Well, maybe he didn't mean anything by it." Or, "Maybe you're just interpreting it the wrong way. I wouldn't think anything of it." Or they'll just get dismissive: "Don't worry about it. Maybe he was having a bad day." But he said it. Most of the time it has been in the context of EJ.

He explained that these aggressions undermine EJ policy implementation by compelling staff who would otherwise champion EJ efforts to find other assignments in the agency or find other employment: "People leave. It gets grinding. People burn out.... Those who are trying to focus on EJ feel a pervasive sense of tension, friction, of personalities clashing. Some of those people have moved on. Some stayed and didn't make any headway. It is a huge emotional drain." Michael feels that he has been blacklisted in the agency because of his EJ work, being passed up for promotions to jobs for which he is more than qualified and not receiving raises when he should. He got visibly upset during this part of our conversation, leaving me with the impression that he has been worn down from such treatment.

Indeed, most of the stories of being bullied and sanctioned came from black EJ staff. (To be clear, not all black staff described such experiences.) In contrast, most white EJ staff expressed frustration and disappointment at the slow pace of their agencies' EJ efforts, but did not report being personally bullied or threatened for their EJ work (with the exception of Peter, who is white). These observations suggest that the experience of working on agencies' EJ efforts varies to some extent along racial lines, where black EJ-supportive staff experience more hostility than their white colleagues do, and underscore the racialized nature of staff resistance to EJ. Research in other settings shows that this is part of a broader pattern whereby, in mostly white institutions whose members stridently identify as race neutral, staff of color who challenge racial narratives or organization practices that reproduce racial inequality are disparaged as "overly focused on race [and] too sensitive" (Evans & Moore, 2015, p. 445; see also Moore, 2008; Moore & Bell, 2017). To avoid sanction and maintain their careers in such settings, staff of color must endure "both everyday racial micro-aggressions and dismissive dominant ideologies that deny the relevance of race and racism" (Evans & Moore, 2015, p. 439).

Racial prejudice is also evident when white EJ staff are taken more seriously than EJ staff of color. For example, Jim, a white EJ staff person who worked at his agency for several decades, told me that, when *he* would talk about EJ reforms with bureaucrats in his agency, they would "take it easier from me" than they did from a particular black EJ staff person he worked with, "because I am white. Whereas, if [the colleague] says it, there is an immediate tension. It's a tense situation of racial issues." Jim asserted that, when EJ staff are all or mostly "minorities," as it was in his agency and is often the case elsewhere, then other bureaucrats think, "Oh, they are just pushing the black agenda." His comments imply that some coworkers perceive black EJ staff as unfairly hoarding resources for black people ("just pushing the black agenda"), whereas a white EJ staff person advocating for the same reforms would be taken more seriously. When EJ staff are persons of color, as is widely the case, such attitudes drive dismissiveness that undermines agencies' EJ efforts.

EJ staff of color emphasized that they do not view their current workplace as uniquely hostile in this regard. For instance, in response to my question about which factors constrain EJ reform at his agency, Malcolm, a black EJ staff person, responded by describing forms of racial discrimination. For half of our recorded interview and a full 20 minutes after I turned off the recorder when I thought our interview had ended, he elaborated about various forms of racial oppression he has experienced in his life in workplaces and interactions with police, forms of discrimination and violence against black men in the United States today, how prevalent racial discrimination and bias are in U.S. society, and how his agency is no different from broader society in this regard. He concluded, "It is all the time. All the time. This is every day."

A few EJ staff conveyed their sense that racial prejudice constrains their work by describing management condoning staff members' racist behavior. For instance, two noted that two senior staff members of California EPA's Department of Toxic Substances Control (DTSC), both white men, were shown to have exchanged numerous emails containing racist and other hate speech about coworkers and communities DTSC serves (Bogado, 2015; McGreevy, 2015). Journalist Aura Bogado and Penny Newman, an EJ advocate she interviewed and who helped unearth these emails through a public records request, emphasize in the article that most of the hazardous waste sites these bureaucrats are responsible for investigating and addressing are located in communities of color (Bogado, 2015). Bogado and Newman argue that these bureaucrats' bigoted attitudes likely lead them to do shoddy work in those communities. Indeed, Newman describes multiple instances in which these two staff members concluded that community concerns about hazardous waste were unwarranted, while subsequent investigations by advocates and other agencies showed that contamination levels exceeded levels that government agencies deem safe.

My research participants who brought up this issue in our interviews noted that DTSC retained these two staff. Thus they feel that the agency's management, by protecting and failing to sufficiently discipline staff who engage in blatantly racist behavior, effectively tolerates such discourse and signals to other staff that management does not value efforts (like EJ reforms) designed to identify racial injustice and improve regulatory practice in communities of color.

CONCLUSION

Environmental regulatory agencies like the EPA are staffed with people who identify as public servants, environmental stewards, and scientists, and who take great pride in protecting the environment and public health. Since their inception, agencies have endured budget cuts and been forced to defend their authority to regulate industry. Staff feel beleaguered by antiregulatory attack and have accomplished much within those constraints. Yet, working-class, racially marginalized, and Native American communities have always been disproportionately exposed to deadly environmental hazards relative to wealthier, white communities. Moreover, as I have shown in this chapter, environmental regulatory agency staff who express deep pride in their work also express disinterest in or even hostile resistance to EJ reforms designed to improve conditions in the nation's most environmentally burdened and vulnerable communities. In so doing, these well-meaning people, widely regarded as politically progressive, engage in practices that undermine prospects for social justice and equity. These findings provide new insights into why agencies have such a hard time implementing EJ reforms that would systematically extend environmental protections to racially marginalized, working-class communities.

In this chapter, I detailed various ways in which staff draw on racial discourses pervasive in U.S. society to denounce proposed EJ reforms as violating what their organization does and should do. I also noted several factors that may motivate these practices. Specifically, some staff resistance to EJ reforms stems from the fact that they interpret bureaucratic neutrality as requiring colorblindness, which leads them to conclude that EJ reforms targeting racial inequalities are unjust. Additionally, bureaucrats' relatively privileged life experiences lead some to believe that racial environmental inequalities are not serious enough to warrant EJ reforms. Also, racist prejudice drives some staff to engage in practices that wound, stall, and otherwise wear down staff of color and those who fight for racial justice. These practices also signal to other staff that racial equity is unimportant, either within the workplace or in terms of environmental inequalities.

Elsewhere (Harrison, 2017, 2019), I elaborate that bureaucrats' pushback against proposed EJ reforms stems in part from numerous other factors. Notably, some bureaucrats resist EJ reforms and EJ staff because they offend their sense of expertise. Additionally, some resistance to EJ reforms is rooted in bureaucrats' principal commitment to conservation of natural resources, which leads them to view programs focused on human health, including EJ reforms, as not a priority. In other cases, staff members are committed to serving industry, which leads them to perceive community-focused EJ reforms as not part of their jobs and/or as unfairly burdening industry. Moreover, some antipathy toward EJ reforms stems from staff members' utilitarian ideas of justice, which lead them to see EJ reforms as irrelevant to what it means for government agencies to do good work. That is, some bureaucrats' resistance to EJ reforms stems in part from the same factor that motivates staff to denounce hostile budget cuts by conservative elites: Both challenge a system they feel is effective, are proud of, and are committed to, and that they feel is fair. While a full discussion of these motivations is beyond the scope of this chapter, I note them to emphasize that, in many cases, staff resist proposed EJ reforms because they see them as violating their organization's identity and what it means to be good public servants.

Nevertheless, bureaucrats' pushback against proposed EJ reforms reinforces a regulatory system that has not protected America's racially marginalized, working-class, and other marginalized groups nearly as well as wealthier, white communities. Justice requires a new environmental regulatory regime, one that does a better job of protecting the health of people in America's most environmentally overburdened and vulnerable communities. To meaningfully support environmental justice, government agencies need committed leadership and resources. Yet agencies must also look within and address their own cultural dynamics through which bureaucrats undermine EJ reform proposals. The practices I showcased here reflect racial ideologies pervasive throughout U.S. society. Reducing racial inequalities requires that we confront these notions—within environmental regulatory agencies and beyond.

DEEPENING OUR UNDERSTANDING

In small groups, choose one state environmental regulatory agency (perhaps the one for the state in which you live now), and scour its web pages and online documents for material pertaining to "environmental justice" or "environmental equity" to describe EJ policies and programs. Consider the following questions to write a short report about the agency you chose:

1. What are its EJ policies and programs?
2. Who is responsible for overseeing them? How do they define "environmental justice"?
3. What evidence do they provide about existing environmental justice issues in their state?
4. Which EJ issues do they *not* discuss?
5. What constraints might influence this agency's EJ efforts?
6. Through which mechanisms can members of the public (like you) contact this agency about its EJ efforts? What could you, as an educated citizen, do to inquire about, support, and help shape this agency's EJ efforts?

References

Banzhaf, H. S. (2011). Regulatory impact analyses of environmental justice effects. *Journal of Land Use and Environmental Law, 27*(1), 1–30.

Bogado, A. (2015, December). What this California department's racist emails could mean for the communities it's supposed to protect. *Grist.* https://grist.org/politics/what-this-california-departments-racist-emails-could-mean-for-the-communities-its-supposed-to-protect/

Bonilla-Silva, E. (2014). *Racism without racists: Color-blind racism and the persistence of racial inequality in America* (4th ed.). Lanham, MD: Rowman and Littlefield.

Bruno, T., & Jepson, W. (2018). Marketisation of environmental justice: U.S. EPA environmental justice showcase communities project in Port Arthur, Texas. *Local Environment, 23*(3), 276–292. doi:10.1080/13549839.2017.1415873

Bullard, R. D., Mohai, P., Saha, R., & Wright, B. (2007). *Toxic wastes and race at twenty 1987–2007.* Cleveland, OH: United Church of Christ Justice and Witness Ministries.

Centeno, M. A., & Cohen, J. N. (2012). The arc of neoliberalism. *Annual Review of Sociology, 38*, 317–340. doi:10.1146/annurev-soc-081309-150235

Coleman-Adebayo, M. (2011). *No fear: A whistle-blower's triumph over corruption and retaliation at the EPA*. Chicago: Chicago Review Press.

Davenport, C. (2017, February). E.P.A. workers try to block Pruitt in show of defiance. *New York Times*. https://www.nytimes.com/2017/02/16/us/politics/scott-pruitt-environmental-protection-agency.html

Dillon, L., Sellers, C., Underhill, V., Shapiro, N., Ohayon, J. L., Sullivan, M., ... Wylie, S. (2018). The EPA in the early Trump administration: Prelude to regulatory capture. *American Journal of Public Health, 108*(S2), S89–S94. doi:10.2105/ajph.2018.304360

Eady, V. (2003). Environmental justice in state policy decisions. In J. Agyeman, R. D. Bullard, & B. Evans (Eds.), *Just sustainabilities: Development in an unequal world* (168–182). Cambridge, MA: MIT Press.

Engelman Lado, M. (2017). Toward civil rights enforcement in the environmental justice context: Step one: Acknowledging the problem. *Fordham Environmental Law Review, 29*, 46–94.

Evans, L., & Moore, W. L. (2015). Impossible burdens: White institutions, emotional labor, and micro-resistance. *Social Problems, 62*, 439–454. doi:10.1093/socpro/spv009

Faber, D. (2008). *Capitalizing on environmental injustice: The polluter-industrial complex in the age of globalization*. Lanham, MD: Rowman and Littlefield.

Foster, S. (2000). Meeting the environmental justice challenge: Evolving norms in environmental decisionmaking. *Environmental Law Reporter, 30*(11), 10992.

Fredrickson, L., Sellers, C., Dillon, L., Ohayon, J. L., Shapiro, N., Sullivan, M., ... Wylie, S. (2018). History of U.S. presidential assaults on modern environmental health protection. *American Journal of Public Health, 108*(S2), S95–S103. doi:10.2105/ajph.2018.304396

Gardner, T. (2017, February). Nearly 800 former EPA officials oppose Trump pick for agency. *Reuters*. https://www.reuters.com/article/usa-epa-pruitt-idUSL1N1G11GK

Gauna, E. (2015). Federal environmental justice policy in permitting. In D. M. Konisky (Ed.), *Failed promises: Evaluating the federal government's response to environmental justice* (pp. 57–83). Cambridge, MA: MIT Press.

Gerber, B. J. (2002). Administering environmental justice: Examining the impact of Executive Order 12898. *Policy and Management Review, 2*(1), 41–61.

Goldberg, D. T. (2002). *The racial state*. Oxford, UK: Blackwell.

Harrison, J. L. (2014). Neoliberal environmental justice: Mainstream ideas of justice in political conflict over agricultural pesticides in the United States. *Environmental Politics, 23*(4), 650–669. doi:10.1080/09644016.2013.877558

Harrison, J. L. (2015). Coopted environmental justice? Activists' roles in shaping EJ policy implementation. *Environmental Sociology, 1*(4), 241–255. doi:10.1080/23251042.2015.1084682

Harrison, J. L. (2016). Bureaucrats' tacit understandings and social movement policy implementation: Unpacking the deviation of agency environmental justice programs from EJ movement priorities. *Social Problems, 63*(4), 534–553. doi:10.1093/socpro/spw024

Harrison, J. L. (2017). "We do ecology, not sociology": Interactions among bureaucrats and the undermining of regulatory agencies' environmental justice efforts. *Environmental Sociology, 3*(3), 197–212. doi:10.1080/23251042.2017.1344918

Harrison, J. L. (2019). *From the inside out: The fight for environmental justice within government agencies*. Cambridge, MA: MIT Press.

Harvey, D. (2005). *A brief history of neoliberalism*. New York: Oxford University Press.

Holifield, R. (2004). Neoliberalism and environmental justice in the United States Environmental Protection Agency: Translating policy into managerial practice in hazardous waste remediation. *Geoforum, 35*, 285–297. doi:10.1016/j.geoforum.2003.11.003

Holifield, R. (2012). The elusive environmental justice area: Three waves of policy in the U.S. Environmental Protection Agency. *Environmental Justice, 5*(6), 293–297. doi:10.1089/env.2012.0029

Holifield, R. (2014). Accounting for diversity in environmental justice screening tools: Toward multiple indices of disproportionate impact. *Environmental Practice*, *16*, 77–86. doi:10.1017/s1466046613000574

Kohl, E. (2015, April). *"People think we're EPA, we can do whatever we have the will to do": Negotiating expectations of environmental justice policies*. Paper presented at the annual meeting of the Association of American Geographers. Chicago, IL.

Kohl, E. (2016, April). *The performance of environmental justice: The Environmental Protection Agency's Collaborative Problem Solving model as cooption*. Paper presented at the annual meeting of the Association of American Geographers. San Francisco, CA.

Konsiky, D. M (Ed.). (2015). *Failed promises: Evaluating the federal government's response to environmental justice*. Cambridge, MA: MIT Press.

Kurtz, H. (2009). Acknowledging the racial state: An agenda for environmental justice research. *Antipode*, *41*(4), 684–704. doi:10.1111/j.1467-8330.2009.00694.x

Lashley, S. E. (2016). Pursuing justice for all: Collaborative problem-solving in the environmental justice context. *Environmental Justice*, *9*(6), 188–194. doi:10.1089/env.2016.0023

Liévanos, R. (2012). Certainty, fairness, and balance: State resonance and environmental justice policy implementation. *Sociological Forum*, *27*(2), 481–503. doi:10.1111/j.1573-7861.2012.01327.x

Liévanos, R. S., London, J. K., & Sze, J. (2011). Uneven transformations and environmental justice: Regulatory science, street science, and pesticide regulation in California. In G. Ottinger & B. R. Cohen (Eds.), *Technoscience and environmental justice: Expert cultures in a grassroots movement* (pp. 201–228. Cambridge, MA: MIT Press.

Lipsitz, G. (1995). The possessive investment in whiteness: Racialized social democracy and the "white" problem in American studies. *American Quarterly*, *47*(3), 369–387. doi:10.2307/2713291

London, J. K., Sze, J., & S, Liévanos, R. (2008). Problems, promise, progress, and perils: Critical reflections on environmental justice policy implementation in California. *UCLA Journal of Environmental Law & Policy*, *26*(2), 255–289.

Mascarenhas, M. (2012). *Where the waters divide: Neoliberalism, white privilege, and environmental racism in Canada*. Lanham, MD: Lexington Books.

Mascarenhas, M. J. (2016). Where the waters divide: Neoliberal racism, white privilege and environmental injustice. *Race, Gender, and Class*, 6–25. 23(3/4),

McGreevy, P. (2015, December). Toxics agency chief condemns racially charged emails. *Los Angeles Times*. http://www.latimes.com/politics/la-me-pc-toxics-agency-chief-condemns-racially-charged-emails-20151209-story.html

Moore, W. (2008). *Reproducing racism: White space, elite law schools, and racial inequality*. Lanham, MD: Rowman & Littlefield.

Moore, W. L., & Bell, J. M. (2017). The right to be racist in college: Racist speech, white institutional space, and the First Amendment. *Law and Policy*, *39*(2), 99–120. doi:10.1111/lapo.12076

Mueller, J. C. (2017). Producing colorblindness: Everyday mechanisms of white ignorance. *Social Problems*, *64*, 219–238.

Noonan, D. S. (2015). Assessing the EPA's experience with equity in standard setting. In D. M. Konisky (Ed.), *Failed promises: Evaluating the federal government's response to environmental justice* (pp. 85–116). Cambridge, MA: MIT Press.

Omi, M., & Winant, H. (2015). *Racial formation in the United States* (3rd ed.). New York: Routledge.

Park, L. S.-H., & Pellow, D. N. (2004). Racial formation, environmental racism, and the emergence of Silicon Valley. *Ethnicities*, *4*(3), 403–424. doi:10.1177/1468796804045241

Payne-Sturges, D., Turner, A., Wignall, J., Rosenbaum, A., Dederick, E., & Dantzker, H. (2012). A review of state-level analytical approaches for evaluating disproportionate environmental health impacts. *Environmental Justice*, *5*(4), 173–187. doi:10.1089/env.2012.0008

Pellow, D. N. (2018). *What is critical environmental justice?* Medford, MA: Polity Press.

Pulido, L. (2000). Rethinking environmental racism: White privilege and urban development in Southern California. *Annals of the Association of American Geographers*, *90*(1), 12–40. doi:10.1111/0004-5608.00182

Pulido, L. (2015). Geographies of race and ethnicity I: White supremacy vs white privilege in environmental racism research. *Progress in Human Geography*, *39*(6), 809–817. doi:10.1177/0309132514563008

Richter, L. (2018). Constructing insignificance: Critical race perspectives on institutional failure in environmental justice communities. *Environmental Sociology*, *4*(1), 107–121. doi:10.1080/23251042.2017.1410988

Save EPA. (2017). Save EPA. Retrieved from http://saveepaalums.info/

Shadbegian, R. J., & Wolverton, A. (2015). Evaluating environmental justice: Analytic lessons from the academic literature and in practice. In D. M. Konisky (Ed.), *Failed promises: Evaluating the federal government's response to environmental justice* (pp. 117–142). Cambridge, MA: MIT Press.

Shilling, F. M., London, J. K., & S, Liévanos, R. (2009). Marginalization by collaboration: Environmental justice as a third party in and beyond Calfed. *Environmental Science and Policy*, *12*(6), 694–709. doi:10.1016/j.envsci.2009.03.003

Targ, N. (2005). The states' comprehensive approach to environmental justice. In D. N. Pellow & R. J. Brulle (Eds.), *Power, justice, and the environment: A critical appraisal of the environmental justice movement* (pp. 171–184). Cambridge, MA: MIT Press.

Taylor, D. (2014). *Toxic communities: Environmental racism, industrial pollution, and residential mobility.* New York: New York University Press.

U.S. Government Accountability Office. (2005). *EPA should devote more attention to environmental justice when developing clean air rules.* https://www.gao.gov/assets/250/247171.pdf

Vajjhala, S. P., Epps, Van., A, Szambelan, S. (2008). *Integrating EJ into federal policies and programs: Examining the role of regulatory impact analyses and environmental impact statements.* Resources for the Future Discussion Paper, 08-45. Washington, DC:Resources for the Future.

Vajjhala, S. P. (2010). Building community capacity? Mapping the scope and impacts of EPA's Environmental Justice Small Grants program. In D. E. Taylor (Ed.), *Environment and social justice: An international perspective (Research in Social Problems and Public Policy 18)* (pp. 353–381). Bingley, UK: Emerald Group.

9

GEOGRAPHIES OF ENVIRONMENTAL RACISM

Capitalism, Pollution, and Public Health in Southern California

Cristina Faiver-Serna

In spring 2013 I visited the home of "Juana," the mother of a young child with asthma, who was a patient at the clinic where I worked. I shadowed and assisted "Antonia," an experienced *promotora de salud* (community health worker) who had been working with Juana and her child to prevent asthma attacks and regular emergency room visits. We sat in Juana's living room as Antonia looked through an illustrated asthma education flip chart. Antonia had assembled this easy-to-use bilingual tool, full of colorful, descriptive images, to educate families in her community about asthma. We explained the biology of asthma and described the sensations of chest restriction that Juana's child experiences during an asthma attack. We also explained that asthma is a lifelong condition, not something that her child would outgrow. But, with proper medication and a clean home environment free of asthma triggers, it can be managed and Juana's child can enjoy a healthy and active life.

Juana gave us a tour of her home, and we pointed out any potential asthma triggers. Stuffed animals and old carpeting harbor dust mites, mold in the bathroom can be harmful, and certain cleaning products have chemicals that will trigger an asthma attack. Juana would be able to eliminate some of these triggers on her own; others would require her landlord's cooperation. We gave Juana a bucket of environmentally-friendly cleaning supplies, and antiallergen mattress and pillow covers for her child. At the end of our visit, Juana reviewed her *asthma action plan* with us. The asthma action plan and the *home environmental assessment* are tools used to measure the fact that Juana had received asthma education and understood her role and responsibility in managing her child's asthma.

The home environmental assessment and the asthma action plan we used are socio-scientific tools developed by the state (see Figures 9.1 and 9.2). The use of these forms to help Juana manage her child's asthma increased her individual accountability to manage an illness not caused by the stuffed animals on her child's bed, but by the air pollution from the *global goods movement* (that is, by the international logistics systems and physical movement of globally imported goods from ports to cities around the country) in her neighborhood (Cowen, 2014; De Lara, 2018). Our use of these forms shifted responsibility to her, the mother, for the cause of her child's suffering. It absolved both the corporations that emitted large amounts of toxic pollution in communities of color and the state that lacked proper regulations to hold corporations responsible for the environmental harm they cause.

CRITICAL PERSPECTIVES ON THE ROLE OF PUBLIC HEALTH IN THE FIGHT FOR ENVIRONMENTAL JUSTICE

From 2010 to 2013 I worked in Long Beach, California, alongside *promotoras de salud* to remediate the life-threatening consequences of industrial air pollution on local community health. I oversaw the operations of an asthma education

---

Photo 9.1
Ben Willardson / Shutterstock

FIGURE 9.1 ■ The asthma action plan is a public domain socio-scientific contract adapted by many organizations and doctor's offices to help families manage the chronic and life-threatening symptoms of asthma.

ASTHMA ACTION PLAN

aafa Asthma and Allergy Foundation of America

Name:	Date:
Doctor:	Medical Record #:
Doctor's Phone #: Day	Night/Weekend
Emergency Contact:	
Doctor's Signature:	

Personal Best Peak Flow: ________

The colors of a traffic light will help you use your asthma medicines.

GREEN means Go Zone! Use preventive medicine.

YELLOW means Caution Zone! Add quick-relief medicine.

RED means Danger Zone! Get help from a doctor.

GO — Use these daily controller medicines:

You have *all* of these:
- Breathing is good
- No cough or wheeze
- Sleep through the night
- Can work & play

Peak flow: from ___ to ___

MEDICINE	HOW MUCH	HOW OFTEN/WHEN

For asthma with exercise, take:

CAUTION — Continue with green zone medicine and add:

You have *any* of these:
- First signs of a cold
- Exposure to known trigger
- Cough
- Mild wheeze
- Tight chest
- Coughing at night

Peak flow: from ___ to ___

MEDICINE	HOW MUCH	HOW OFTEN/ WHEN

CALL YOUR ASTHMA CARE PROVIDER.

DANGER — Take these medicines and call your doctor now.

Your asthma is getting worse fast:
- Medicine is not helping
- Breathing is hard & fast
- Nose opens wide
- Trouble speaking
- Ribs show (in children)

Peak flow: reading below ___

MEDICINE	HOW MUCH	HOW OFTEN/WHEN

GET HELP FROM A DOCTOR NOW! Your doctor will want to see you right away. It's important! If you cannot contact your doctor, go directly to the emergency room. DO NOT WAIT.

Make an appointment with your asthma care provider within two days of an ER visit or hospitalization.

Source: Asthma and Allergy Foundation of America (https://www.aafa.org/asthma-treatment-action-plan/).

program that was funded by federal money and a community air pollution mitigation grant from the Port of Long Beach. The asthma education program, "Bridge to Health," was centered around the *promotora de salud*, who acted as a "bridge" between the patient and family, and the healthcare system, that is, insurance providers, doctors, nursing staff, and pharmacies. The ideology of the

FIGURE 9.2 ■ First page of the home environmental assessment for asthma developed by federal agencies and available for public use. The document is 12 pages. It lists questions about the structure and conditions of the building and home environment that indicate if it is safe or unsafe for someone with asthma, and suggests potential action steps to get rid of conditions that would be triggering for someone with asthma.

ASTHMA HOME ENVIRONMENT

☑ CHECKLIST

Home visits provide an opportunity to educate and equip asthma patients with the tools to effectively manage their disease in concert with a physician's care. This checklist—designed for home care visitors—provides a list of questions and action steps to assist in the identification and mitigation of environmental asthma triggers commonly found in and around the home. The checklist is organized into three sections—building information, home interior and room interior. The room interior is further subdivided by categories (such as bedding and sleeping arrangements, flooring, window treatments, and moisture control). This will allow the home care visitor to focus on the specific activities or things in a room—in particular the asthma patient's sleeping area—that might produce or harbor environmental triggers. The activities recommended in this checklist are generally simple and low cost. Information on outdoor air pollution follows the checklist. The last page includes information on U.S. Environmental Protection Agency (EPA) resources and an area for the home care visitor to record a home visit summary.

If the patient's sensitivities to allergens (such as dust mites, pests, warm-blooded pets and mold) and irritants (such as secondhand smoke and nitrogen dioxide) are known, the home care visitor should begin by focusing on relevant areas. This checklist covers the following allergens and irritants, which are commonly found in homes. Information is also provided on chemical irritants—found in some scented and unscented consumer products—which may worsen asthma symptoms.

Dust Mites

Triggers: Body parts and droppings.

Where Found: Highest levels found in mattresses and bedding. Also found in carpeting, curtains and draperies, upholstered furniture, and stuffed toys. Dust mites are too small to be seen with the naked eye and are found in almost every home.

Pests (such as cockroaches and rodents)

Triggers: Cockroaches – Body parts, secretions, and droppings.
Rodents – Hair, skin flakes, urine, and saliva.

Where Found: Often found in areas with food and water such as kitchens, bathrooms, and basements.

Warm-Blooded Pets (such as cats and dogs)

Triggers: Skin flakes, urine, and saliva.

Where Found: Throughout entire house, if allowed inside.

Mold

Triggers: Mold and mold spores which may begin growing indoors when they land on damp or wet surfaces.

Where Found: Often found in areas with excess moisture such as kitchens, bathrooms, and basements. There are many types of mold and they can be found in any climate.

Secondhand Smoke

Trigger: Secondhand smoke – Mixture of smoke from the burning end of a cigarette, pipe or cigar and the smoke exhaled by a smoker.

Where Found: Home or car where smoking is allowed.

Nitrogen Dioxide (combustion by-product)

Trigger: Nitrogen dioxide – An odorless gas that can irritate your eyes, nose, and throat and may cause shortness of breath.

Where Found: Associated with gas cooking appliances, fireplaces, woodstoves, and unvented kerosene and gas space heaters.

1

Source: U.S. EPA https://www.epa.gov/sites/production/files/2018-05/documents/asthma_home_environment_checklist.pdf

work was rooted in an individualized approach to address community health problems caused by a much larger environmental problem. The health education and outreach programs I oversaw through the federally qualified health center (FQHC) where I worked were vital to serving medically underserved areas and populations.

Populations that FQHCs serve are defined according to geographic, demographic, and socioeconomic markers of non-white, working class, and poor. I began to wonder: Why does the state (that is, local, state, and federal governmental institutions, networks, and systems) simultaneously allow for the poisoning of racialized and marginalized communities, yet also set aside funds and develop programs to pay for the medical treatment of residents of those same communities? And why is the poisoning of certain communities justified and legitimated according to the growth of local industry, for the sake of local, national, and global economic development? As I came to understand the limited capacity that these programs, and the role I played in them, had to achieve social justice, this understanding generated larger questions about public health, racial capitalism, and geographies of environmental racism.

In recent years, environmental justice scholars have become increasingly critical of the state as a partner in the fight for environmental justice. Much of the field's critique has been directed at the failure of the U.S. Environmental Protection Agency (EPA) to effectively implement Executive Order 12898, "Federal Actions to Address Environmental Justice in Minority Populations and Low-Income Populations," and its failure to properly regulate polluting industry (Harrison, 2015; Konisky, 2015; Kurtz, 2009; Liévanos, 2012; Pellow, 2018; Pulido, 2016a, 2016b; Pulido, Kohl, & Cotton, 2016). And although demands for the state to provide public health services have always been a part of the call for environmental justice (Brulle & Pellow, 2006; Bullard, 2000; Cole & Foster, 2001; Pulido, 2016a; Schlosberg, 2007; Sze, 2006), critical consideration for the role that public health plays in the perpetuation of injustice is an area of much needed scholarship.

The recent emergence of *critical environmental justice* has conceptualized the state beyond the paradigm of regulator and protector. Pellow (2018) argues that the state is "authoritarian, coercive, racist, patriarchal, exclusionary, militaristic, and anti-ecological" (p. 23). Pulido (2016a) and Pellow (2018) both argue that continued environmental racism and injustice is perpetuated by the state and should be understood as "state-sanctioned violence." Further, Pulido (2016a, 2016b) contends that the state's deep entanglements with *racial capitalism*, a global economic system dominant in the United States, are precisely what makes the state a site of contestation for achieving true environmental justice.

Racial capitalism has always been entangled with the goals, operations, and expansion of the United States (Robinson, 2000). Native peoples were violently dispossessed of their land and culture in order for settlers to establish the United States as a nation-state and profit from the land as a "natural resource." Africans were dehumanized, imported as slave labor, and used by wealthy white landowners to accumulate capital and develop a racial caste system still very much alive today. Racial capitalism has played a key role in the rise of the production of cheap goods produced by racialized groups of low-wage workers, and it drives geographies of global economic trade (De Lara, 2018; Ferguson, 2001). In the United States, and specifically in Southern California, the import and movement of goods depends on both the racialized marginalization of people of color and the state's active endorsement in order to produce profit.

Moving forward in this chapter, we will use racial capitalism as a framework for mapping geographies of environmental racism in Southern California so as to better understand the role of the state, and specifically public health, in relation to environmental racism, and as a framework to discuss the spaces of contestation and contradiction that racial capitalism produces within and beside itself.

MAPPING GEOGRAPHIES OF ENVIRONMENTAL RACISM

The goal for profit and state hegemony has been a driving force for profound changes in the natural landscape of Southern California since U.S. settler westward expansion, and it has influenced the geography of the built environment of today (De Lara, 2018). The City of Long Beach was officially incorporated in 1897 and is the second largest city in Los Angeles County. Seven miles down the road from the Port of Los Angeles, the Port of Long Beach was founded in 1911 and built on 800 acres of coastal wetlands (Cunningham & Cunningham, 2015). In 1931 the Long Beach City Charter was amended to create the Harbor District, with public commissioners appointed by the city council to manage it. Then, in 1953, the harbor commissioners voted to use more than $12 million of public funds to complete Interstate 710 as a main transport route for the port (Estrada, 2014).

The 710 Freeway runs 23 miles from the Port of Long Beach along the Los Angeles River. Its designated purpose has always been to serve the Harbor District and facilitate the movement of goods through Los Angeles County for nationwide distribution. Today the Port of Long Beach, along with the Port of Los Angeles, brings in 43 percent of imported goods into the continental United States (Khouri, 2015).

Prior to the I-710 completion project, racist private and public lending had produced a segregated landscape in Los Angeles County. The National Housing Act of 1934 operationalized racist and segregationist agendas by public and private lenders. In the 1930s the Home Owners Loan Corporation in Los Angeles made "predictions" about the urban and suburban landscape of the Los Angeles Basin, designating many of the neighborhoods adjacent to the Los Angeles River as "definitely declining" or redlining them as "hazardous" (Figures 9.3 and 9.4). These designations made for fast sells by white homeowners fleeing to the suburbs, and made property values drop fast so that they were worth less than what black and Latinx people paid for them once they moved into the neighborhoods and owned their own homes (Lipsitz, 2011; Rothstein, 2017). In the 1950s the completion of I-710 along the Los Angeles River and through the "definitely declining" and "hazardous" neighborhoods cemented their status as "undesirable" real estate, as the neighborhoods now had a noisy, dirty freeway running through them.

The system and incentives of racial capitalism are deeply embedded in how the state operates. The state's decisions enable capitalist profit from the political, social, and environmental marginalization of poor and working-class people of color. In 2009 the Port of Long Beach commissioned an environmental impact report (EIR) to review the potential environmental impact of an expansion project in the port's middle harbor. The Port of Long Beach was keen to capitalize on the anticipated widening of the Panama Canal by investing in

FIGURE 9.3 ■ 1939 Home Owners Loan Corporation redlined map of Los Angeles County, Los Angeles River.

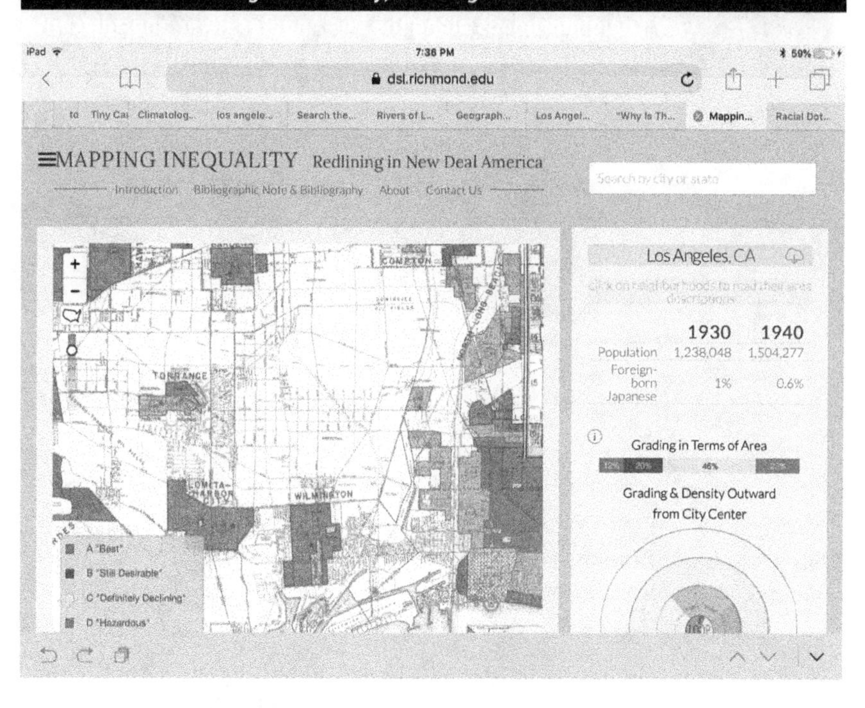

Source: University of Richmond,Mapping inequality: Redlining in New Deal America dsl.richmond.edu/panorama/redlining

infrastructure to handle more goods imported to Southern California. The middle harbor expansion project promised 14,000 new permanent jobs. The fallout from the 2008 Great Recession had hit the local economy hard. Unemployment rates in Los Angeles County were up to 12 percent at this time, and the expansion of the Port of Long Beach promised to alleviate some of this hardship. Port executives, as well as harbor commissioners, who called the expansion project their own "stimulus package," wanted it approved (Sahagun, 2009).

The California Environmental Quality Act of 1977 stipulates that an EIR must be provided to the public with detailed information about the effects that any proposed project is likely to have on the environment. EIRs must also list the ways in which significant negative effects can be minimized and provide possible alternatives to the project (Hildreth, 1977). The Port of Long Beach submitted a 1,500-page EIR final draft to the harbor commissioners, and the public, 10 days before they were to vote on its approval.

Despite the limited opportunity to review and respond to the EIR, environmental justice activists in the community organized and pushed back against the commissioners' immediate approval of the project. Community leaders cited heavy truck traffic along I-710 and through predominantly poor and working-class neighborhoods of color. They also cited a disproportionate excess of noise and light pollution, increased volume of toxic air pollution, and anticipated higher rates of respiratory disease and cancer in low-income neighborhoods and communities of color. Their efforts put pressure on the commissioners to require

FIGURE 9.4 ■ Redlining in the 1930s worked to solidify a segregated landscape in Southern California in the 2010s. The neighborhoods along the Los Angeles River and adjacent to the 710 Freeway are majority Latinx and people of color.

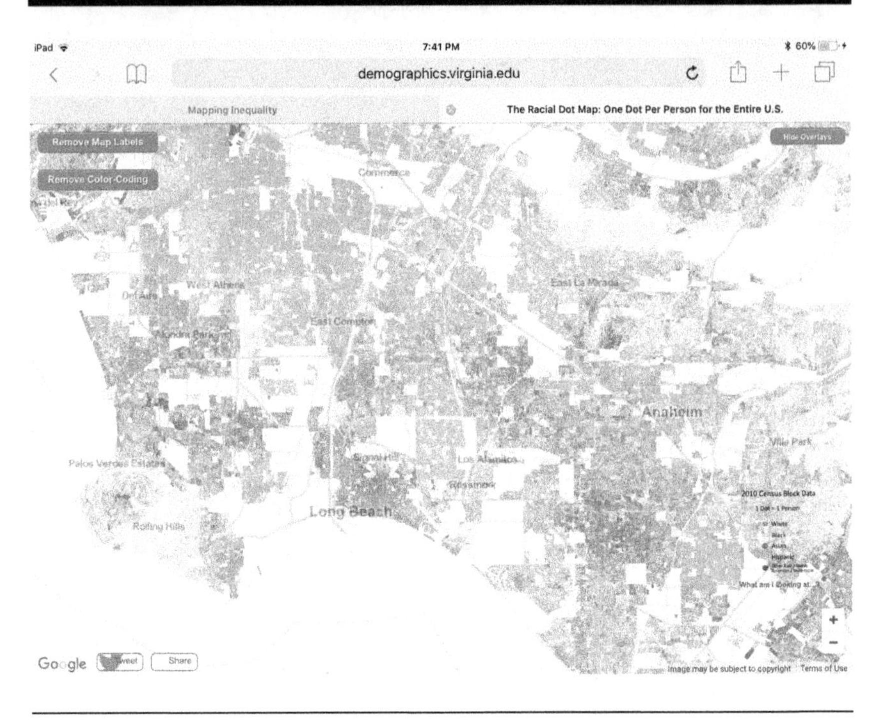

Source: University of Virginia, Racial dot map, 2010 U.S. Census (https://demographics.virginia.edu/DotMap/).

the Port of Long Beach to make some concessions to the community. The harbor commissioners thus required the port to provide an environmental mitigation fund as part of the expansion budget. The Port of Long Beach was directed to allocate $15 million, or 2 percent, of the $750 million expansion budget to mitigate the anticipated environmental impact on the local community (Daily News Wire Services, 2009; Port of Long Beach, 2009; Sahagun, 2009).

The geography of the Port of Long Beach heavily, and disproportionately, impacts majority Latinx neighborhoods in Los Angeles County. The routes its trucks take deposit toxic pollution in poor and working-class communities of color. In this way, the Harbor District's investment in the capitalist projects of the Port of Long Beach inflict violence on the local community and produce geographies of environmental injustice and racism.

Residents along I-710 experience high rates of asthma and other respiratory illnesses (Parvini, 2014; Perez et al., 2009; Polidori & Fine, 2012). Birth defects, asthma, cardiopulmonary disease, stroke, cancer, and chronic respiratory symptoms are linked to elevated and prolonged exposure to the pollutant particulates in diesel exhaust (Li et al., 2003; Perez et al., 2009; Sioutas et al., 2005). Particulate matter is a physical by-product from the burning of diesel fuel.

Particulates have uneven, sticky surfaces that toxic chemical traces easily latch onto. Fine particulate matter measures between 2.5 and 0.1 micrometers

FIGURE 9.5 ■ Particle pollution is measured in micrometers (µm). The California Environmental Protection Agency measures and regulates PM 10 and PM 2.5 but does not regulate PM < 0.1 in its own separate category despite research demonstrating the unique dangers of PM < 0.1 to human health.

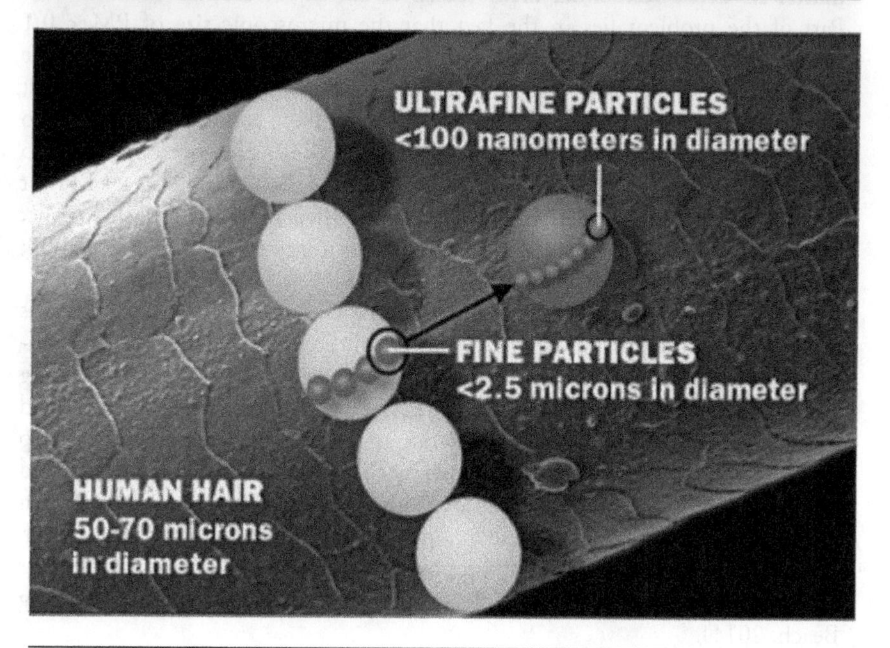

Source: State of California EPA (https://airnow.gov/index.cfm?action=aqibasics.particle).

in diameter, designated PM 2.5 (Figure 9.5). Studies that have captured and studied fine particulates in Southern California have found chemical traces of arsenic, sulfates, and nitrates, among other chemicals that are dangerous to human health (Li et al., 2003; Sioutas et al., 2005). The microscopic size and aerodynamics of PM 2.5 make it easy to travel far and wide, but it largely settles in the region due to the dry and hot climate, largely stagnant airflow, and bowl-like structure of the Los Angeles Basin (Littman & Magill, 1953; Schmool et al., 2013).

Ultrafine particulate matter is also a toxic by-product of diesel fuel but is even smaller than PM 2.5. Ultrafine particulates are any particulates less than 0.1 micrometers in diameter, designated PM < 0.1. These tiny, invisible objects quickly and easily find their way through a body's porous surface. Ultrafine particles can penetrate the mitochondria of individual cells, altering cellular function and inducing disease, especially in a child's growing and developing body. Concentrated exposures to ultrafine particulates are strongly linked to birth defects in utero, childhood asthma, as well as adult-onset asthma, heart disease, and cancer (Li et al., 2003; Ostro et al., 2015).

The State of California Environmental Protection Agency regulates air pollution according to geographic region. The South Coast Air Quality Management District monitors air quality for the Los Angeles Basin and measures the amount of PM 2.5 in the region. Although there have been decades of research on the dangerous effects of PM < 0.1 (see Li et al., 2003; Ostro et al., 2015; Sioutas et al., 2005), it is not monitored or regulated by the state of California, or federally. Academic and community partnerships have monitored and studied the high concentrations of ultrafine particulate matter in Long Beach and along I-710, finding high concentrations of this toxic matter in dense residential areas (Perez et al., 2009; Sioutas et al., 2005). Part of the problem lies in the fact that the microscopic size of PM < 0.1 makes it difficult, if not impossible, to filter and trap. The other part of the problem lies in the fact that the state is invested in the success of capitalist projects that continue to use dirty diesel fuel, and there isn't the political will to abandon its use. While data on the dangers of PM < 0.1 have been presented to regulatory agencies like the California Air Resources Board at the state's EPA, it continues to go unregulated and, thus, is less studied and known to the public, which compounds the danger it poses to community health (Li et al., 2003; Ostro et al., 2015).

In 2009 the Port of Long Beach's middle harbor expansion project was approved, and the port designated 2 percent of the project's budget to environmental mitigation as directed to by the harbor commissioners. The Port of Long Beach assigned $5 million per three categories: greenhouse gases, schools, and healthcare. In the first round of mitigation grant funding, the port drew a map of the areas that would be most affected by the expansion. The map designated communities along the I-710 corridor as high priority. The Port of Long Beach awarded the largest single grant to a nonprofit, federally qualified health center, The Children's Clinic. The clinic proposed to see and serve the largest number of community residents made sick by the port's air pollution (Port of Long Beach, 2011).

Federally qualified health centers are community health centers that meet a strict list of requirements set by the federal government in order to receive Medicaid funds, and they qualify to apply for additional grants to provide medical and public health services. FQHCs serve medically underserved areas (MUAs) and medically underserved populations (MUPs). These are geographic and demographic federal designations that identify communities and populations with limited access to primary health care (Figure 9.6). MUAs are determined according to state geographic designations, such as a county, a group of neighboring counties, census tract(s), or census county or civil divisions such as townships or precincts. MUPs are determined according to U.S. Census data by homelessness, low-income status, Medicaid-eligible, Medicare recipients, tribal enrollment, and/or migrant farmworker employment. MUA/P status is determined according to low-medical-provider to high-MUP-qualified-patient ratio, the percentage of a population below the federal poverty level, the percentage of a population above the age 65, and/or the infant mortality rate (Health Resources & Services Administration, n.d.).

FQHCs grew out of President Lyndon B. Johnson's War on Poverty initiative. They have their roots in the neighborhood health centers established under the 1964 Economic Opportunity Act. The FQHC status is unique and not easy to attain. Section 330 of the Public Health Service Act of 1996 designates special grant funding for FQHCs, in addition to the Medicaid and Medicare reimbursement for which they are eligible. FQHCs must routinely prove to the federal government that their comprehensive health care is accessible to, and serving, as many MUA/Ps as possible (NACHC, 2010). A unique component

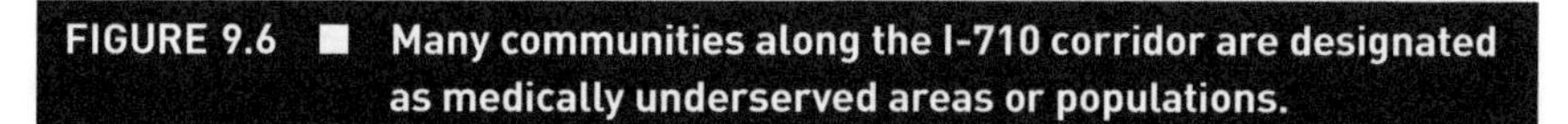

FIGURE 9.6 ■ Many communities along the I-710 corridor are designated as medically underserved areas or populations.

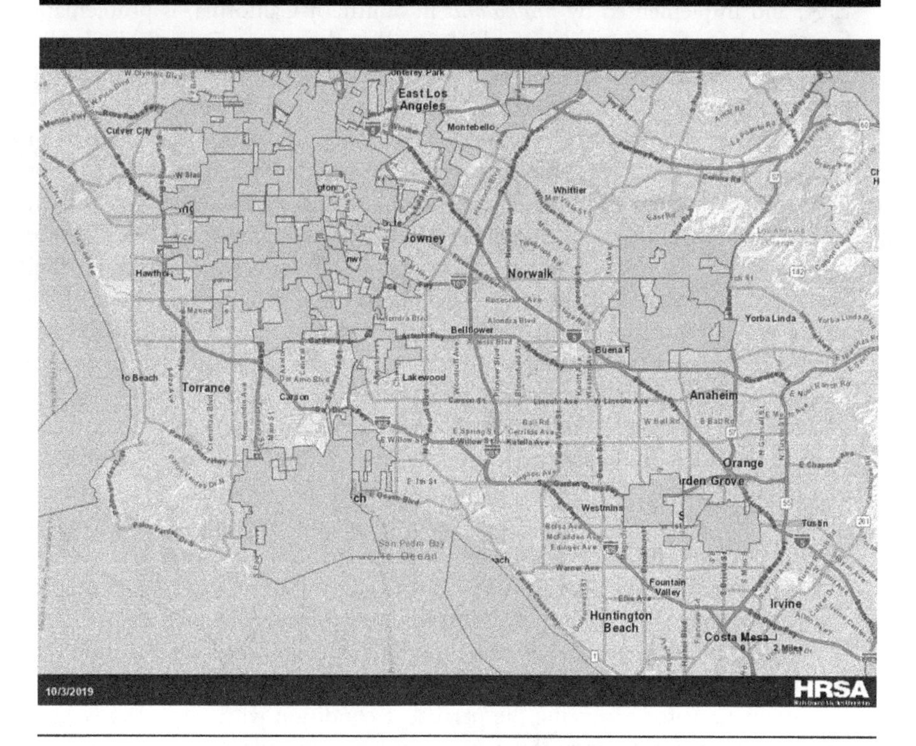

Source: Health Resources & Services Administration (https://data.hrsa.gov/hdw/tools/MapTool.aspx).

of FQHCs is the health education and outreach services provided by health educators and community health workers. These nonmedical staff are well versed in the healthcare world but are focused on helping the community access services.

The Port of Long Beach awarded The Children's Clinic more than $800,000 over three years to provide asthma care and management services to the communities most affected by the port expansion project. As part of the project, The Children's Clinic expanded an asthma care program that centered around management, guidance, and education provided by a *promotora de salud*. The *promotoras* worked with the medical care team to assess patients' current asthma management by taking into account how frequently patients were using their emergency inhaler, how frequently they were having asthma attacks, how often they would visit the emergency department, and how conducive their home environment was to keeping their asthma controlled.

Promotoras working with parents of children with asthma would simulate the feeling of chest restriction by having the parents breathe through a straw. Part of the *promotoras* job was to win over adults with asthma and parents of children with asthma, to ensure their participation in the 12-month program. The *promotoras* would visit patients' homes at least once to conduct the initial environmental assessment and have the adults sign an asthma action plan (see Figures 9.1 and 9.2). Sometimes the *promotora* would do a second home environmental assessment to ensure changes were made that removed home environmental triggers that made it easier to breathe.

As environmental justice scholarship is beginning to observe (see Bruno & Jepson, 2018), the home environmental assessment recommended by the EPA, and implemented by *promotoras* in Southern California, is problematic for a number of reasons, two of which I will explore here. First, public health programs that address environmental injustice and racism without holding polluters responsible are downstream solutions. Instead of working to prevent asthma in the community, the *promotora* program is a makeshift solution to an existing, ongoing, and expanding problem created by the state and the Port of Long Beach. Asthma and other chronic conditions are widespread in Southern California in communities along the I-710 corridor, and certainly *promotoras* are providing a much-needed, lifesaving service. Whereas they are not preventing asthma in the community, they are working to teach patients and parents how to prevent asthma attacks and save lives. Nevertheless, the Port of Long Beach funded these *promotora* programs precisely as a means to continue, and expand, diesel (and other) pollution in the region.

Second, but relatedly, *promotoras* are teaching patients and parents to mitigate triggers in their homes. Asthma patients live in various housing conditions in communities along I-710. A few own their homes, but many are renters. Much of the housing along the freeway corridor lacks adequate air filtration to prevent outside pollutants from entering homes. Additionally, the presence of mold and old carpeting that a landlord refuses to replace can harbor asthma triggers despite the best cleaning efforts. Asthma attacks lead to missed school and/or work days, and missed income, which can make the difference between whether one eats or pays the electricity bill. The home visit and environmental assessment are important tools for managing the persistent condition when you already have asthma; however, they do nothing to prevent more people from getting asthma and do not address the environmental triggers beyond the confines of the home.

These tools put the burden of asthma entirely on the family, which often has little control or power to change their situation. They also mitigate political will in the community to demand greater accountability by the Port of Long Beach, since it is funding public health solutions. The use of these tools works to absolve the port and other industry that is responsible for the region's poor and toxic air quality. They also absolve the state of having to regulate the port and other corporate polluters more closely, and of its refusal to regulate ultrafine particulate matter. The Children's Clinic, the *promotoras*, and the tools they use, like the home environmental assessment and the asthma action plan, put pressure on families to meticulously manage the air quality of their homes, even though the air quality of their larger community environment is unmanageably toxic.

SPACES OF CONTESTATION AND SPEAKING TRUTH TO POWER

Racial capitalist projects produce in excess of what is needed to make a profit. In Southern California, the goods movement and the coordination between the Port of Long Beach and the state have produced geographies of environmental racism. The system of racial capitalism has built spaces of excess: racialized marginalization and environmental harm. Within these spaces, and because of racial capitalist projects, asthma has become a widespread problem among poor and working-class Latinx communities and *promotoras de salud* are relied-upon community leaders. However, the excesses of racial capitalism are both within and beside itself.

Racial capitalism also produces the possibilities for, and spaces of, contestation and contradiction (Melamed, 2015). The Long Beach Alliance for Children with Asthma is another Long Beach–based nonprofit organization, funded by the Port of Long Beach, that supplements asthma medical care in the community by providing an asthma management care program similar to the one provided by The Children's Clinic. In addition to providing asthma education, a home environmental assessment, and check-ins with the patient and/or the family for up to 12 months, the Alliance was funded by the Port of Long Beach to do policy advocacy on behalf of the I-710 corridor community.

The Alliance held bimonthly community meetings where they invited local and state policymakers, public health officials, university researchers, and port officials to discuss the congestion of diesel trucks on the roads in their neighborhoods and the problem of air pollution in their communities. The meetings have historically been a unique opportunity for local community leaders (including *promotoras* from The Children's Clinic) and community members to speak directly to the power brokers of the air pollution in their neighborhoods: the politicians, experts, and industry leaders. The meetings are always interpreted in Spanish and English, and often in Khmer as well, as Long Beach has the largest Cambodian refugee population in the United States. The meetings provide space both for the historically disenfranchised and for those with immense political power to sit, more or less, at the same table (Nicholas et al., 2006).

Community members often speak publicly about their own struggles with the poor air quality at the bimonthly meetings. They engage in *testimonio* as a means to speak truth to power and harness truths evident in their daily experience of breathing toxic air, for social change. *Testimonio*, or giving public testimony, is a Latina feminist methodology that politicizes one's personal experience for the sake of the larger group. *Testimonio* has been used in movements for liberation across Latin America for generations and is as much a truth-telling practice as one that is rooted in the tradition of fighting for radical social and political transformation across multiple scales (Cahuas & Levkoe, 2017; Latina Feminist Group, 2001).

The Alliance has by no means rejected funding by capitalist projects that continue to pollute the community. However, the meeting space and the collective action of noneducated immigrant Latinas has commanded the attention of local and state policymakers, scientific and medical researchers, and Port of Long Beach officials. By engaging in *testimonio*, and telling their stories—in their own languages—in a community forum, they have politicized the individual experiences of struggling to breathe and developed a collective political consciousness about their daily encounters with toxic particulate matter.

Promotoras from both The Children's Clinic and the Alliance have also educated the mothers of children with asthma about the benefits of giving *testimonio* and getting involved in community efforts against environmental injustice and racism. The Alliance runs a group of community volunteers of mothers of children with asthma, or the "A-Team." The A-Team routinely count port trucks moving through their neighborhood, they educate themselves on the science of particulate matter, and they are vocal advocates against pollution in their communities at harbor commissioners' meetings and city council meetings, and they have even taken part in actions in Sacramento, California. The Port of Long Beach's community mitigation grants funded much of this advocacy work in 2010 and beyond, despite the deep contradictions. The mitigation funds were about making amends, while still enabling the port to expand its operations, move more goods through Southern California, and ultimately increase levels of pollution in the region.

CONCLUSION

Racial capitalism produces in excess of itself, and within its geographies are spaces of contestation. In Southern California, the goods movement is a racial capitalist project that has profited off the movement of imported goods through majority Latinx communities. Working-class communities of color have been politically, economically, and environmentally marginalized in the toxic wake of diesel trucks moving through their neighborhoods.

Racial capitalism is a global economic system rooted very much in the history of Indigenous land dispossession and chattel slavery in the United States. Racial capitalism is deeply embedded in the operation of the state. Historically the state has required racial capitalism in order to expand its territory and economy. This is particularly true in Southern California, and in Los Angeles County, where 43 percent of imported goods are brought into the continental United States.

Toxic particulate matter pollutes and poisons majority Latinx communities, and environmental injustice is met by the state with individualized public health solutions that shift responsibility and burden from the corporate polluters and state regulators onto individuals and families. Public health is an arm of the state that requires deeper critical consideration for how it contributes to geographies of environmental racism. The need for public health services has historically always been included in calls for environmental justice. Federally qualified health centers were created as a healthcare safety net for the most socially vulnerable communities. Their community outreach arm is what makes them special, but in Southern California *promotoras de salud* do lifesaving work that simultaneously shifts responsibility for the consequences of environmental racism from the polluters to the community they are there to serve.

Promotoras de salud address the individual consequences and experiences of environmental injustice in Southern California. They do so through strategies that both sustain, and contest, geographies of environmental racism. Environmental justice scholarship has begun to critically assess the state, and its entanglements with racial capitalism, as a perpetrator of environmental violence. However, more research still needs to be done to understand the complex role public health can play in both fighting against, and reinforcing, geographies of environmental violence, injustice, and racism.

DEEPENING OUR UNDERSTANDING

1. Visit and explore *Mapping Inequality* (dsl.richmond.edu/panorama/redlining) to learn more about redlining in the United States (see Figure 9.3). Search "Los Angeles" to zoom in specifically on Long Beach and the Los Angeles Basin. Look for the Los Angeles River and notice the redlined areas near the Port of Long Beach along the coast (Downtown Long Beach), and the yellow-lined areas along either side

of the Los Angeles River. These maps from 1939 were produced by the Home Owners Loan Corporation in Los Angeles, California. Ten years later, in 1949, construction would be in full swing for the 710 Freeway to follow the pathway of the Los Angeles River.

2. a) Visit and explore the *Racial Dot Map* (https://demographics.virginia.edu/DotMap/) to examine the racial/ethnic demographics of regions across the continental United States from U.S. 2010 Census Data (see Figure 9.4). Make sure the Color Coding in on, and add Map Labels in the upper left-hand corner. Zoom in on Southern California. Look along the southern coast and find the Port of Long Beach. Follow from the port and find the main road leading inland directly from the coast. Even though it is not labeled as such, this is the 710 Freeway. You will also see it labeled as the Los Angeles River further north—remember that the 710 Freeway was built to follow the pathway of the Los Angeles River in the early 20th century. Follow along the 710 Freeway.

   b) What communities along the 710 Freeway are racially diverse? Where along the freeway is there more explicit racial segregation?

   c) How do the redlined ("hazardous") and yellow-lined ("declining") areas that flank either side of the Los Angeles River in 1939 compare to the racial demographics of those same areas from the 2010 census? What is the historical and geographic relationship between redlined and yellow-lined communities in 1939 and the racial and ethnic composition of those same neighborhoods in 2010?

3. a) Based on the descriptions of asthma triggers commonly found in homes (see Figure 9.1), what triggers do you identify in your own home?

   b) How easy would it be for you to remove these asthma triggers from your home, either on a routine basis or permanently? To answer this question, use a scale from 1 (it would take no time at all) to 5 (it would be very time consuming).

   c) Does the degree of difficulty of managing these asthma triggers increase or decrease depending on renting status?

Visit https://www.epa.gov/sites/production/files/2018-05/documents/asthma_home_environment_checklist.pdf to view the entire home environmental assessment.

References

Brulle, R. J., & Pellow, D. N. (2006). Environmental justice: Human health and environmental inequalities. *Annual Review of Public Health, 27*, 103–124. doi:10.1146/annurev.publhealth.27.021405.102124

Bruno, T., & Jepson, W. (2018). Marketization of environmental justice: US EPA Environmental Justice Showcase Communities Project in Port Arthur, Texas. *Local Environment, 23*(3), 276–292. doi:10.1080/13549839.2017.1415873

Bullard, R. D. (2000). *Dumping in Dixie: Race, class, and environmental quality*. Boulder, CO: Westview Press.

Cahuas, M. C., & Levkoe, C. Z. (2017). Towards a critical service learning in geography education: Exploring challenges and possibilities through testimonio. *Journal of Geography in Higher Education, 41* (2), 246–263.

Cole, L. W., & Foster, S. R. (2001). *From the ground up: Environmental racism and the rise of the environmental justice movement.* New York: New York University Press.

Cowen, D. 2014. *The deadly life of logistics: Mapping violence in global trade*. Minneapolis: University of Minnesota Press.

Cunningham, G., & Cunningham, C. (2015). *Port town: How the people of Long Beach built and defended their port.* Long Beach, CA: City of Long Beach.

Daily News Wire Services. (2009). Long Beach Port's environmental impacts to be released. *Long Beach Press Telegram*. http://www.presstelegram.com/article/ZZ/20090402/NEWS/904029976

De Lara, J. (2018). *Inland shift: Race, space, and capital in Southern California*. Berkeley: University of California Press.

Estrada, G. (2014). *The 710 freeway: A history of America's most important Freeway.* KCET. https://www.kcet.org/shows/departures/the-710-long-beach-freeway-a-history-of-americas-most-important-freeway

Ferguson, R. (2001). *Aberrations in black: Toward a queer of color critique*. Minneapolis: University of Minnesota Press.

Harrison, J. L. (2015). Coopted environmental justice? Activists' roles in shaping EJ policy implementation. *Environmental Sociology, 1*(4), 241–255. doi:10.1080/23251042.2015.1084682

Health Resources & Services Administration. (n.d.) Medically underserved areas and populations. https://bhw.hrsa.gov/shortage-designation/muap

Hildreth, R. G. (1977). Environmental impact reports under the California Quality Act. The new legal framework. *Santa Clara Law Review, 17*(4), 805. http://digitalcommons.law.scu.edu/lawreview/vol17/iss4/3

Khouri, A. (2015, January). Ports of L.A. and Long Beach see best year yet since recession. *Los Angeles Times*. http://www.latimes.com/business/la-fi-ports-traffic-20150122-story.html

Konisky, D. (2015). *Failed promises: Evaluating the federal government's response to environmental justice*. Cambridge, MA: MIT Press.

Kurtz, H. (2009). Acknowledging the racial state: An agenda for environmental justice research. *Antipode, 41* (4), 684–704.

Latina Feminist Group. (2001).*Telling to live: Latina feminist testimonios*. Durham, NC:Duke University Press.

Li, N., Sioutas, C., Cho, A., Schmitz, D., Misra, C., Sempf, J., ... Nel, A. (2003). Ultrafine particulate pollutants induce oxidative stress and mitochondrial damage. *Environmental Health Perspectives, 111*(4), 455–460. https://www.ncbi.nlm.nih.gov/pmc/articles/PMC1241427/

Liévanos, R. (2012). Certainty, fairness, and balance: State resonance and environmental justice policy implementation. *Sociological Forum, 27*(2), 481–503.

Lipsitz, G. (2011). *How racism takes place*. Philadelphia, PA: Temple University Press.

Littman, F. E., & Magill, P. L. (1953). Some unique aspects of air pollution in Los Angeles. *Air Repair, 3*(1), 29–34. doi:10.1080/00966665.1953.10467586

Melamed, J. (2015). Racial capitalism. *Critical Ethnic Studies, 1* (1), 76–85. doi:10.5749/jcritethnstud.1.1.0076.

Nicholas, E. A., Dwyer, M., & Murphy, A. (2006). Coalition-based approaches for addressing environmental issues in childhood asthma. *Health Promotion Practice, 7*(2), 108S–116S.

Ostro, B., Hu, J., Goldberg, D., Reynolds, P., Hertz, A., Bernstein, L., & Kleeman, M. J. (2015). Associations of mortality with long-term exposures to fine and ultrafine particles, species and sources: Results from the California Teachers Study Cohort. *Environmental Health Perspectives, 123* (6), 549–556. doi:10.1289/ehp.1408565

Parvini, S. (2014). *Breathing uneasy: Living along the 710 corridor*. KCET. https://www.kcet.org/shows/departures/breathing-uneasy-living-along-the-710-freeway-corridor

Pellow, D. N. (2018). *What is critical environmental justice?* Cambridge, UK: Polity Press.

Perez, L., Künzli, N., Avol, E., Hricko, A. M., Lurmann, F., Nicholas, E., ... , McConnell, R. (2009). Global goods movement and the local burden of childhood asthma in Southern California. *American Journal of Public Health,99* (3), 622–628. doi:10.2105/AJPH.2008.154955

Polidori, A., Fine, P. M. (2012). *Ambient concentrations of criteria and air toxic pollutants in close proximity to a freeway with heavy-duty diesel traffic*. South Coast Air Quality Management District. http://www.aqmd.gov/docs/default-source/air-quality/air-quality-monitoring-studies/near-roadway-study.pdf

Port of Long Beach. (2009). Port grant programs to begin this fall: Funds available to offset impacts of cargo operations (news release). http://www.polb.com/civica/filebank/blobdload.asp?BlobID=6767

Port of Long Beach. (2011). Port awards $5 million to hospitals, clinics: Grants to aid respiratory health programs for local residents (news release). http://www.polb.com/civica/filebank/blobdload.asp?BlobID=8313

Pulido, L. (2016a). Geographies of race and ethnicity II: Environmental racism, racial capitalism and state-sanctioned violence. *Progress in Human Geography*, 1–10.

Pulido, L. (2016b). Flint, environmental racism, and racial capitalism. *Capitalism Nature Socialism*. doi:10.1080/10455752.2016.1213013

Pulido, L., Kohl, E., & Cotton, N.-M. (2016). State regulation and environmental justice: The need for strategy reassessment. *Capital Nature Socialism*. doi:10.1080/ 10455752.2016.1146782

Robinson, C. (2000). *Black Marxism: The making of the black radical tradition*. London: Verso.

Rothstein, R. (2017). *The color of law: A forgotten history of how our government segregated America*. New York: Liveright.

Sahagun, L. (2009, April). Long Beach port expansion moves closer to package. *Los Angeles Times*. http://articles.latimes.com/2009/apr/02/business/fi-long-beach-port2

Schlosberg, D. (2007). *Defining environmental justice: Theories, movements, and nature*. New York: Oxford University Press.

Schmool, J. L. C. (2013). Saturation sampling for spatial variation in multiple air pollutants across an inversion-prone metropolitan area of complex terrain. *Environmental Health,13*, 28. https://ehjournal.biomedcentral.com/articles/10.1186/1476-069X-13-28

Sioutas, C., Delfino, R. J., & Singh, M. (2005). Exposure assessment for atmospheric ultrafine particles and implications in epidemiologic research. *Environmental Health Perspectives,113* (8), 947–955. https://www.ncbi.nlm.nih.gov/pmc/articles/PMC1280332/

Sze, J. (2006). *Noxious New York: The racial politics of urban health and environmental justice*. Cambridge, MA: MIT Press.

10

ENVIRONMENTAL JUSTICE AND THE LAW

Oday Salim

THE CASE STUDY OF FLINT, MICHIGAN

Anyone who wants to know what happened during the Flint water crisis can simply search the Internet for the terms *Flint* and *timeline,* and they will have pages and pages of answers. The existence of so many timelines that describe the Flint water crisis in painstaking detail is almost unsettling (Brush, Williams, Smith, & Scullen, 2015; Michigan Civil Rights Commission, 2017; MSNBC, 2016). Some timelines start a few years ago with the decision to switch the drinking water source from Lake Huron and the Detroit River to the Flint River. Others go back much further to white flight and the beginning of disinvestment in a city with increasingly more concentrated minority and low-income populations. No matter where the timelines start, we arguably know more about the Flint water crisis than any other environmental crisis in American history. Most observers agree that beyond the technical causes of corrosive water and lead pipes, the real cause is environmental injustice or environmental racism.

Water distribution starts with source water, which can be a lake, a river, or an aquifer. A drinking water plant takes the source water, treats it, and conveys it through large-diameter water lines until it reaches points of consumption such as commercial corridors and residential neighborhoods. Once there, the water lines often run underneath a road perpendicular to the businesses and homes they serve. At each home, a lateral service line comes off the main line and feeds into that home's plumbing.

When cities first began installing water distribution plumbing in the 19th century, lead was the metal of choice largely because lead pipes are more malleable and resistant to corrosion than pipes made from other metals. Some city building codes actually required the use of lead. In Flint, as in many other older cities, the vast majority of lateral service lines were made of lead, hence, the term *lead service lines.*

The Safe Drinking Water Act (SDWA) was enacted in the 1980s to ensure that public water suppliers were delivering safe, potable water to Americans. The SDWA works mainly by regulating how much of a particular contaminant is allowed in drinking water. As a result of those contaminant limits, public water suppliers must treat their water at the plant before distribution to the public. For example, the maximum level for barium is 2 milligrams per liter. That means before the water leaves the plant, it should contain no more than 2 milligrams of barium per liter of water.

The problem with lead is that, unlike most other contaminants that arise in the source water or at the plant and can be treated there, lead typically enters drinking water by leaching into it from the inside of lead pipes. Therefore, public water supplies cannot really treat lead at the plant. As a result, the U.S. Environmental Protection Agengy (EPA) promulgated the Lead and Copper Rule in 1991. The rule purported to address lead pollution through an action level and a treatment technique. The main treatment technique is corrosion control, which consists of adding chemicals like orthophosphates to finished drinking water. The chemicals coat lead service lines and create a physical barrier to mitigate the risk of lead leaching. The action level is the level of lead pollution

---

PHOTO 10.1: Istockphoto.com/PierreDesrosiers

at which water systems must take additional steps. In a water system's service territory, the system is responsible for collecting and analyzing samples taken from homes that have lead service lines; that way, a water system can understand whether the pipes are leaching lead and at what concentrations. If the 90th percentile value of sampled homes exceeds the lead action level of 15 parts per billion, then the system has to address lead pollution through techniques such as source water evaluation, public education, and lead service line replacement. Replacement of lead service lines is often the costliest form of treatment but is by far the most effective as it eliminates most lead risk.

The story of Flint is the story of structural racism told against the backdrop of lead in the drinking water. In 2002 Michigan's governor declared a financial emergency in Flint based on Flint's fiscal deficit (caused in large part, many argue, by decisions by prior governments to reduce revenue sharing to cities). As a result, the governor appointed an emergency manager to make decisions for the city on the assumption that the elected city leaders were incapable of doing so. Various people challenged the emergency manager law in federal court, arguing that it discriminated against communities of color (McVicar, 2017). In spite of a decades-old relationship with Detroit's water system, there was interest from the appointed city leadership in connecting Flint to a new pipeline called the KWA pipeline that would obtain source water from Lake Huron. In the interim, to save money, rather than continue to buy water from Detroit's system, it was decided in 2014, in large part by the emergency manager, that Flint's retail water system would simply take water from the Flint River, treat it, and distribute it to customers.

Flint's water system was in no way ready for a source water switch. The Flint River was contaminated with corrosive agents from historic pollution. Even General Motors, in early 2015, stopped using Flint River water because it was corroding materials in its engine plant (Fonger, 2014). After the switch, there was a failure to properly evaluate the consequences, to implement badly needed corrosion control, to measure the lead action level, and to perform a number of other necessary tasks. Flint, a city composed mainly of African American residents, for years lacked local governance due to emergency management. For even longer than that, ever since white flight, there was increasingly less investment in education, public health, and infrastructure. Now, many Flint residents who drank water that passed through lead service lines were drinking water poisoned with lead.

For years, I had been practicing mainly straightforward environmental law in Pennsylvania. I represented large national environmental organizations to enforce the Clean Air Act and Clean Water Act against facilities violating their permits. I worked with suburban communities to address the ever-approaching shale gas drilling (commonly known as fracking) operations. I challenged environmental approvals on behalf of statewide coalitions. In 2015, the Flint water crisis changed my outlook. As a person of color from an immigrant community, I was afforded, through the crisis, an opportunity to see environmental protection through the lenses of race and class. I returned to my home state of Michigan soon after to help advance environmental justice throughout the state. Upon arriving, I asked myself: *Where was the law during the Flint water crisis? What can I and others do with the law to prevent another crisis?*

WHAT COUNTS AS ENVIRONMENTAL JUSTICE LAW?

Environmental justice is often described as comprising three kinds of justice: procedural, distributive, and corrective (Kuehn, 2000).[1] Procedural justice is the right to full participation and meaningful involvement in environmental decision-making. Distributive justice is the right of every person and community to be treated equally such that the burden and benefits of activity that impacts the environment do not fall disproportionately on historically disadvantaged populations. Corrective justice addresses harm already done by focusing on fair punishment of wrongdoers and helping an affected community repair what is broken.

The law of environmental justice is not easy to define. On one end of the spectrum, there are laws that themselves have nothing to do with environmental justice but can be used to advance it. On the other end of the spectrum, there are laws that are expressly about environmental justice. As the concept is relatively new, laws that expressly address environmental justice are newer and less plentiful.

Generally Applicable Laws That Advance Environmental Justice

Toxic Torts

Most people are familiar with tort law even if they have never heard it called that. According to *Black's Law Dictionary* (Garner, 2014), torts are wrongs committed against persons that are civil, not criminal, in nature. Common torts that should be familiar to laypersons are negligence, nuisance, and trespass. Depending on the circumstances, tortious conduct can be intentional, negligent, or even accidental.

For the most part, tort law is state law, not federal law. Although state legislatures are free to create, modify, or eliminate torts, they have existed in English law for centuries and have been part of the American common law tradition since the country's founding.

For a claim to be successful, the claimant typically must establish that the person accused of tortious conduct owed a duty to the claimant, that the duty was breached, that there was an injury, and that the breach of duty caused the injury. Claimants typically seek forward-looking remedies in the form of prevention of the wrongful conduct and backward-looking remedies in the form of monetary compensation for harm already done.

There is a category of tort law called toxic torts that has sometimes been used to try to advance environmental justice. Toxic torts are simply torts that involve toxic substances, such as pesticides, solid waste, and air contaminants (Garner, 2014). Because toxic torts often involve harm to multiple people, groups of injured persons can file claims or class actions in an attempt to more efficiently use finite litigation resources and to exert more pressure on defendants.

Toxic tort claims can be brought to advance environmental justice in the sense that a successful claim's remedies can help alleviate the material harm suffered by members of that community, thus addressing in some form the environmental injustice suffered. For example, if someone purchases property that is contaminated, learns of the contamination, fails to do anything about it, and allows it to spread and injure others, that landowner may be liable of torts like nuisance and negligence.[2] If the person whose property was contaminated is in an economically disadvantaged African American neighborhood, then it is not difficult to see how a successful toxic tort claim can be considered to advance environmental justice if, to benefit that community, the claim yields remedies such as financial compensation for the victim's increased medical expenses and loss of property value, as well as cleanup activity.

Environmental justice communities may turn to toxic tort claims because the environmental regulatory system is unable to address their problems. Environmental justice communities are often frustrated by the obstacles to participation in environmental decision-making and the failure of environmental laws to address harms that can be traced to discriminatory conduct (Lazarus, 1997; Yang, 2002). Additionally, environmental laws are designed to benefit the public at large, which is why the financial penalties imposed on violators go to the government and not the injured community. Toxic torts are private claims and can produce remedies that are more tailored to the communities that bring them.

There are also various drawbacks to using toxic tort claims to achieve environmental justice. First, there is something inherently unsatisfying about achieving a material remedy without also obtaining a formal finding that the cause of the harm was bigotry against that community. It may be good for the community to achieve monetary compensation, prevention of further harm, and environmental remediation, but it is unclear whether that sort of victory can deter future instances of discriminatory conduct.

Second, winning a toxic tort case is difficult and expensive. Toxic tort cases often become a "battle of the experts." Good expert opinion can be expensive. Additionally, many toxic tort claims require proof of causation, and it can be challenging for claimants to obtain the quantity and quality of proof that courts require (Browne, Keeley, & Hiers, 1998; Humphreys, 1990). How a substance affects health and well-being will vary depending on the person, and the manifestations of toxic exposure will not occur at the same time for all involved (Roisman, Judy, & Stein, 2004).

For example, the case of *Nonnon et al. v. City of New York*[3] was settled for $12 million after more than 20 years of litigation. *Nonnon* involved the 81-acre Pelham Bay Landfill that was closed in 1978 after allegations of illegal hazardous dumping. After the city sued the private companies responsible for the dumping, local residents discovered that the city had allowed hazardous waste to be dumped there and had allowed landfill wastewater to enter the surface water and groundwater to which the plaintiffs' children were exposed (Hill, 2018). The plaintiffs were a group of Bronx residents who lived near the landfill and who claimed that the toxic chemicals released from the landfill caused their children to develop various cancers. One can imagine the families' pain for having to wait too long for any kind of justice. In the end, each plaintiff received approximately $750,000 to settle the case. Is $750,000 enough to compensate a family for a child's cancer? Was the result worth the 22-year wait? Would the families have been more satisfied with a court ruling that the city was at fault? Did this lawsuit change the city's decision-making with regard to siting landfills and addressing landfill pollution? These are the inevitable questions that arise when plaintiffs choose toxic tort litigation.

Civil Rights Law: Private Causes of Action

In some ways, Title VI of the Civil Rights Act of 1964[4] provides what many environmental laws cannot when it comes to environmental justice (Outka, 2005). Environmental laws are universally applicable; one can apply them to pollution in rich or poor neighborhoods and in white or minority neighborhoods. If anything, environmental laws are underapplied and under-enforced in communities of color (Lazarus, 1997). Although Title VI does not expressly address the environment or public health, it specifically applies to discrimination based on race, color, or national origin (notably, Title VI does not protect against discrimination based on income status). Therefore, when you succeed in bringing a Title VI claim, you not only obtain remedies like monetary damages and injunctions, but you also can obtain a court ruling that makes environmental and public health impacts part of a wider narrative of discrimination.

Title VI was enacted as part of the broader Civil Rights Act, with 11 titles that address voting rights, discrimination in housing and public accommodations, desegregation of schools, and employment discrimination. Title VI prohibits discrimination on the basis of race, national origin, color, or religion, by the federal government and others who receive federal funds.[5] Among others, state environmental, transportation, and natural resource agencies receive federal funding and so Title VI's prohibition applies to them. Sections 601 and 602 are the principal components of the law. Section 601 states that no person can "on the grounds of race, color, or national origin, be excluded from participation in, or denied the benefits of, or be subjected to discrimination under any program or activity receiving Federal financial assistance."[6] Section 602 authorizes the federal government to enforce Title VI.

The courts have always interpreted Section 601 to allow a private cause of action. That is, if an individual or community believed that a state agency receiving federal funding had violated Section 601's prohibition against discrimination, they could sue that agency in court. If successful, they could receive monetary damages as well as an injunction correcting the agency's behavior moving forward.

For a while, there was a question as to whether Section 601 required a showing of disparate treatment or disparate impact. *Disparate treatment*, which is the more difficult of the two to demonstrate, requires proof of discriminatory animus, motivation, and bias. *Disparate impact* can apply where the law or practice on its face is neutral and there is no direct evidence of intentional discrimination, yet the negative effects of the law or practice fall disproportionately on vulnerable populations, such as communities of color. The bar is lower to prove disparate impact because you do not have to prove intention, which is often challenging to establish with the evidence available.

The Supreme Court ruled in 1983 that a Section 601 claim requires proof of disparate treatment, adding to the burden of winning a Title VI claim.[7] While affected communities continued to bring Section 601 claims, many environmental justice advocates turned also to Section 602. With the express authority of Section 602, federal agencies promulgated regulations that implemented the Section 601 discrimination prohibition. Federal agencies believed they could enforce their antidiscrimination regulations using the disparate impact standard, and various courts upheld that interpretation. The idea behind Section 602 lawsuits was that aggrieved communities could themselves go to court and enforce those regulations against federally funded state and local agencies and in doing so could take advantage of the disparate impact standard. However, in 2001, the Supreme Court cut off this avenue and ruled that Section 602 does not create a private cause of action.[8] Although the Court questioned the notion that federal agencies could apply disparate impact when writing and enforcing their own Title VI regulations, it did not disturb it. After 2001, then, environmental justice advocates could still sue state and local government agencies under Section 601 so long as they established disparate treatment with proof of intentional discrimination, but they could not directly enforce Section 602. At best, as described more fully below, they can file administrative complaints with the federal agency, alleging that a local or state government had violated Title VI, and hope that the federal agency follows up with enforcement.

Settlements and Supplemental Environmental Projects

A negotiated settlement to address violations of an environmental law can advance environmental justice. Although any settlement can do so in theory, a specific negotiated settlement tool called a supplemental environmental project or SEP is a way to advance environmental justice by ensuring that the violators do not just halt their violations but also do something to improve the community they negatively impacted (U.S. Environmental Protection Agency, 2015).

Not all pollution is illegal. Environmental laws like the Clean Air Act and Clean Water Act impose standards on polluters that allow them to pollute but only in certain ways and only in prescribed amounts. When polluters violate those standards, there are numerous ways to address the violations. As environmental laws are public laws, the government can unilaterally issue administrative orders to the polluter, or can sue the polluter in court. Because many of our federal environmental laws allow for citizen enforcement, citizens can also take polluters to court.

Enforcement of environmental laws is often insufficient to fully advance environmental justice. The end result of enforcement typically comprises monetary penalties and a requirement that the polluter change its behavior moving forward to prevent further violations. Monetary penalties function in theory to deter further violations by eliminating or at least reducing the economic benefit of noncompliance with the law. The money itself, however, does not go to the injured community but instead to the government treasury.[9] Therefore, both monetary penalties and behavior change are positive but only address the future. Neither is intended to specifically address the local affected community's needs or address harm already done.

In addition to imposing penalties and altering behavior, enforcement of environmental laws can also yield SEPs. A SEP is an environmentally beneficial project that is not required by law but that a polluter agrees to undertake as part of an overall agreement that settles an enforcement action (U.S. Environmental Protection Agency, 2015). The EPA has a comprehensive SEP policy.[10] Although states are not required to have SEP policies as a condition of implementing federal environmental programs, most do have one (Bonorris, Holloway, Lo, & Yang, 2005). The federal policy allows polluters to engage in SEPs to mitigate the monetary penalties they might otherwise have to pay. There is supposed to be a robust process to involve the affected community, which is not always effective, particularly in environmental justice communities (Simms, 2017). Also, there must be a significant connection between the SEP and the violation; for example, a polluter cannot simply provide computers to the local elementary school to address being over its air quality permit's emissions limits (Kristl, 2007).[11]

The EPA and the Department of Justice recognize SEPs as a way to advance environmental justice (U.S. Department of Justice, 2014). When the EPA evaluates whether to include a SEP in a settlement agreement, environmental justice is one of the six factors it considers (U.S. Environmental Protection Agency, 2015). If there is robust community engagement to ensure the affected community has a say in what the SEP will be, and if the SEP provides the affected community a project that materially improves environment and health, then a SEP becomes a way to address harm already done.

In 1997, to resolve environmental violations, the Sherwin-Williams paint manufacturing facility in southeastern Chicago agreed to a SEP requiring it to remediate contaminated property located nearby in a neighborhood consisting largely of African Americans (U.S. Environmental Protection Agency, 2001). In another case, the federal government in conjunction with the state of Arkansas settled with the City of Fort Smith.[12] Fort Smith operated a sewage treatment plant and had been illegally discharging untreated sewage to surface waters. Fort Smith agreed to conduct a series of remedial measures to return to compliance with the law, including assessments of its sewer system and necessary repair. It agreed to pay $300,000 in civil penalties. Finally, it agreed to a SEP involving private sewer lines. The city agreed to spend approximately $400,000 to replace the defective private sewer lines of certain eligible low-income residents, which would help the residents but also would reduce the number of times that defective sewer lines exacerbated the raw sewage discharge problem.

A recent consent decree illustrates more obviously how a SEP can address environmental justice. In 2015, the EPA and the Michigan Department of Environmental Quality sued the AK Steel Corporation for violations of the Clean Air Act. AK Steel and predecessor companies for decades have operated a steel manufacturing facility in Dearborn, southwest of Detroit, a community that consists mainly of low-income minority residents, including African Americans and Arab Americans. The lawsuit alleged that AK Steel was emitting air contaminants in excess of state law and its permit. In the settlement agreement, the company agreed to update its pollution control equipment in order to return to compliance, and to pay $1,353,126 in civil penalties.[13] It also agreed to pay a local school district approximately $375,000 to install indoor air filters at two nearby schools. The filters would remove organic chemicals and reduce odor.

Clearly, SEPs can be an instrument to advance environmental justice. They can address distributive justice by reducing pollution-related impacts, and corrective justice by providing a community-based environmental benefit. However, the issue with many SEPs is that they lack procedural justice. Not enough is done to bring the local community into the SEP development process to ensure the projects reflect community needs and achieve local objectives.

Environmental Justice Laws

Laws and policies come from several sources. In terms of layers of governance, there is federal, state, and local law. At whatever level, elected legislatures enact laws that are called statutes at the federal and state levels and ordinances at the local level. These laws tend to be general. Legislatures, through the general laws they pass, often delegate to executive branch agencies the authority to provide the details. For example, Congress may pass a law like the Clean Water Act that says a permit is required to discharge pollutants into bodies of water, but it delegates to the EPA the authority to provide the details on what those permits look like, how they will be written, how water quality monitoring will be performed, and so on. Apart from regulations, agencies also have the ability to publish nonbinding guidance documents. Finally, the head of any executive branch—the president of the United States, state governors, county executives, and city mayors—can issue orders directing the agencies under their control to conduct themselves in certain ways.

Generally speaking, laws that expressly address environmental justice have taken the form of executive orders, regulations, and agency guidance. More recently, there have been a handful of state and local laws.

Executive Order 12898 and Federal Agency Regulations

The U.S. General Accounting Office delivered its report in 1983 on the siting of landfills in poor and minority communities (U.S. General Accounting Office, 1983). A few years later the Commission for Racial Justice (1987) of the United Church of Christ (UCC) released a similar study (see also Bullard, Mohai, Saha, & Wright, 2007). Both reports concluded that there *was* a connection between race, economic status, and pollution. Before, between, and after those studies, there had been numerous demonstrations and lawsuits complaining about what we now label *environmental injustice*.

The federal government finally began to seriously respond in the 1990s. In September 1993, the National Environmental Justice Advisory Council or NEJAC was created as an advisory committee to the EPA. Shortly after that, the Clinton administration issued Executive Order 12898, titled "Federal Actions to Address Environmental Justice in Minority Populations and Low-Income Populations,"[14] which required that "each Federal agency shall make achieving environmental justice part of its mission by identifying and addressing, as appropriate, disproportionately high and adverse human health or environmental effects of its programs, policies, and activities on minority and low-income populations."[15] Title VI of the Civil Rights Act and the National Environmental Policy Act (NEPA) were the principal legal bases for the Executive Order (Lado, 2017). Though not enforceable, the executive order required agencies to satisfy their obligation under Title VI by implementing environmental justice in a manner that reflected their work and focus.

The Civil Rights Act, enacted in 1964, has 11 titles that address voting rights, discrimination in housing and public accommodations, desegregation of schools, and employment discrimination. Title VI of the Civil Right Act prohibits discrimination on the basis of race, national origin, color, or religion, by government agencies that receive federal funds.[16] Although the Department of Justice has the primary responsibility to implement Title VI, in practice, each agency that receives federal funding takes on the responsibility of enforcing it (Dana & Tuerkheimer, 2017).[17]

With regard to federal agencies, many of them have addressed environmental justice through regulations and policies. Not surprisingly, the EPA has the most robust scheme.

Other federal agencies also implement the Title VI principles. The Department of Transportation, for example, has a specific environmental justice strategy that applies to itself and its subdepartments (U.S. Department of Transportation, 2016). Pursuant to that strategy document, the Federal Highway Administration has issued orders and directives that require an assessment of whether an authorized project may disproportionately impact minority populations and imposes mitigation measures to avoid such impacts.[18]

Numerous state agencies receive federal financial assistance, so each of those state agencies is responsible for abiding by Title VI's nondiscrimination standard, and the assistance-providing federal agency is responsible for enforcing those standards. For example, the EPA has an Office of Civil Rights that ensures compliance with all relevant civil rights laws. Because the EPA provides financial assistance to state environmental agencies, the EPA is tasked with oversight and enforcement responsibility to ensure those state environmental agencies comply. Currently, the EPA division that oversees non-EPA financial assistance recipients is the External Civil Rights Compliance Office.

For example, in 2014, numerous environmental justice organizations in North Carolina filed a complaint with the EPA's civil rights division. In it, they claimed that the North Carolina Department of Environmental Quality's renewal of a solid waste general permit for hog farms disproportionately impacted African Americans, Latinos, and Native Americans. Among other things, the groups alleged that the permitting allowed the hog farms to mostly disrupt communities of color, that the state agency ignored or moved too slowly to address resident complaints about odor and contamination, and that the state agency did not do enough to prevent certain individuals from the hog industry from harassing and intimidating them.[19] The EPA investigated and confirmed many of the allegations. In 2018, the EPA and the state agency entered into a settlement that developed a framework for improving the general permit process.

It has to be said that victories like the one in North Carolina have been few and far between. In 2009, the U.S. Court of Appeals for the Ninth Circuit ruled that the U.S. EPA's Office of Civil Rights had systematically failed to address environmental justice complaints. The plaintiff, Rosemere Neighborhood Association (located in Clark County, Washington), filed its administrative complaint in 2003. Far exceeding the deadlines, it took the EPA two years simply to accept the complaint. When the EPA exceeded deadlines again by taking more than two years to investigate, Rosemere filed suit. The ruling highlighted the fact that the EPA, when it moved at all on environmental justice complaints, moved so slowly as to render the complaint process almost meaningless. The ruling demonstrates that sometimes, however well environmental justice laws are written, they require agencies that have the political will to implement and enforce them (see Chapter 8 in this volume).

State Laws

Although many states have environmental, transportation, and other agencies with environmental justice policies, few states have passed legislation that expressly addresses environmental justice. For the states that do, the legislative focus is often on development of a commission or similar body that can evaluate the state's ability to advance environmental justice and to make recommendations for improvements. Some legislation substantively focuses on specific environmental injustices, such as lack of meaningful involvement in decision making and the siting of polluting activity.

Arkansas passed a statute in 1993 (even before President Clinton's Executive Order 12898), called the Environmental Equity Act (University of California Hastings College of the Law, Public Law Research Institute, 2010).[20] It focused on the siting of solid waste disposal facilities. The law had as its purpose to "prevent communities from being involuntary hosts to a proliferation of high impact" facilities. The law created a rebuttable presumption against permitting solid waste management facilities to be located within 12 miles of any existing facility. The facility can rebut the presumption by establishing that there are no other suitable sites or that the host community is receiving sufficient benefits from locating the facility there.

Illinois passed its Environmental Justice Act in 2011.[21] The law acknowledged that there can be disproportionate environmental impacts on certain communities with regard to their race, national origin, age, or income. The law established a Commission on Environmental Justice that consists of 24 members, of whom 10 represent the legislature or state agencies, and the remaining 14 represent the affected communities themselves, environmental organizations, business organizations, local government units, and labor organizations. The purpose of the commission is to generally advise state government on environmental justice issues, review and analyze the impact of state laws and policies on affected communities, and develop criteria to evaluate impacts onenvironmental justice communities. The law tasks the Illinois Environmental Protection Agency with providing logistical support to the commission.

In 2007, Oregon passed Senate Bill 420 with the main purpose of creating an intragovernmental body called the Environmental Justice Task Force (University of California Hastings College of the Law, Public Law Research Institute, 2010). The task force would consist of 12 governor-appointed members, all of whom have experience with environmental justice and who "to the greatest extent practicable, represent minority communities, low-income communities, environmental interests, industry groups and geographically diverse areas of the state." The purposes of the task force included advising the governor and state agencies on environmental justice issues, identifying and meeting with environmental justice communities, and making recommendations to the state with regard to how to advance environmental justice. Aside from the task force, the law called on agencies to improve public participation in decision making and avoid disproportionate impacts by conducting an environmental justice analysis before taking action, holding hearings in ways that maximize participation by environmental justice communities, and creating a citizen advocate position focused on resolving environmental justice issues.

Local Ordinances

Local ordinances that address environmental justice are fairly new. In 2009, Cincinnati was arguably the first major American city to pass an environmental justice ordinance (Bergeron, 2018). With regard to impacts on the environment and human health, local governments often regulate through land use and zoning, which can be broadly defined as the practice of dividing land in a city and defining which activities can take place in which divisions. The federal and state governments regulate to a limited extent the siting of activities, but for the most part, *where* an activity can take place is addressed through local land-use and zoning practices. For years, one of the principal environmental injustices has been siting facilities that cause pollution adjacent to or near communities of color as well as other vulnerable populations such as the elderly and children (Maantay, 2002). Local law, then, clearly has a role to play in advancing environmental justice by better regulating the siting of potentially harmful land uses.

Apart from regulation of siting and incompatible adjacent land uses, local ordinances can address environmental justice in various ways. Like their federal and state counterparts, they can increase public participation opportunities for local decision-making. They can also set up institutional bodies to collect data about impacts on affected communities and provide advice to elected officials and city departments. They can try to regulate a polluter's conduct by limiting air emissions or defining how they will operate a waste management facility. However, cities that try to regulate conduct too much may face preemption. Preemption occurs when a higher authority of law displaces a lower authority of law when there is a conflict between the two. Whereas many cities are motivated to significantly reduce pollution risk in environmental justice communities by regulating conduct, federal and state law is often capable of displacing local law should local law replicate or otherwise conflict.

In 2016, Los Angeles, California, revised its zoning code to create Clean Up Green Up (CUGU) Supplemental Use Districts in three communities[22] with the expressed purpose of reducing "cumulative health impacts resulting from land uses including, but not limited to, concentrated industrial land use, on-road vehicle travel, and heavily freight-dominated transportation corridors, which are incompatible with the sensitive uses to which they are in close proximity," such as homes and schools.[23] The communities to which the current district designations apply are low-income minority communities that are considered toxic hotspots (Barboza, 2016). The ordinance places restrictions on development in the CUGU districts in order to mitigate the impacts on vulnerable populations from land uses that may generate various forms of pollution. The ordinance applies more scrutiny during permitting and requires different forms of landscaping, noise control, and setbacks to protect nearby communities. The ordinance also requires signage that serves as a warning label. One kind of signage discourages vehicle idling, which can significantly affect local air quality. Another kind requires municipal buildings located near freeways to post a notice that roadway emissions such as exhaust and particulate matter can lead to health problems, and that being within 500 feet of a roadway can make one more vulnerable to those issues.

Camden, New Jersey, also has an ordinance that consequentially addresses environmental justice but takes a different approach.[24] The Camden City Sustainability Ordinance requires that those interested in building new developments submit an evaluation of potential impacts and benefits of the proposed development with regard to the environment, public health, and general welfare. The evaluation is called an environmental impact and benefits assessment and is reviewed by Camden's planning and zoning board. In that regard, the ordinance is similar to the National Environmental Policy Act. NEPA does not regulate conduct or require certain substantive outcomes, but it does require that, for certain kinds of federal projects, the responsible agency fully evaluate alternatives to and environmental impacts of the project. The public is invited to contribute to the evaluation while it is being drafted and then to comment on the final version. The idea behind NEPA and laws like it are to enhance public participation in government decision-making that will impact the environment and public health. Like NEPA, Camden's ordinance requires an impact assessment that will address impacts from activities such as chemical storage, traffic, and increasing impervious surface area. Some of the considerations that relate expressly to environmental justice include "Are affected residents involved in the planning process?" and "Are disadvantaged populations at greater risk of exposure to environmental hazards?" (Drewes, 2015). Additionally, once impacts are identified, Camden requires the applicant to address opportunities to minimize the impacts through initiatives such as tree planting, green infrastructure, and riparian buffers. It is not clear whether the city can reject a development proposal as a result of the evaluation, but certainly the reviewing board has the ability to shape the development proposal to reduce its impact.

New York City passed two ordinances in 2017 that are specifically focused on advancing environmental justice. The Environmental Justice Study Bill[25] requires the mayor to create an interagency working group, composed of representatives of various city agencies, that will create a comprehensive environmental justice plan to guide the city's implementation of environmental justice practices. The ordinance also establishes an advisory board for the interagency working group that consists of appointees with environmental justice expertise. The second ordinance is the Environmental Justice Policy Bill,[26] which details the environmental justice plan that the interagency working group must author. In part, this ordinance requires the creation of a study that delineates environmental justice communities, describes the environmental injustices in those communities, and provides recommendations for improving the situation. The ordinance also requires the creation of an environmental justice website that, among other things, must contain an interactive map that highlights affected communities and the potential sources of problems in them.

Local environmental justice is important, but there is often tremendous opposition to it. That opposition is usually based on arguments that it hinders economic development, which makes achieving local action difficult. Cincinnati may be the prime example. The Regional Chamber of Commerce vigorously opposed the Cincinnati ordinance,[27] which requires an environmental justice permit from the Office of Environmental Quality to expand or construct certain kinds of polluting facilities such as landfills (Trinity Consultants, 2009). The permit application would provide for public notice and comment, and the Environmental Justice Board, created to conduct the review, would consist in part of experts and community members (Bergeron, 2018). Cincinnati's approach to addressing environmental justice through local law has been purposefully public. The city has a webpage devoted to its efforts, which says in part, "Environmental justice refers to efforts to prevent harms caused by pollution from occurring disproportionately in poor and minority populations," and that the city is "conscious of the tendency for polluting industries to be located in predominantly poor and minority communities."[28] However, the 2009 ordinance was enacted with the caveat that it would take effect only if there was adequate funding. That funding appears never to have materialized, and so the ordinance, however good on paper, remains unused (Bergeron, 2018).[29]

THE FEDERAL HOUSING ACT AND FLINT'S WATER CRISIS

The Fair Housing Act (FHA), otherwise known as Title VIII of the Civil Rights Act of 1964, contains Section 804(b), which makes it unlawful for a person "to discriminate against any person in the terms, conditions, or privileges of sale or rental of a dwelling, or in the provision of services or facilities in connection therewith, because of race, color, religion, sex, familial status, or national origin." In one case, residents of a predominantly African American community in Chicago sued the Illinois Sports Facilities Authority, claiming that, based on the race of the residents, the authority selected a nearby site for the new Chicago White Sox baseball stadium, which had negatively impacted the neighborhood by causing more noise, subjecting residents to potentially dangerous interactions with stadium visitors, forcing small businesses to move to another neighborhood, and the like.[30] In another case, members of the Fundamentalist Church of Jesus Christ of Latter-Day Saints filed a lawsuit claiming the town in which they lived had denied them, based on their religion, water, electricity, and sewer connections.[31]

Many environmental justice advocates have posed two questions about the FHA. First, in a civil lawsuit, would a plaintiff have to prove disparate treatment or disparate impact? Second, in what senses would "the provision of services or facilities in connection" with "the sale or rental of a dwelling" apply to the context of environmental protection?

Although groups continue to file Title VI claims and have managed to succeed in some, the Supreme Court certainly made Title VI claims more difficult when deciding in 2001 that litigants would have to leap over the higher disparate treatment hurdle with Section 601 claims, and that Section 602's lower-hurdle, disparate impact standard had to be utilized by the government in its enforcement actions, not by private citizens. However, environmental justice advocates saw a ray of hope in a 2015 Supreme Court opinion about Texas housing policy.[32]

In a case known as *Inclusive Communities*, the Supreme Court held that citizens bringing private causes of action under the FHA could succeed by proving disparate impact. *Inclusive Communities* is about housing tax credits. The federal government provides tax credits to states to encourage affordable housing. State housing departments, such as the Texas Department of Housing and Community Affairs, distribute those credits to developers based on various factors. The Inclusive Communities Project is a nonprofit that assists families with obtaining affordable housing. They brought a disparate impact lawsuit based on the FHA, claiming that the Texas agency exacerbated segregated housing patterns by allocating too many tax credits to urban, predominantly African American neighborhoods and too few in suburban, predominantly white neighborhoods. In that sense, the agency was subsidizing the segregated status quo rather than allocating credits that would go toward suburban affordable housing that might attract low-income African Americans to those suburban white neighborhoods. The Supreme Court held that the FHA could support disparate impact claims but sent the case back to the trial court so that it could apply the correct standard to the facts at issue.

Clearly, it helps environmental justice advocates to have access to a disparate impact standard in a statute that can be used in the environmental context.[33] Before the Supreme Court confirmed the standard in 2015, environmental justice advocates had already pursued Section 804(b) claims using the disparate impact standard. In one case, a California city decided to facilitate significant expansion and redevelopment of a commercial area in a predominantly African American neighborhood.[34] The plaintiffs claimed that the city had engaged in a discriminatory pattern of conduct over years that caused their neighborhood to be deemed blighted and therefore appropriate for commercial redevelopment, and that the redevelopment would affect the availability of municipal services, thus forcing them out (Brown & Lyskowski, 1995). In another case, a group of low-income African Americans in Cocoa, Florida, sued the city, alleging that its zoning code and zoning permits had allowed incompatible industrial uses to be located too near the residential areas, subjecting residents to loud noises, bad smells, and increased traffic.[35] The court required protective rezoning of the area to address the problem (Brown & Lyskowski, 1995).

The question on many advocates' minds today is this: Can the FHA's Section 804(b) be used to prevent another Flint water crisis by addressing lead service lines? If it is attempted, whom would the lawsuit target? Water delivery infrastructure is almost exclusively a local issue, not a state or federal one. Would the defendant be the water system? In some areas, a single water system obtains water from a source such as a river or lake, treats it, and delivers it to consumers. However, in other areas, regional wholesale water systems do much of the water acquisition and treatment, selling that water at wholesale rates to a retail water system that is typically a municipal department. That local system sells water at retail rates to residents. Usually, the local water system will have the most direct access to the lead service lines, but it is the regional wholesale systems that likely provide the orthophosphate treatment.

In the context of lead service lines, what are the "services" or "facilities" and what is the related discriminatory conduct? In some of the cases described earlier, municipalities permitted or otherwise facilitated pollution in a neighborhood, and the affected community sued before the statute of limitations period was over. However, lead service lines are by definition decades old and it is not clear whether a lawsuit for constructing the lines in the first place would be timely in most communities. In other cases, the allegation was that a service necessary to a community was withheld or eliminated or made more difficult to access. Can one argue that failure to replace the lead service lines constitutes the withholding or elimination or reduction of services where the service is defined as providing healthy water?

In some communities, the water system will replace only the portion of the lead service line that runs beneath private property if the private property owner pays for it, which can cost thousands of dollars. Partial lead service line replacement often makes matters worse and has been criticized by the EPA because the risk of lead pollution continues to exist in the unreplaced portion of the line, and the partial replacement may cause more lead to enter the water supply during the replacement process (U.S. Environmental Protection Agency, 2011). One can imagine a community in which homeowners in more affluent neighborhoods can afford to pay for replacement of the portion of the line on their property, but homeowners in a poorer neighborhood (often minority due to practices that lead to the enactment of the FHA in the first place) cannot afford to pay and so are more vulnerable to lead pollution. Can the FHA address that? Even if it could, would a court order lead service line replacement or simply better application of orthophosphate along with other risk-mitigation measures?

DEEPENING OUR UNDERSTANDING

1. Find a local situation in which it is alleged that a community is exposed some kind of environmental injustice.
2. Develop a Supplemental Environmental Project (SEP) that could advance environmental justice in that community.
3. Based on what you now know about environmental law, how would you propose to deliver procedural, substantive and corrective justice?

References

Barboza, T. (2016, April). L.A. City Council adopts rules to ease health hazards in polluted neighborhoods. *Los Angeles Times*. https://www.latimes.com/local/lanow/la-me-pollution-protection-20160412-story.html

Bergeron, E. (2018). Local justice: How cities can protect and promote environmental justice in a hostile environment. *Natural Resources & Environment, 32*, 8.

Bonorris,S., Holloway, C., Lo, A., & Yang, G. (2005). Environmental enforcement in the fifty states: The promise and pitfalls of supplemental environmental projects. *Hastings Environmental Law Journal,11*, 185 -222

Brown, A. L., & Lyskowski, K. (1995). Environmental justice and Title VIII of the Civil Rights Act of 1968 (The Fair Housing Act). *Virginia Environmental Law Journal, 14*, 741–756.

Browne, M. N., Keeley, T. J., & Hiers, W. J. (1998). The epistemological role of expert witnesses and toxic torts. *American Business Law Journal, 36*, 1–72.

Brush, M., Williams, R., Smith, L., & Scullen, L. (2015) TIMELINE: Here's how the Flint water crisis unfolded. Michigan Radio. https://www.michiganradio.org/post/timeline-heres-how-flint-water-crisis-unfolded

Bullard, R. D., Mohai, P., Saha, R., & Wright, B. (2007). *Toxic wastes and race at twenty 1987–2007.* Cleveland, OH: United Church of Christ Justice and Witness Ministries.

Commission for Racial Justice. (1987). *Toxic wastes and race in the United States: A national report on the racial and socio-economic characteristics of communities with hazardous waste sites.* New York: United Church of Christ.

Dana, D. A., & Tuerkheimer, D. (2017). After Flint: Environmental justice as equal protection. *Northwestern University Law Review, 111*, 879.

Drewes, D. (2015). Camden passes New Jersey's first sustainable site plan ordinance. *The Dodge Blog*. http://blog.grdodge.org/2015/02/18/camden-passes-new-jerseys-first-sustainable-site-plan-ordinance/

Fonger, R. (2014). General Motors shutting off Flint River water at engine plant over corrosion worries. *M Live*. https://www.mlive.com/news/flint/index.ssf/2014/10/general_motors_wont_use_flint.html

Garner, B. (Ed.) (2014). *Black's law dictionary* (10th ed.). Eagan, MN: Thomson Reuters Legal.

Hill, B. E. (2018). *Environmental justice: Legal theory and practice* (4th ed.). Washington, DC: Environmental Law Institute Press.

Humphreys, S. L. (1990). Comment: An enemy of the people: Prosecuting the corporate polluter as a common law criminal. *American University Law Review, 39*(2), 311–341.

Kristl, K. T. (2007). Making a good idea even better: Rethinking the limits on supplemental environmental projects. *Vermont Law Review, 31*, 217–271.

Kuehn, R. (2000). A taxonomy of environmental justice. *Environment Reporter, 30*, 10681–10700.

Lado, M. E. (2017). Toward civil rights enforcement in the environmental justice context: Step one: Acknowledging the problem. *Fordham Environmental Law Review, 29*, 19.

Lazarus, R. (1997). Fairness in environmental law. *Environmental Law, 27*, 705–739.

Maantay, J. (2002). Zoning law, health, and environmental justice: What's the connection? *Journal of Law, Medicine and Ethics, 4*, 572–593.

McVicar, B. (2017). Michigan's emergency manager law is racist, lawsuit argues. *M Live.* https://www.mlive.com/news/index.ssf/2017/12/michigans_emergency_manager_la.html

Michigan Civil Rights Commission. (2017). *Flint water crisis: Systemic racism through the lens of Flint.* https://www.michigan.gov/mdcr/0,4613,7-138--405318--a,00.html

MSNBC. (2016). *Flint water crisis: A timeline.* http://www.msnbc.com/msnbc/flint-water-crisis-timeline

Outka, U. (2005). Environmental injustice and the problem of the law (Comment). *Maine Law Review, 57*, 209–258.

Roisman, A., Judy, M., & Stein, D. (2004). Preserving justice: Defending toxic tort litigation. *Fordham Environmental Law Review, 15*, 191–208.

Simms, P. L. (2017). Leveraging supplemental environmental projects: Toward an integrated strategy for empowering environmental justice communities. *Environmental Law Reports News & Analysis, 47*, 10511.

Trinity Consultants. (2009). *Cincinnati environmental justice ordinance summary.* https://www.trinityconsultants.com/news/states/ohio/cincinnati-environmental-justice-ordinance-summary

U.S. Department of Justice. (2014). *Environmental justice plan: DOJ guidance concerning EJ.* Washington, DC: Author. https://www.justice.gov/sites/default/files/ej/pages/attachments/2014/12/19/doj_guidance_concerning_ej.pdf

U.S. Department of Transportation. (2016). *Environmental justice strategy.* https://www.transportation.gov/transportation-policy/environmental-justice/environmental-justice-strategy

U.S. Environmental Protection Agency. (2001). *Beyond compliance: Supplemental environmental projects.* EPA-325-R-01-001139. Washington, DC: Author.

U.S. Environmental Protection Agency. (2011). *Science Advisory Board evaluation of the effectiveness of partial lead service line replacements.* EPA-SAB-11-015. Washington, DC: Author. https://www.epa.gov/sites/production/files/2015-09/documents/sab_evaluation_partial_lead_service_lines_epa-sab-11-015.pdf

U.S. Environmental Protection Agency. (2015). *Supplemental environmental projects policy, 2015 update.* https://www.epa.gov/sites/production/files/2015-04/documents/sepupdatedpolicy15.pdf

U.S. Government Accountability Office. (1983). *Siting of hazardous waste landfills and their correlation with racial and economic status of surrounding communities. GAO/RCED 83-168 B-211461.* Washington, DC: Author. https://www.gao.gov/assets/150/140159.pdf

University of California Hastings College of the Law, Public Law Research Institute. (2010). *Environmental justice for all: A fifty state survey of legislation, policies and cases* (4th ed.). http://gov.uchastings.edu/public-law/docs/ejreport-fourthedition1.pdf. Third edition available to public at https://www.issuelab.org/resource/environmental-justice-for-all-a-fifty-state-survey-of-legislation-policies-and-cases-3rd-ed.html

Yang, T. (2002). Environmental regulation, tort law and environmental justice: What could have been. *Washburn Law Journal, 41*, 607–628.

Notes

1. Kuehn also describes environmental justice as including social justice, though for the sake of simplicity and because social justice has been considered part of the broadly accepted definition for quite as long, this chapter will focus on the other three.
2. *New York v. Shore Realty Corp.*, 759 F.2d 1032 (2d Cir 1985)
3. Nos. 8576/91, 12648/91, 15687/92, 23354/92, 14920/92, and 7957/06, settlement announced (N.Y. Sup. Ct., Bronx County Aug. 12, 2013).
4. This chapter highlights the Civil Rights Act of 1964. There are other civil rights laws that can be applied to environmental injustices. The main alternative is the equal protection clause in the 14th Amendment of the U.S. Constitution, which states that no state may "deny to any person within its jurisdiction the equal protection of the laws." The Fifth Amendment contains a similar prohibition against discrimination that applies to the federal government. The Supreme Court in the 1970s held that an equal protection claim under the U.S. Constitution applies only to address government action based on "discriminatory purpose." Because private claims brought under Title VI face the same hurdle requiring proof of discriminatory motive or intent, the chapter does not elaborate on the equal protection cases.
5. 42 U.S.C.A. § 2000d et seq.
6. 42 U.S.C.A. § 2000d.
7. *Guardians Association v. Civil Service Commission*, 463 U.S. 582 (1983).
8. *Alexander v. Sandoval*, 532 U.S. 275 (2001)
9. Whether the penalty monies go to a general fund or to a particular agency program varies state by state.
10. In a recent memorandum, the Department of Justice announced that federal SEPs can no longer include payments made to third-party nongovernmental organizations, who in many instances have helped to implement the SEP in a local community. Memorandum, The Attorney General to All Component Heads and United States Attorneys, Prohibition on Settlement Payments to Third Parties (June 5, 2017), https://www.justice.gov/opa/press-release/file/971826/download.
11. Some observers argue that the nexus requirement can be too restrictive and that communities should have more freedom to achieve the beneficial project that is most meaningful to them.
12. *United States of America and State of Arkansas v. City of Fort Smith, Arkansas*. https://www.epa.gov/sites/production/files/2015-01/documents/ftsmith-cd.pdf.
13. Summary of Consent Decree, *United States and Michigan Department of Environmental Quality v. AK Steel Corporation*, Civil Action No. 2:15-cv-11804 (E.D. Mich.). https://archive.epa.gov/region5/swdetroit/web/pdf/ak-steel-cd-summary-eng.pdf.
14. 59 Fed. Reg. 32 (1994).
15. 40 C.F.R. Part 7—Nondiscrimination in Programs or Activities Receiving Federal Assistance from the Environmental Protection Agency.
16. 42 U.S.C.A. § 2000d et seq.
17. 28 C.F.R. § 42.401.
18. Federal Highway Administration Order 6640.23A. https://www.fhwa.dot.gov/legsregs/directives/orders/664023a.cfm.
19. Letter from Environmental Protection Agency to North Carolina Department of Environmental Quality (2017). Retrieved from http://waterkeeper.org/wp-content/uploads/2017/01/Letter-to-Complainants-in-Case-11R-14-R4-Forwarding-Letter-of-Concern-to-NC-DEQ-1-12-2017.pdf.
20. Ark. Code Ann. §§ 8-6-1501 et seq.
21. 415 ILCS 155/.
22. Ordinance No. 184246 (2016).
23. Sec. 13.18.

24. Ordinance Approving Sustainability Requirements for the City of Camden(Jan. 13, 2015).
25. Ordinance No. 2017/064.
26. Ordinance No. 2017/060.
27. Title X, Chapter 1041, Environmental Justice, Ordinance 210-2009.
28. https://www.cincinnati-oh.gov/oes/citywide-efforts/environmental-justice/
29. As of the date of publication, the ordinance was not part of the official online municipal code, https://library.municode.com/oh/cincinnati/codes/code_of_ordinances, and the link to the ordinance on the city's environmental justice webpage was broken. https://www.cincinnati-oh.gov/oes/linkservid/7C8F0FDF-ABBF-6DA7-9A95F3DC99D13A12/showMeta/0/. The purported text of the ordinance can be found on an unofficial site: http://cincinnati-oh.elaws.us/code/coor_titlex_ch1041.
30. *Laramore v. Illinois Sports Facilities Authority*, 722 F. Supp. 443 (N.D. Illinois 1989).
31. *Cooke v. Town of Colorado City, Ariz.*, 934 F. Supp. 2d 1097 (D. Ariz. 2013).
32. *Texas Dept. of Community Affairs v. Inclusive Communities Project*, 135 S. Ct. 2507 (2015).
33. It should be noted that the Supreme Court rendered its opinion based on claims brought under Sections 804(a) and 805(a). The language about facilities and services that is most often used by environmental justice advocates is located in Section 804(b), though there is no reason to believe that the disparate impact standard should not also apply to claims brought under Section 804(b).
34. *Oliver v. City of Indio, CA*, No. SA CV 90-0097 (C.D. Cal. filed Feb. 19, 1990).
35. *Houston v. City of Cocoa, FL*, No. 89-82-CIV-ORL-19 (M.D. Fla. Dec. 22, 1989).

PART IV

ENVIRONMENTS OF (IN)JUSTICE AND ACTIVISM

11

FAIR HOUSING AND HEALTH

A Social Ecology Framework

George Lipzitz

HOUSING, HEALTH, AND RACE

Housing discrimination plays a key role in enabling and exacerbating environmental racism. The concentration of communities of color in discrete spatial locations makes them vulnerable to experiencing disproportionate exposure to polluted water, air, and land; to being isolated from access to health care and healthful food options; and to suffering the most from the collateral health consequences of predatory policing and mass incarceration. The Fair Housing Act of 1968 is not a perfect remedy for housing discrimination because the law was written deliberately to be weak and unenforceable. Yet 50 years of grassroots struggle by activist individuals and organizations has generated forms of fair housing mobilization, litigation, and legislation that provide potentially promising weapons in the battle against environmental racism.

In Zanesville, Ohio, and Modesto, California, social movements have employed the resources offered by the Fair Housing Act to identify and redress instances of systematic environmental racism against black and Latinx individuals and communities. These campaigns evidence a transformation in ideas about law and environmental health. They go beyond the tort model of injury in law and the biomedical model of illness in public health to produce a framework rooted in the concept of social ecology. A tort is an act of omission that gives rise to injury (the invasion of any legal right) or harm (loss or detriment an individual suffers) (see Chapter 10 in this volume). Thus, the tort model of injury in fair housing law treats discrimination as a finite private and personal act aimed at impeding access to property. The social ecology framework connects individual acts of discrimination to the collective, cumulative, and continuing opportunity hoarding that is intrinsic to the structured preferences and advantages produced by the possessive investment in whiteness (Lipsitz, 1998). The biomedical model of illness focuses on the individual level: genetic makeup, health habits, lifestyle choices, and degree of access to medical care. The social ecology framework when applied to illness focuses on the collective experiences of groups with threats to health, created by structural social conditions, the economy, the law, and politics.

The collateral health consequences of racial discrimination in housing compel millions of people of color to experience the racial identities ascribed to them through heightened susceptibility to premature death. Relegating people of different races to different places produces racial wealth gaps that are also racial health gaps. Members of aggrieved communities of color are routinely relegated to urban and rural sacrifice zones filled with infectious agents and environmental pollutants but devoid of physicians, pharmacies, and fresh food. Instead of parks and playgrounds, they reside near polluting incinerators, garbage dumps, and metal-plating shops. Children raised in these places are exposed to lead-based paint on inside walls and playground equipment (Gee & Ford, 2011). Moreover, being the object of racist treatment by itself can do damage to physical and mental health (Burris, Kawachi, & Sarat, 2002).

PHOTO 11.1
HenryKSadura/Shutterstock

White people in the United States are only half as likely as people of color to live in proximity to chemical waste facilities (Crowder & Downey, 2010). Moreover, the toxic waste sites that are routinely situated in communities of color are almost twice as likely to be a source of hazardous chemical incidents as those located in majority-white communities (Starbuck & White, 2016). Scholarly research reveals that, even when adjusted for income and education, racial minorities encounter higher levels of exposure to toxic substances than white people face. A 2007 study found that more than half of the residents of neighborhoods with at least one hazardous waste site are people of color, compared to only 30 percent of residents in neighborhoods without any such sites. Studies have found that decisions about placement of commercial hazardous waste facilities hinge largely on the racial composition of the surrounding area (Bullard, Mohai, Saha, & Wright, 2007, p. xii). Racial considerations not reducible to class position predict exposure to most environmental dangers, including polluted air, lead poisoning, and contaminated water and food (Bullard, 2016).

Native American men who reside on the Pine Ridge Reservation in South Dakota have a life expectancy of only 48 years (Ansell, 2017, p. 39). On average, white women nationwide can expect to live four years longer than black women (Acevedo-Garcia, Osypuk, & McArdle, 2010, p. 132). Black children are three times more likely than white children to be hospitalized for asthma and four times more likely to die from the disease (Sze, 2007, p. 95). The asthma hospitalization rate for Latinx children living in National City, California, in 2010 exceeded the county average by 50 percent (Environmental Health Coalition, 2017). A comprehensive 2017 study of neighborhood race effects in Chicago conducted by the Institute for Research on Race and Public Policy discovered that physicians' offices are concentrated in white neighborhoods. These areas experience almost no incidents of childhood lead poisoning, while eight predominantly black neighborhoods report elevated lead levels in children that are twice the city average. All 17 of the city's neighborhoods with the highest levels of premature death have a majority of residents who are black, and 16 of the neighborhoods have a population that is at least 85 percent black. Four out of the five neighborhoods with the highest rates of breast cancer have black populations exceeding 90 percent, and each of the 13 communities in the city with rates of low-birth-weight babies above 15 percent have a population that is at least 90 percent black (Institute for Research on Race and Public Policy, 2017, pp. 148, 150, 152).

THE FAIR HOUSING ACT AND RESIDENTIAL SEGREGATION

Pervasive but unprosecuted violations of the 1968 Fair Housing Act combine with the enduring effects of powerful patterns of racial segregation, which were firmly in place for nearly a century when that law was passed, to render the nation's racial order a quintessentially spatial order. The law came into being in response to insurrections that broke out in more than 100 cities in response to the murder of Martin Luther King Jr. The death of Dr. King brought new life to legislation that had previously been rejected repeatedly by legislators, guided by popular opinion among whites that overt racial discrimination in housing rentals and sales constituted a white property right that should never be abridged. The proposed law entailed two significant innovations in civil rights law. The first innovative feature banned many of the discriminatory practices that previously played key roles in skewing opportunities and life chances along racial lines, in particular, deed restrictions, racial zoning, mortgage redlining, real estate agent blockbusting, and racially motivated refusals to engage in housing transactions. The second novel aspect of the law entailed a statement that the creation of racially integrated communities should be considered the highest priority of Congress and a goal to be pursued through a government responsibility to "affirmatively further fair housing." Like other civil rights laws, the Fair Housing Act rendered illegal many forms of overtly racist impediments to full citizenship and social membership. Unlike most other civil rights laws, however, the act also expressed an obligation to create new democratic practices and opportunities (Smith & Cloud, 2010).

In the wake of King's death and the disorder it provoked, the enemies of fair housing saw a need to make concessions. They moved from a stance of outright opposition to any changes in the hierarchical racial order of the property system, to support a bill that proclaimed the right to fair housing in theory but lacked the enforcement mechanisms needed to bring it about in practice (Smith & Cloud, 2010). Signed into law only seven days after King's murder, the act gave little oversight authority to the Department of Justice or to the Department of Housing and Urban Development. It required citizens to make complaints quickly, to assemble evidence on their own without subpoena power, and to prove not merely an individual unjust act, but a pattern and practice of discrimination. If a plaintiff somehow overcame all those hurdles, the first remedy the law provided was "conference, conciliation, and persuasion," that is, asking if the perpetrator of the discrimination wished to talk about it with the victim. Under rare conditions, the government could go to court to punish unfair housing offenders, but even in those cases damages were capped at $1,000, which was much less than the cost of litigation and certainly less than the monetary harm inflicted by most racially motivated denials of shelter. Decades later, Sen. Edward Kennedy referred to the original law as "a toothless tiger," a law that proclaimed a commitment to equality but one that contained no attendant statutory power of enforcement (Seng & Caruso, 2010, p. 54).

Yet the very weaknesses built into the Fair Housing Act curiously made it possible for citizen activists to transform the law into an effective tool for social justice. In 1970, Paul J. Trafficante, a white man, and Dorothy M. Carr, a black woman, filed suit against their landlord in the Parkmerced apartments in San Francisco. They charged that the complex's managers worked systematically to deny housing to African Americans, and that as a result they suffered injuries because this policy (a) denied them the opportunity to interact as neighbors in a diverse environment, (b) artificially limited their circle of associates in ways that impeded their ability to engage in commerce with the broadest possible market, and (c) damaged their public reputations by making it appear as if they preferred to live in a "white ghetto" by residing in a complex with a black population that was less than 1 percent. In the early stages of litigation, courts rejected these claims on the grounds that since Trafficante and Carr were in fact residents of the Parkmerced apartments they could not claim that they had been denied housing for racial reasons and therefore did not have standing to bring claims under the aegis of the Fair Housing Act.

When the litigation reached the Supreme Court, however, the case brought an unusual and extremely generative ruling. All nine justices agreed that the intent of the Fair Housing Act was to help "any person who claims to have been injured by a discriminatory housing practice," and therefore the law had to be construed in broad terms. The Court ruled that Trafficante and Carr indeed had standing to pursue litigation and were in fact injured by the policies of their landlord. Even more important, writing for the unanimous court, Justice William O. Douglas argued that the very weaknesses that were written into the design and implementation of the act compelled ordinary citizens like Trafficante and Carr to take the actions they had. Citizen enforcement of the law through private litigation was necessary, Douglas declared, because (a) the law gave no enforcement power to the Department of Housing and Urban Development, (b) the litigation authorized by the act was confined to the Department of Justice and then only if it discerned a systematic pattern and practice of discrimination, and (c) the housing section of the Justice Department employed only 24 attorneys and was not set up to enforce the law in any meaningful way. Under these conditions, "the main generating force" for enforcing this law, Justice Douglas declared, must be private lawsuits through which citizens act as "private attorneys general in vindicating a policy that Congress considered to be a high priority" (*Trafficante,* 1972).

In *Trafficante v. Metropolitan Life Insurance Company*, the Court not only endorsed citizen enforcement but also offered a broad reading of who is injured by housing discrimination and the nature of that injury. Citing the debates that took place on the floor of the U.S. Senate when the bill was being considered, Justice Douglas held that in this case "the person on the landlord's blacklist is not the only victim of discriminatory housing practices" but, rather, that the entire community is harmed by a case like this because the discriminatory action undermines the clear intention of Congress "to replace the ghettos" and promote truly integrated living patterns (*Trafficante,* 1972).

The recognition by the Supreme Court in the Trafficante case of the necessity for citizens to act as private attorneys general and the identification of housing discrimination as detrimental to the free interactions required for both democratic citizenship and robust and open market opportunities set the stage for decades of litigation, legislation, and administration grounded in an expansive understanding of the reach and scope of fair housing law. Over the years, the Fair Housing Act has provided a basis for securing justice by a wide range of injured parties, including immigrant victims of hate crimes, women harassed and assaulted by landlords, entrepreneurs denied legitimate business opportunities by racially motivated ordinances designed to prevent blacks and Latinx home seekers from moving into previously majority-white areas of residence, and city governments burdened by the collateral consequences of waves of mortgage foreclosures instigated by predatory lending practices (*U.S. v. Piekarsky & Donchak, United States v. Southeastern Community and Family Services, Lozano v. Hazelton, Garrett v. Escondido, Greater New Orleans Fair Housing Action Center v. St. Bernard Parish, Baltimore v. Wells Fargo*).

ZANESVILLE AND MODESTO

In 2002 nearly 60 residents of the Coal Run community near Zanesville, Ohio, filed a complaint with the state civil rights commission. They alleged that for more than 50 years city and county officials had deliberately denied the plaintiffs access to water and sewer connections because they resided in a predominantly black community. The population of Coal Run was 85 percent black in a county that was 90 percent white. This concentration stemmed in part from real estate agents and mortgage lenders confining African Americans to an artificially constrained segment of the local housing market as well as from the harassment and violence that blacks encountered when trying to rent or purchase homes in predominantly white neighborhoods.

The water connections from the city stopped at exactly the place where houses inhabited by blacks began. Some pipelines that delivered water to distant areas of the county inhabited by whites were located right next to houses where blacks lived that had been refused water service. All of the plaintiffs lived approximately one mile from the city of Zanesville's water plant or from public water lines that were already in place, yet they found themselves forced to purchase water in bulk, at 10 times the cost of water that flowed through the pipelines, and haul it themselves to their homes (*Kennedy v. City of Zanesville*, 2008).

The refusal to provide Coal Run residents with water created an ongoing public health crisis in the community. In an area historically focused on coal mining, acid residues permeated ground and well water, giving it a peculiar red-orange hue and making it foul smelling and impossible to drink. The water that Coal Run residents purchased or collected in rain barrels had to be stored in cisterns that frequently became stagnant and consequently vulnerable to being fouled by parasites, worms, mosquitoes, snails, rats, snakes, and crawfish (Johnson, 2008). Water stored in cisterns would freeze in winter months, compelling residents to chip away at the ice. Capturing water in rain barrels and melting snow did not produce enough supply to enable regular baths or showers. Water for washing dishes was reused several times before being discarded. Children had to take baths in tubs that contained the water in which their siblings had already bathed. The quality of the water Coal Run residents could access was so bad that they could not wash clothes in it. The absence of a secure water supply also made the area a fire hazard, leading to higher premiums on homeowners' insurance.

Over a period of nearly 50 years, Coal Run residents repeatedly requested access to the municipal water system. Some of them even paid deposits and tap fees for water connections that were never set up. The racial animus propelling the denial of service was made evident to them again and again. One plaintiff remembers overhearing a water authority administrator vow that "those niggers will never have running water." Another testified that county commissioner Dorothy Montgomery "joked" that Coal Run would not get water unless "the President dropped spiral bombs and hopefully hit deep enough to hit good water." Shortly after they filed their complaint with the Ohio Civil Rights Commission, African American residents of Coal Run began to receive threats. A severed pig's head was left in the driveway of Cynthia Hairston's home in late 2002. Slightly more than six months later, a co-worker taunted Hairston by saying, "You people out on Coal Run act like the Klan don't exist out here any more" (*Kennedy v. Zanesville*, 2008).

The Ohio Civil Rights Commission ruled that the treatment of Coal Run residents violated the state's civil rights laws. When that judgment produced no remedial action, however, community residents joined forces with a fair housing agency to start litigation that accused the city and county of violating the Fair Housing Act by denying water to the community. The defendants countered that the denial of water to Coal Run residents had nothing to do with racism. They contended that it would simply have been too expensive to extend the pipelines to the neighborhood. Yet trial testimony revealed that the decision to locate the pipeline sending water to white neighborhoods in the county *around* rather than *through* Coal Run made the project *more* rather than less expensive. The defendants also claimed that the plaintiffs had never requested water service, but evidence in the trial showed that they had spoken up at meetings, made oral requests, and even started paperwork and paid fees for service that never materialized. In July 2008, a federal jury awarded the Zanesville plaintiffs nearly $11 million in damages. Moreover, as explained by Vincent Curry, the executive director of the Fair Housing Advocates Associates that was party to the litigation, the victory sent a warning to other municipal agencies that racially discriminatory policies could make them vulnerable to claims filed under the aegis of the Fair Housing Act (Johnson, 2008).

In 2002, the same year when the Zanesville case began, a parallel fair housing and environmental justice case emerged in Modesto, California, when the Committee Concerning Community Improvement and the South United Neighbors association joined with seven individual plaintiffs to charge the City of Modesto, Stanislaus County, the county sheriff, and other officials with violating the Fair Housing Act. The plaintiffs claimed that the city's practice of annexing only areas inhabited mostly by whites secured for those neighborhoods an array of amenities and services that were denied systematically to the largely Latinx neighborhoods left out of the annexation process. The Modesto case revealed the existence and import of the practice of "underbounding," a new form of racial subordination. Underbounding enacts but hides racial discrimination by drawing municipal boundary lines in such a way as to exclude communities of color from municipal services. A key mechanism for underbounding in Modesto was the Modesto-Stanislaus master tax-sharing agreement that identified the unincorporated areas inhabited by whites as eligible for sharing property tax revenues while excluding areas inhabited largely by Latinx from securing access to those same funds (Molina, 2014, pp. 183–184).

Court filings revealed that Latinx in the Modesto area were three times more likely than non-Latinx to dwell in an unincorporated area. They showed that county officials treated the unincorporated areas within their jurisdiction that were populated largely by Latinx as places of last resort, denying them necessary health, safety, and police services. The litigation revealed that residents of the areas known as "Bret Harte," "the Garden," "No Man's Land," and "Robertson Road" lived in isolated, underserved pockets of land, on poorly maintained streets in neighborhoods without proper sanitation, sidewalks, storm drainage systems, or adequate street lighting. The percentage of Latinx in these four aggrieved neighborhoods ranged from 63 percent in the Garden to 76 percent in Bret Harte.

The areas' impoverished infrastructures meant that water and sewage collected in the streets. Pools of water stood stagnant for weeks. Streets that children traversed on their way to schools became filled with mud. Placing these neighborhoods under the jurisdiction of the distant county sheriff's office rather than the nearby city police department deprived them of adequate law enforcement. The absence of patrols and surveillance opened the way for city and county residents to dump household garbage, mattresses, appliances, and abandoned cars in vacant lots and alleys in the unincorporated areas. The presence of these objects in the alleyways endangered the health and safety of the residents by posing potential impediments to fire and police vehicles. The garbage also provided a nesting ground for flies, mosquitoes, mice, and rats. On occasions when residents implored the county to clean up refuse in these neighborhoods, they found themselves receiving warnings to remove the debris themselves, threatening them with fines for failing to remove garbage they had no role in dumping.

The Modesto area's main sewage treatment plant was positioned right on the border of the Garden and Robertson Road communities. Residents of these neighborhoods had to endure the noxious smells that the facility emitted, but because they lived outside the city limits, they were not served by it. Instead, they had to rely on septic tanks out of which sewage rose to the surface during rainstorms, overflowing onto streets that flooded frequently because of their inadequate drainage (*Committee Concerning Community Improvement*, 2004).

Like the Zanesville case, residential racial segregation in Modesto rendered the mistreatment of particular places a *de facto* mistreatment of non-whites. The city and county gave favored treatment and full municipal services to the white neighborhoods located northeast of Highway 99, while denying services to those on the wrong side of what has historically been the color line in the area. Annexation policies followed prior patterns of housing discrimination, enhancing the value of assets secured in the past through restrictive covenants, mortgage redlining, racial zoning, real estate agent blockbusting, and outright refusal to integrate neighborhoods.

In a manner that paralleled the litigation history of the Trafficante case, the plaintiffs in Modesto suffered an initial setback when the U.S. District Court for the Eastern District of California dismissed their lawsuit, ruling that the protections of the Fair Housing Act applied exclusively to the initial moment of rental or purchase, not to discriminatory actions after occupancy. An appeal to the U.S. Court of Appeals for the Ninth Circuit successfully challenged that opinion. The Ninth Circuit reversed the initial verdict and sent the case back to the district court for trial, declaring that Fair Housing Act protections bar discriminatory actions throughout the time of occupancy. The Ninth Circuit also ruled that in the Modesto case the refusal to spend city and county tax revenues in unincorporated areas evidenced clear and intentional racial discrimination. In 2011, attorneys for the plaintiffs and defendants negotiated a settlement that secured commitments from the city and the county to establish priorities for infrastructure improvements on an objective and nondiscriminatory basis, to make sewer projects in Latinx neighborhoods a high priority, and to support future annexation of the four aggrieved neighborhoods (Murthy, 2011). Like the Zanesville and Trafficante cases, the litigation in Modesto secured justice for the plaintiffs, established legal precedents that expanded the reach and scope of the Fair Housing Act, and created a disincentive for future discriminatory acts by establishing them as clearly illegal and potentially quite costly.

THE SOCIAL ECOLOGY MODEL OF FAIR HOUSING

Cases like the ones protesting segregation in the Parkmerced apartments in San Francisco, denial of water provision to the residents of Coal Run near Zanesville, and the underbounding of Latinx neighborhoods in Modesto apply a social ecology approach to housing that can serve as the basis for legal strategies aimed at securing appropriate damage awards for successful plaintiffs.

The social ecology framework challenges the spatial and temporal parameters of the tort model of injury. By showing how housing discrimination concentrates poverty and damages the social cohesion of communities, it explains how every act of discrimination against individual victims has collateral consequences for the physical and mental health of people living nearby. One foreclosed home on a block can be a fire hazard, a place that squatters use to sell drugs or promote prostitution, and an attraction to vandals who strip the dwelling of copper pipes and other items they can sell. Each foreclosure reduces the equity of everyone in the neighborhood and deprives municipalities of tax revenue needed for fire and health inspectors. The act of discrimination is itself a health hazard that can lead to hypertension and other physical illness, while producing excessive psychological introversion and extroversion in its victims. These practices undermine social cohesion and fray social networks, leading to more physical and mental health problems (Burris et al., 2002; Gee & Ford, 2011).

The cases of *Trafficante v. Metropolitan Life*, *Kennedy v. Zanesville*, and *Committee Concerning Community Improvement v. City of Modesto* all evidence dimensions of a long struggle within fair housing litigation to move beyond the tort model of injury. When applied to housing discrimination, this model imagines the standard fair housing problem to entail a discrete, individual, and intentional biased action that disrupts a previously just status quo. In actual practice, however, housing discrimination is collective, cumulative, and continuing, not so much the disruption of a just baseline norm, but the continuation of a system of racialized capitalism that has never been just. Wealth secured from past acts of discrimination has been passed down across generations, giving the descendants of discriminators unearned advantages. Passing the Fair Housing Act made it illegal to continue to use restrictive covenants, racial zoning, and mortgage redlining. Yet the effects of those practices did not end simply because they were to be considered illegal from that point on. The assets that appreciated in value and were passed down across generations continued to shape the wealth portfolios of white families, while the asset poverty imposed on communities of color impeded their entry into the housing market. Each present act of discrimination adds to the collective and cumulative deprivation of aggrieved groups and increases the unearned rewards of whiteness (Lipsitz, 1998). The tort model of injury as applied to fair housing law leads to egregious underestimation of the mental and physical health injuries created by racially motivated housing discrimination. It does not recognize that the economic, mental, and physical damage done to people by racial discrimination can function like a time bomb, manifesting itself long after the act has taken place, and even after it has seemingly been adjudicated (Burris et al., 2002).

The presumption of a previously just social order and the emphasis on individual injury and liability in the tort model of civil rights law grievously distorts how racism takes place. It puts beyond the reach of normal legal adjudication most of the practices that produce and preserve racial subordination. As Scott Burris astutely observes, civil rights law "does not so much prohibit as regulate discrimination," because it "prohibits a narrow range of behavior driven by provable racial animus," thus, in effect, protecting the innumerable other practices that harm non-whites because of past and present forms of racial subordination. Without being pushed to a higher level by activist litigation, legislation, and mobilization, civil rights law can function as a handbook for the perpetuation of racism, a body of information that authorizes many forms of discrimination by declaring only a select few of them to be illegal (Burris 2002, p. 517). It would not be an exaggeration to estimate that 90 percent of the harmful actions and policies that perpetuate racial subordination fall *outside* the purview of civil rights laws (Crenshaw, 1995).

The Trafficante, Zanesville, and Modesto cases reveal links that connect housing to other social ills. Most people understand that using race as a basis to exclude people from housing imposes artificial, arbitrary, and irrational impediments to opportunities for asset accumulation. Yet what is not as widely understood is that housing discrimination hurts more than its direct victims. In *Trafficante*, the Court ruled that the entire community was injured by the ways in which exclusionary housing policies constrained potentially vibrant social and economic interactions. In the Zanesville and Modesto cases, courts held that administering space in a racially biased way produced health and safety problems for both the plaintiffs and broader communities. These rulings demonstrate how nearly every obstacle to social justice in this society has a fair housing dimension.

Because of residential racial segregation, place becomes a proxy for race and it becomes easy for communities of color to be targeted for systematic voter suppression, educational inequality, environmental racism, and predatory policing. Such practices entail a range of consequences that include disproportionate vulnerability to paying municipal fines and fees, going into debt, and being targeted for mass incarceration, which researchers have shown is itself a public health problem. David Ansell (2017) notes that HIV is five times more prevalent in prisons than on the outside and that hepatitis C is nine times more prevalent. Forty percent of adults in jails and prisons have more than one chronic health condition. Mandatory minimum sentences and "three strikes" policies that mete out life sentences for minor crimes lead to an aging among inmates that increases their health risks. These conditions on the inside of penal institutions have health consequences outside them, especially in the urban and rural sacrifice zones from which most prisoners come and to which they return. The mortality rate is 3.5 times greater among returning ex-offenders than in the general population, and 12.7 times greater during the first two weeks of reentry. This imposes a hidden burden on ex-offenders' families. Ansell (2017) finds that women caregivers of returning ex-offenders are especially susceptible to chronic health problems, especially hypertension and depression (pp. 93–94).

Because neighborhoods inhabited by members of aggrieved racial groups are targeted for surveillance and harassment by "broken windows policing" that relentlessly punishes small crimes of condition as if they were crimes of conduct, policing constitutes a health hazard, even short of the much publicized shootings of black and brown civilians by officers. Sewell, Jefferson, and Lee (2016) have shown that "stop and frisk" policies damage mental and physical health by making residents in targeted areas feel insecure in place. She reveals that a 9 percent rise in the risk of being frisked results in an equal increase in reporting ill health, an 11 percent increase in reports of diabetes, and a 17 percent rise in reports of high blood pressure (Sewell et al., 2016, pp. 5–6).

Moreover, racism itself has been shown to be a health hazard, exposing its targets to physical and psychological damage. Studies show that people who have been subjected to racism are likely to have longer periods of unemployment and underemployment, of illness and incarceration, and shorter lives, shorter careers, and fewer retirement years (Gee, Walsermann, & Brondolo, 2012). Unfair treatment can produce depression and anxiety (Chae, Nuru-Jeter, Lincoln, & Francis, 2011, p. 70). Nancy Krieger notes that experiencing and anticipating racial discrimination "mobilizes lipids and glucose to increase energy supplies and sensory vigilance and also produces transient elevations in blood pressure that can lead to sustained hypertension" (quoted in Burris, 2002, p. 514).

In much the same way that public health researchers and environmental justice advocates and activists have recognized the need to embrace a social ecology framework to transcend the limits of the biomedical model of human health, fair housing litigators have labored to find ways to address discrimination as collective, cumulative, and continuing, as the product of complex interactions among diverse practices, processes, institutions, ideologies, structures, and systems. Comparing fair housing injuries to incidents of environmental destruction can be a useful point of entry to the development of a social ecology perspective in fair housing practice. Thompson, Fullilove, and other social scientists often illustrate the reach and scope of racist injuries by comparing them to living downstream from a factory that pollutes a river. The chemicals released into the waterways poison the water; make fish, birds, animals, and plant life sick; and facilitate the movement of diseases. People living downstream may not know about the source of the pollution upstream, but they will see sick fish floating down the river. Motivated by compassion and concern for their own health, they may remove sick fish from the stream, bring them to the shore, and nurture them back to health. The fish may get better, but they will not be completely well, especially because to survive they must be returned to the very waters that made them ill in the first place. Someone needs to go upstream and do something about the source of the pollution, or new groups of poisoned fish will appear in the river every day.

The traditional practices of fair housing law grounded in the tort model of injury resemble efforts to rescue and treat fish from a polluted river. Only a miniscule number of the victims of housing discrimination get to the point of filing a complaint. Some are discouraged, demoralized, and defeated by the initial discriminatory act. Others do not know they can file complaints or lack the resources to do so. Overworked and underfunded fair housing agencies and attorneys can pursue only a small number of the cases they encounter. When victories are won and the victorious plaintiffs step outside the courtroom, they return to the same artificially constrained housing market that injured them in the first place. The social ecology framework looks upstream and attempts to identify and correct the sources of the pollution that poisons lives in the present. This "upstream" may be defined by the origins, effects, and continuing consequences of now outlawed but once legal forms of overt housing discrimination, or it may be the complex collateral consequences of decisions about things like water provision in Zanesville and underbounding in Modesto.

EQUITY-ORIENTED, COMMUNITY-BASED COLLABORATIVE RESEARCH

Struggles to contest the biomedical model of illness in public health and the tort model of injury in law require challenges to established paradigms by both credentialed experts and community groups. Fights for fair housing and environmental justice require not only new pieces of research but also new relations of research. The ideas and actions of communities in struggle produce new forms of knowledge that enable augmented awareness and understanding of aspects of social structures and social relations that are absent from the perspectives of dominant groups. Project-based learning that is useful to the struggles of social movement groups deepens and expands the sources of data from which evidence is derived and from which research questions are formulated (Hale 2008, p. 23).

CONCLUSION

The times we live in are no ordinary times. In the midst of the systemic breakdown of the economy, the environment, and the educational and electoral systems, we face daunting challenges. Developing the clarity, conviction, and courage we need for the difficult days that lie ahead requires us to know more about how power works, and to challenge received knowledge and conventional wisdom. In a neoliberal society, experts and power brokers offer top-down technical and managerial solutions to what are, at root, political problems. Employing a social ecology framework connects individual ills to discrimination that is collective, cumulative, and continuing, the product of complex interactions among diverse practices, processes, institutions, ideologies, structures, and systems. Scholars can learn from and with people in struggle. We can look "upstream" to identify and attempt to correct the sources of "pollution" that poisons lives in the present. We can learn with those who have been "poisoned," developing new knowledge paradigms on campuses in communities that serve the interests not of foundation funders, philanthropists, or finance capital, but that instead fuel what Dr. King called "the bitter but beautiful struggle" for a new and better world.

DEEPENING OUR UNDERSTANDING

Project choice 1:

1. Access the 1939 Home Owners Loan Corporation Secret City Survey files on eight California cities at http://salt.umd.edu/T-RACES/ (or another location if a similar database exists).

   a. Your group will be assigned a city and instructed to compare and contrast one neighborhood that has an A rating marked in green on the map with an area that has a D rating marked in red on the map. These designations were used to parcel out federally secured home mortgage loans on an expressly racist basis: The A neighborhoods that got loans most easily and with the best conditions were exclusively white (usually because of restrictive covenants), and the D neighborhoods invariably contained black residents or other aggrieved communities of color.

   b. Using the data in these files, especially the area descriptions, identify expressly racist statements that influenced the area designations and predict the future of the neighborhood with respect to asset appreciation and depreciation, amenities and hazards, and economic opportunities and impediments.

   c. Examine the contemporary condition of each neighborhood based on census data and other sources, and report your findings.

Project choice 2:
An instance of racial discrimination has denied a family of four a rental dwelling in a ZIP code with many advantages and resources. (Find a ZIP code near you that fits this description.) The family then finds an apartment for the exact same rent in a ZIP code plagued by poverty.

a. Your group will take on the role of an expert witness in this hypothetical fair housing case. Keep in mind that despite finding an apartment with the same rent, your group needs to compare the conditions in the two ZIP codes: the quality of local schools; the incidence of crime; insurance rates; possibilities of advantageous local social networks; the number and kinds of stores, especially fresh food outlets, pharmacies, and banks; the quality of the local air, water, and land; and the availability of transportation, medical care, and employment opportunities.

b. Write a report recommending the damages that should be paid to the family in reparation for their treatment.

   For example, a family denied housing in the 94563 zip code in Orinda, California, that finds lodging in ZIP code 94607 in Oakland for the same amount of rent incurs new costs. In Oakland, the general pollution burden is worse than in 70 percent of the census tracts in the state, while only 5 percent of California's census tracts have less pollution than Orinda (CalEnviro Screen 2.0). In Orinda, 98 percent of local residents possess a high school diploma and 78 percent possess a bachelor's degree or higher. Miramonte High School in Orinda has one teacher for every 18 students, while Ralph Bunche High School in Oakland has one teacher for every 20 students (Zillow). The projected annual household income in Oakland is $52,962; in Orinda, that figure is $166,856. Yet auto insurance rates average $585–$1,583 per year in Orinda and $672–$1,884 in Oakland. In 2016, only 18 homes were foreclosed in Orinda, compared to 350 in Oakland (Realty Trak).

References

Acevedo-Garcia, D., Osypuk, T. L., & McArdle, N. (2010). Racial/ethnic integration and child health disparities. In C. Hartman & G. D. Squires (Eds.), *The integration debate: Competing futures for American cities* (pp. 131–152). New York: Routledge.

Ansell, D. (2017). *The death gap: How inequality kills*. Chicago, IL: University of Chicago Press.

Bullard, R. (2016). Environmental racism and the environmental justice movement. In S. Nicholson & P. Wapner (Eds.), *Global environmental politics: From person to planet* (pp. 238–245). New York: Routledge.

Bullard, R., Mohai, P., Saha, R., & Wright, B. (2007). *Toxic wastes and race at twenty: 1987-2007*. Cleveland, OH: United Church of Christ Justice and Witness Ministries.

Burris, S. (2002). Introduction: Merging law, human rights, and social epidemiology. *Journal of Law, Medicine, & Ethics, 30*, 498–509.

Burris, S., Kawachi, I., & Sarat, A. (2002). Integrating law and social epidemiology. *Journal of Law, Medicine & Ethics, 30*, 510–521.

Chae, D. H., Nuru-Jeter, A. M., Lincoln, K. D., & Francis, D. D. (2011). Conceptualizing racial disparities in health: Advancement of a socio-psychobiological approach. *Du Bois Review, 8*, 63–77.

Committee Concerning Community Improvement et al. v. City of Modesto. CIV-F-04-6121 (2004).

Crenshaw, K. (1995). Race, reform and retrenchment: Transformation and legitimation in anti-discrimination law. In K. Crenshaw, N. Gotanda, G. Peller, & K. Thomas (Eds.), *Critical race theory: The key writings that formed the movement* (pp. 103–122). New York: New Press.

Crowder, K., & Downey, L. (2010). Inter-neighborhood migration, race and environmental hazards: Modeling micro-level processes of environmental inequality. *American Journal of Sociology, 115*, 1110–1149

Environmental Health Coalition. (2017, February). New data show San Diego low-income communities of color remain among most heavily polluted in California (press release). https://www.environmentalhealth.org/images/releases/2017/CalEnviroScreen_Release_FINAL.pdf

Gee, G. & Ford, C. L. (2011). Structural racism and health inequities. *Du Bois Review, 8*(1), 115–132

Gee, G., Walsemann, K., & Brondolo, E. (2012). A life course perspective on how racism may be related to health inequities. *American Journal of Public Health, 102*, 967–974.

Hale, C. R. (Ed.) (2008). *Engaging contradictions: Theory, politics, and methods of activist scholarship*. Berkeley: University of California Press.

Institute for Research on Race and Public Policy. (2017). *A tale of three cities: The state of racial justice in Chicago report*. Chicago, IL: Author.

Johnson, D. (2008, August 12). For a recently plumbed neighborhood, validation in a verdict. *New York Times*.

Kennedy et al. v. City of Zanesville (2008).

Lipsitz, G. (1998). *The possessive investment in whiteness: How white people profit from identity politics*. Philadelphia, PA: Temple University Press.

Molina, E. T. (2014). Race, municipal underbounding and coalitional politics in Modesto, California and Moore County, North Carolina. *Kalfou, 1*, 180–187.

Murthy, S. (2011, June 30). Latino residents reach settlement with City of Modesto and Stanislaus County on equal access to municipal services (press release). California Rural Legal Assistance. http://www.crla.org/110630-pr-residents-reach-settlement

Seng, M., & Caruso, F. W. (2010). Achieving integration through private litigation. In C. Hartman & G. Squires (Eds.), *The integration debate: Competing futures for American cities* (pp. 53–66). New York: Routledge.

Sewell, A., Jefferson, K., & Lee, H. (2016). Living under surveillance: Gender, psychological distress, and stop-question-and-frisk policing in New York City. *Social Science Medicine, 159*, 1–13. doi:10.1016/j.socscimed.2016.04.024

Smith, S. L., & Cloud, C. (2010). Welcome to the neighborhood? The persistence of discrimination and segregation. In C. Hartman & G. Squires (Eds.), *The integration debate: Competing futures for American cities* (pp. 9–22). New York: Routledge.

Starbuck, A., & White, R. (2016). *Living in the shadow of danger: Poverty, race, and unequal chemical facility hazards*. Washington, DC: Center for Effective Government.

Sze, J. (2007). *Noxious New York: The racial politics of health and environmental justice*. Cambridge, MA: MIT Press.

Trafficante v. Metropolitan Life Ins. Co. (1972).

12

FOR TRIBAL PEOPLES, FOOD JUSTICE REQUIRES ENVIRONMENTAL JUSTICE

Elizabeth Hoover

The term *environmental racism* describes the disproportionate way in which communities of color shoulder the burden of environmentally harmful projects. This is especially salient for Indigenous people who not only continue to suffer from the alienation from much of their tribal homeland due to settler colonial policies, but who also are fighting against the contamination of land to which they do have access. Sites ranging from industries to mines, military bases, oil wells, and pipelines negatively affect not only the surrounding environment but also the health and cultures of the Indigenous communities they border (Hoover et al., 2012; LaDuke, 1999).

Environmental issues impacting Indian Country need to be considered through the unique colonial history and relationship that tribes have with the United States, and their status as sovereign nations. Mvskoke Creek scholar Daniel Wildcat refers to the dislocation of Native peoples from the environment as a result of contamination as the "fourth removal," following relocation from tribal homelands through forced removal, the compulsory attendance of Indigenous children in boarding schools, and the removal of tribal identity through assimilationist programs (Wildcat, 2009). The ongoing project of settler colonialism entails removing Indigenous people from the land through theft, genocide, assimilation, and increasingly through making it difficult for traditional people to harvest food from the land because of concerns about environmental contamination (Hoover, 2017a). Ongoing research across the United States has documented the negative health impacts to Indigenous people of consuming contaminated traditional foods (Adamson, 2011; Donatuto, Satterfield, & Gregory, 2011; Hoover, 2013, 2017b; Miller et al., 2013).

As such, much of the organizing around environmental justice issues in Indigenous communities is in part to protect traditional food sources. The notion that Indigenous peoples have a sovereign "right to food" has been affirmed in an array of documents and declarations created by international governmental and nongovernmental alliances. This includes the affirmation by the United Nations Declaration on the Rights of Indigenous Peoples of the range of rights required for the full exercise of food sovereignty, paying particular attention to the connections between the right to cultural self-determination and the right to maintain and protect seeds and land (Adamson, 2011). But as Potawatomi philosopher Kyle Powys Whyte (2015) notes, "[M]any of the more visible theories of environmental justice have not explicitly referenced the relationship between food and environmental justice" (p. 144). For many people, being fed by today's corporate food regime—dominated by global monopolies and global supply chains—means that the average person does not see their local environment as a source of food.

PHOTO 12.1: Mohawk midwife Katsi Cook at a fundraiser for the Akwesasne Freedom School, August 2019.
Photo by Elizabeth Hoover.

FOOD JUSTICE

In their book *Food Justice*, Robert Gottlieb and Anupama Joshi (2010) trace how the environmental justice conceptualization of the environment as "where we live, work, and play" could be extended to "where, what, and how we eat," to reflect the growing food justice movement to transform where, what, and how food is being grown, produced, transported, accessed, and eaten (p. 5). Continuing this extension of the environmental justice movement, Gottlieb and Joshi (2010) define food justice as "ensuring that the benefits and risks of where, what, and how food is grown and produced, transported and distributed, and accessed and eaten are shared fairly" (p. 6).

The organization Just Food defines food justice as "communities exercising their right to grow, sell, and eat healthy food. Healthy food is fresh, nutritious, affordable, culturally-appropriate, and grown locally with care for the well-being of the land, workers, and animals. People practicing food justice leads to a strong local food system, self-reliant communities, and a healthy environment."[1] The three dimensions of food justice include equal access to safe, healthy, and culturally appropriate foods; assurances that everyone working within the food system is paid fair wages and is working in safe conditions; and accounting for the value of food in relation to the self-determination of communities, especially "urban communities of color, Indigenous peoples and migrant farm workers among many other groups"—that is, those for whom their food systems are often challenged (Whyte, 2016, p. 123). In their assessment of the growing field of food justice, Charlotte Glennie and Alison Hope Alkon (2018) note that food justice encompasses many issues, including the opportunity to grow or purchase healthy food; diet-related health disparities; access to land; and wages and working conditions in agriculture, food processing, and restaurant work. The movement also focuses on "just sustainability"—the fusion of concerns for ecological sustainability and social justice.

Food justice advocates recognize that the problems of the food system are connected to other systemic social, economic, and racial injustices and, thus, that effective food-related initiatives could be used as tools to develop a set of community-based solutions that might help transform the very political and economic systems that have historically oppressed low-income and ethnic minority communities around the world (Broad, 2016). Conversely, food injustice occurs when at least one human group systematically dominates one or more other human groups through their connections to and interactions with one another in local and global food systems (Whyte, 2018, p. 345). Food injustice also occurs when environmental conditions do not allow for traditional levels of harvesting foods for Indigenous peoples—through the fault of polluting industries but also due to environmental policies that dictate levels of acceptable discharge or cleanup based on the needs of the average American population that does not subsist off the land.

As such, there is increasingly a convergence in many communities between environmental justice and food justice, as people work to have their environments cleaned up and to make safe food accessible to everyone regardless of social standing. As opposed to life in many urban environments, for rural and Indigenous communities the immediate environment is often the source of food. Both the environmental justice and food justice movements have cultivated (a) a place-based and health-related focus that is critical of corporate dominance of the food system, (b) the empowerment of community members, and (c) the development of sustainable and livable communities (Gottlieb & Fisher, 2000). Proponents of both environmental justice and food justice seek to alter power relations at the root of social and ecological problems. In Indigenous communities, understandings of environmental injustice are often tied to notions of the wrongful disruption of Indigenous food systems. Specifically, as Whyte (2015) notes, "Indigenous food systems refer to specific *collective capacities* of particular Indigenous peoples to cultivate and tend, produce, distribute, and consume their own foods, recirculate refuse, and acquire trusted foods and ingredients from other populations" (p. 145). Indigenous peoples often describe food injustice as a violation of their collective self-determination over their food systems, and a disruption of these collective capacities, in many cases by environmental contamination (Whyte, 2016).

This chapter includes three case studies that attempt to deepen our understanding of food justice for Indigenous peoples. The first case highlights how the Akwesasne Mohawk people have fought against the contamination of fish as a threat to environmental reproductive justice, arguing that risk-avoidant protective measures like fish advisories, made necessary by environmental contamination and followed in order to make breastfeeding safe, disrupt the community's ability to safely engage in culturally important activities. The second case examines how tribal scientists and their allies have pushed for the enactment of different measures of traditional food consumption in order to shift the amount of permitted industrial discharge in waterways. And the third case examines the ways in which tribal communities are using the fight for food sovereignty as another means of preventing polluting infrastructure such as mines and pipelines.

FIRST FOODS: HOW CONTAMINATED FISH AND BREAST MILK ARE A THREAT TO ENVIRONMENTAL REPRODUCTIVE JUSTICE

Akwesasne is a Mohawk community located at the nexus of New York, Ontario, and Quebec. The community is also downwind and downriver from three Superfund sites, all of which leached polychlorinated biphenyls (PCBs) into the waterways from which Mohawks once drew much of their diet. In the early 1980s, shortly after the discovery of the extent of the industrial contamination leaching into her community, Mohawk midwife Katsi Cook went to the State of New York and demanded that the Mohawk community be assisted in determining the safety of its food sources—namely, the fish coming from the river being directly polluted by industry, and the breast milk of new mothers.

As a midwife, Katsi was most concerned about community members at the top of the food chain—breastfeeding infants. She developed a research field team of Mohawk women and named it the First Environment Research Project, in acknowledgment that the womb is the first environment that we are exposed to and that the contamination of this environment is a significant human transgression, impacting the sacred relationship between mother and child, and the ability for a Nation to properly reproduce itself. Studies conducted in collaboration between the First Environment Research Project, the Akwesasne Task Force on the Environment, and the State University of New York (SUNY) at Albany, through a Superfund Basic Research Program (SBRP) grant, connected levels of PCBs in participants' breast milk and blood to fish consumption, which decreased as community members began heeding fish advisories published by the tribal government (Hoover, 2013).

The research determined that traditional food practices—which entail a web of language, culture, family, and interspecies relations, in addition to the nutrition provided—were delivering contamination from the industrial site to the bodies of Akwesasne. PCBs leached from the General Motors site into the river were taken up by fish and then eaten by Mohawk people. From dinner plates, these PCBs made their way into Mohawk bodies, into their fat, into women's breast milk, which was then passed on to the most sensitive members of the community (Hoover, 2017a). In her book *Tainted Milk*, Maia Boswell-Penc (2006) notes that breast milk is "symbolic of the most essential human connection," and when breast milk is contaminated, that "should be a wakeup call" (p. 12). Boswell-Penc (2006) demonstrates "how child sustenance becomes a figure for the oppressive frameworks at work in any historical juncture," and this is seen in Akwesasne, where Indigenous women were instructed to abandon their traditional food practices in order to prevent accumulating more PCBs that would then be transferred to their children (p. 169). In communities impacted by environmental contamination, residents are often encouraged to avoid the source of contamination—to practice risk avoidance to make up for the regulatory agency's and industry's lack of risk reduction—in lieu of cleaning up the source.

Fish advisories are often issued in an effort to protect human health from exposure to contaminants, but many Native American communities suffer from unintended health, social, and cultural consequences as a result of warnings against eating local fish (Hoover, 2013). Prior to the fish advisories that were issued in the 1980s to protect people in Akwesasne from exposure to the industrial contaminants, fishing had been a central part of the diet, economy, and social culture of their community for centuries. More than just a means of acquiring a dietary mainstay, fishing was considered a livelihood, a lifestyle, and a culture. Almost everyone I spoke to in this Mohawk community had a connection in some way to fish or fishing. The process of catching and cooking fish out of the river was at the root of many people's childhoods and something that connected them to their ancestors (Hoover, 2017a).

As part of a broader project in which I interviewed Akwesasne community members in 2008 and 2009, I asked participants about the extent to which they had changed their diets as a result of the fish advisories, knowledge about the industrial sites, or changes they noticed in the fish. Three-quarters of the 50 Akwesasne Mohawks interviewed had ceased or significantly curtailed their local fish consumption due to the issuance of fish advisories or witnessing or hearing about deformities on fish. Many of these respondents turned to outside sources of fish, from other communities or from grocery stores. This change in fish consumption concerned many residents because cultural and social connections developed around fishing were being lost and because fish was replaced with high-fat, high-carbohydrate processed foods, which has led to other health complications. One-quarter of the 50 interviewees still eat local fish, but these are generally middle-aged or older residents; fish consumption no longer occurs in the multigenerational social context it once did (Hoover, 2013).

When these fish advisories were issued, calling for people to diminish or eliminate fish from their diet, many residents felt they had lost more than just omega-3 fatty acids from their diet; another part of their culture was being eroded by outside influences. A cessation in fishing gradually diminished Mohawk culture in several ways. As Henry Lickers, who worked for the Mohawk Council of Akwesasne department of environment describes, the language and culture around tying knots in nets as well as the social interactions that occurred around the process of creating these nets are lost when there is no longer a use for those nets:

> People forget, in their own culture, what you call the knot that you tie in a net. And so, a whole section of your language and culture is lost because no one is tying those nets anymore. The interrelation between men and women, when they tied nets, the relationship between adults or elders and young people, as they tied nets together, the stories . . . that whole social infrastructure that was around the fabrication of that net disappeared. (Hoover 2017, p. 175)

Similarly, the language around the names and descriptions of certain fish was lost. As one older man described to me, "A lot of that has been forgotten, the fish names in our language. Because a lot of the fishermen when they go fishing they talk about their Indian names to them, there is no English part of it, but that has been sort of forgotten now."

A series of health studies at Akwesasne demonstrated that PCB exposure had negative impacts on residents' health (Hoover, 2017b). But people's health was also impacted by the environmental contamination even without the ingestion of fish: Fear of exposure led to the replacement of this low-fat source of protein and other important nutrients with high-fat and high-carbohydrate sources of food. Akwesasne community members Alice Tarbell and Mary Arquette (2000) posit, "Diabetes is on the rise because more people no longer eat traditional foods and no longer participate in cultural activities that once provided healthy forms of exercise" (p. 102). SUNY Albany researchers conducting environmental health studies with Mohawk adolescents also noted that while the community might have decreased their exposure to fish-borne contamination, they have lost a primary source of protein and other important nutrients such as calcium, iron, zinc, and omega-3 fatty acids (Fitzgerald et al., 2004). The replacement of fish with cheap foods has had the effect of "further exacerbating chronic, diet-related health problems in the community, such as diabetes and cardiovascular disease" (Schell et al., 2003, p. 961).

Human health in Native American communities such as Akwesasne is intimately tied to the health of the environment. Avoiding contact with that environment through risk-avoidance policies—which put the onus on the fish consumers to prevent contamination exposure, rather than risk-reduction policies, which would force the polluting industries to clean up their messes—puts an unfair burden on an innocent public who is then faced with a conundrum of culture and nutrient loss or exposure to contamination. Fighting for a clean environment that supports the ability to birth and then breastfeed healthy babies, and the ability to pass on the community's fishing culture, is part of a broader movement for environmental reproductive justice.

Mohawk midwife Katsi Cook (2007), who originated the term *environmental reproductive justice*, said that "[e]nvironmental justice and reproductive justice intersect at the very center of woman's role in the processes and patterns of continuous creation" (p. 32). She notes that in her community of Akwesasne the struggles for environmental justice and reproductive justice coevolved and included understanding the impact of environmental contaminants on women's health as well as on language and culture issues (Cook, 2007). The concept of environmental reproductive justice involves expanding reproductive justice to include a deeper focus on the environment, and to include the reproduction of language and culture as concerns, in addition to the reproduction of human beings. Environmental reproductive justice also aims to expand the framework of environmental justice to more closely consider the impact of environmental contaminants on physical and cultural reproduction. In Akwesasne, the fight for environmental justice, reproductive justice, and food justice are intimately intertwined, as the community has worked hard to push for the most stringent cleanup possible; for the crafting of new up-to-date fish advisories; and for the development of social and cultural programming to support the transfer of knowledge about traditional food production methods—for everything from fishing, farming, and plant gathering, to breastfeeding.

SHIFTING THE STANDARD: HEIRLOOM CONSUMPTION RATES AND PROTECTING TRADITIONAL FOODS CONSUMERS

Because of subsistence lifestyles, spiritual practices, and other cultural behaviors, Indigenous people often suffer multiple exposures from resource use that result in environmental health impacts disproportionate to those seen in the general population (Hoover et al., 2012). "Exposure scenarios designed for suburban activities and lifestyles are not suitable for tribal communities," note Stuart Harris and Barbara Harper (1997, p. 789), who have conducted extensive work around developing Native American exposure scenarios and risk-assessment tools (for example, see Harris & Harper 2000, 2011). For this reason, they argue, risk needs to be calculated differently in Indigenous communities. Indigenous scholars Jaclyn Johnson and Darren Ranco (2011) have also demonstrated how the risk-assessment process used by the U.S. Environmental Protection Agency (EPA) to determine how much of a pollutant can be emitted into or remain in the environment is not equipped to accurately assess and confront the risks facing tribal communities and specific tribal needs. They point out that the process used by the EPA "essentially ignores tribal cultural differences (and really any other kind of difference that deviates from the suburban lifeway). It is much easier for the EPA to apply one already established method used for all Americans, rather than to create a different process each time a tribal culture is impacted by a source of pollution" (Johnson & Ranco, 2011, p. 187). Because of the unique history and political relationships between Native communities and the settler government, achieving environmental justice and food justice for tribes necessitates going beyond "equal protection" and, instead, demands environmental standards that are supportive of cultural differences.

To help tribes fight for standards that will be more protective of the traditional foods that members of their communities consume, scientists like Barbara Harper and Stewart Harris have been working for decades to help tribes document actual and traditional rates of consumption and exposure. They note that it is not useful for the EPA to use the average suburban resident as a standard, because "[i]t appears that an average subsistence lifestyle would be equivalent to at least a 90th percentile of the average suburban exposure. Initial sensitivity analyses show that the difference between means of the two types of life styles ranges from two- to 100-fold" (Harris & Harper, 1997, p. 794). They note that this magnitude of difference is due to the fact that "the traditional way of life as it is currently practiced is more than just a suburban lifestyle with extra fish consumption" but also includes other types of exposures associated with obtaining and processing traditional foods (Harris & Harper, 1997, p. 794).

Working with the Confederated Tribes of the Umatilla Indian Reservation (CTUIR) in northeastern Oregon, Harris and Harper (1997) found a fish consumption rate of 540 grams/day, rather than the EPA value of 70 grams/day, demonstrating how inaccurate EPA risk assessments would be for this community. They note that because individual and collective well-being for these tribal members is derived from having a healthy community with access to ancestral lands and resources in order to satisfy responsibilities to participate in traditional community activities and pass along knowledge between generations, ecosystem integrity of those specific landscapes is imperative (Harper & Harris, 1997).

Barbara Harper also worked with Darren Ranco and Maine tribes to develop the "Wabanaki Cultural Lifeways Exposure Scenario," which is "a numerical representation of the environmental contact, diet, and exposure pathways present in traditional cultural lifeways in Maine" (Harper & Ranco, 2009, p. 2). The intent of the report was to reflect "the lifeways of people fully using natural resources and pursuing traditional cultural lifeways, not lifeways of people with semi-suburban or hybrid lifestyles and grocery-store diets" (p. 2). So while present-day environmental conditions may not allow many people to fully engage in a traditional lifestyle until resources are restored, this report demanded that heritage rates of consumption be taken into account—the amount of fish and other traditional foods that people would be eating if afforded the opportunity to do so safely. The goal of the project was to ensure "that exposure pathway information that is collected for the Tribes in Maine will not be biased by contemporary consumption rates," since tribes in Maine have only recently begun to regain a sufficient land base that would support the utilization of natural resources (p. 2). The report notes that "social oppression, economic factors, and current fears about the safety of consuming natural resources (i.e. fish advisories, etc), severely compromises the effectiveness of a consumption survey to accurately characterize the exposures that would occur during fully traditional use" (p. 14). As such, they do not want current consumption data to be misinterpreted by regulatory agencies to make decisions about acceptable levels of pollutants in the environment. "Specifically, a regulatory agency might assume that Tribes want to use natural resources only as much as they are doing so today. In other words, evaluating only today's tribal exposures and risks, and then setting standards or risk levels based on today's risks could prevent safe uses at higher usage rates (i.e., as restoration improves resource availability, resource cleanliness must increase in parallel)" (p. 14). As such, developing heritage rates—through archival, archeological, historical, and ethnographic research—better represents the goals regulators should be aiming for. A similar report has been prepared with the Spokane Tribe in Washington (Harper, Flett, Harris, Abeyta, & Kirschner, 2002) and is in process with the Narragansett Tribe in Rhode Island, who are also seeking to protect and restore traditional food consumption.

Johnson and Ranco (2011) note that two approaches have been developed to address the problems of cross-cultural risk assessment. One approach is to make science more responsive to the ways that Native people actually use resources in the environment. So, rather than using aggregate models from the entire American population for something like fish consumption, the EPA and tribal scientists measure the actual fish consumption of a tribal community—either currently or the historical consumption rates mentioned earlier. Along these lines, at the urging of the Columbia River Inter-Tribal Fish Commission and other advocates and scientists, in 1997 the EPA finally set a subsistence consumption rate of fish for Native Americans at 70 grams/day—which, while still lower than many tribal fish consumers, was three and a half times the mean ingestion rate of the overall U.S. population (p. 187). But this still does not necessarily cover the exposures to pollution in all aspects of a subsistence lifestyle—including "exposures during the gathering of plants for medicinal and other uses, higher water and air ingestion rates during hunting and fishing, and the exposures to soil, air, and water in other cultural activities like sweat lodges" (p. 187).

The second approach noted by Johnson and Ranco, which is more difficult to define from a scientific perspective, is defined as the "health and well-being" model, which would allow for tribal nations to "redefine health itself in culturally relevant terms" (Johnson & Ranco, 2011, p. 180). In this model, "a risk assessment would not just emphasize the potential deaths to a population caused by cancer, but would look at the risks to a healthy, culturally defined lifestyle in each Tribal Nation. This approach would potentially allow Tribal Nations to redefine health in much broader terms than cancer death rates and would include cultural indicators such as access to a healthy traditional diet, ability to participate in ceremonies, the passing down of traditional knowledge, and so on" (p. 180). Unlike a classic risk-assessment paradigm, a health and well-being paradigm is driven by tribal priorities rather than EPA regulations and measurements of risk; allows communities to define health and establish priorities and relationships accordingly; and, rather than focusing on one aspect of the environment, takes a holistic approach "so that the interconnectedness of all aspects of a community is respected" (p. 192). It is only through allowing tribal communities to define health for themselves; considering the myriad environmental exposures that can occur through the activities necessary to achieve and maintain health; and then ensuring that the environment is clean enough to safely support all of these activities that true environmental justice and food justice will be achieved in tribal communities.

FIGHTING DESTRUCTIVE INFRASTRUCTURE IN ORDER TO PROTECT TRADITIONAL FOODS

Photo by Elizabeth Hoover.

PHOTO 12.2: Melvin Gasper runs his hands over freshly roasted, hulled, and winnowed wild rice at the Red Lake food summit, September 2016.

The preceding two sections have been about tribes contending with existing contamination. In addition, tribes across the nation have been fighting against allowing the types of projects that could create these kinds of exposures—especially pipeline and mining projects. These battles include against the Keystone XL pipeline, which will threaten water sources across the Midwest and Great Plains; the Dakota Access Pipeline (DAPL), which stands to threaten the water supply for the Standing Rock Sioux Tribe and others, while carrying oil whose extraction has poisoned First Nations in Alberta; the Sandpiper Pipeline, which would have threatened wild rice beds in Minnesota; the ongoing fight against the expansion of Line 3, also in Minnesota; and others. Similarly, mining projects across the Great Lakes region threaten wild rice beds—the sacred food of Anishinaabe people—with the possibility of contaminated runoff.

The Kakagon and Bad River Sloughs are an important source of food for tribal communities like the Bad River Band of the Lake Superior Tribe of Chippewa Indians, located on a 125,000-acre reservation in northern Wisconsin, on the south shore of Lake Superior. Together comprising more than 10,000 acres, the sloughs were recognized as a wetland of international importance, or a Ramsar site. Often called the Everglades of the North, the sloughs mark the first Ramsar site to be owned by a tribe. When asked about the importance of the sloughs and reaching this milestone, Tribal Chairman Mike Wiggins Jr. commented:

> The Kakagon and Bad River Sloughs wetland complex represent everything our Tribal People hold dear and sacred on many different levels. Spiritually, the "place" and everything it has, the clean water, the winged, the seasons, the rice and fish, connects us with our ancestors and the Creator. The Sloughs sustain the physical well-being of our community with foods such as wild rice, fish, cranberries, waterfowl, venison, and medicines. From an Anishinaabe (Chippewa) world-view perspective, the wetlands ecosystem is a tangible representation of our values of caring for the environment. The international Ramsar recognition is an honor for the Bad River Band and maybe even more importantly, the recognition sends a message about the importance and critical need for biologically productive and water rich areas such as the Kakagon and Bad River Sloughs wetland complex. There is water purification, ecological harmony, and people who are interwoven into this "place" where the Bad River Reservation dovetails with Lake Superior. (Bad River Band of Lake Superior Chippewa Tribe, 2012)

One of the main environmental challenges that Bad River and other local Ojibwe communities were facing at the time of my visit in 2014, that could have posed a threat to their precious wild rice, was the world's largest open-pit taconite mine that was proposed to be located near the head waters of the Bad River. Taconite is a low-grade iron ore, originally discarded as waste when higher-grade iron was available. But in the early 21st century, there was a renewed interest in even low-grade iron ore to feed industrial growth in China and India (Grossman, 2017). Because wild rice is sensitive to hydrological changes from development and to sulfates that can be released by taconite mines and tailing piles, this potential mine posed a huge threat to the Ojibwe communities downstream.

In 2011 Gogebic Taconite (GTAC) began leasing land in the Penokee Range, upstream from Bad River, with plans to spend $1.5 billion in constructing the largest mine in Wisconsin history, using millions of tons of explosives to dig a 4.5-mile-long open pit in sulfide-laden, asbestos-type rock as part of a mountaintop removal operation that could extend 22 miles (Grossman, 2017). In an effort to assuage concerns, GTAC representatives appeared before Bad River tribal citizens and assured them that their project would follow existing law and be environmentally responsible. But quietly the company had already hired a law firm to write a new bill exempting iron mines from these state regulatory standards (Langston, 2017). In 2013, the Republican-controlled house and senate in Wisconsin passed a new iron mining bill that exempted taconite mining from many state water quality standards and established the expansion of the iron mining industry as state policy—meaning if any provision of the iron mining laws were to contradict other state environmental laws, the conflict would be resolved in favor of the former. But this state legislation began to face stiff local opposition and federal hangups. For example, the U.S. Army Corp of Engineers said that they would not work with the Wisconsin state Department of Natural Resources on an environmental impact statement, one of the most important parts of getting a permit to mine, because the differences in state and federal environmental requirements would not allow the two governments to work together. In addition, an earlier battle over the Mole Lake mine led to the landmark 2002 Supreme Court decision that affirmed the right of Indian nations to work with the EPA to set and enforce their own clean air and water standards. But when the Wisconsin mining bill was drafted, the state senate majority leader refused to consult with the Bad River Tribe. After the bill passed, six tribes joined forces to create the Wisconsin Chippewa Federation to protect ceded territories against environmental threats.

Community members mobilized to educate the public about the potential impact of the mine on their environment and food systems. In June 2011 they founded the Penokee Hills Education Project, which gave tours of the site and hosted potlucks to bring together local residents, Native and non-Native (Grossman, 2017). The site of the proposed mine fell within the ceded territory that Ojibwe people have access to for traditional gathering, and so the Lac Courte Oreilles Ojibwe community and allies set up a Harvest Education and Learning Project (HELP) camp, starting in April 2013, calling it a "living demonstration of treaty rights" (Seeley, 2013). Mining companies owned 22,000 acres in the Penokee Hills, which they enrolled as managed forestlands, in order to pay fewer taxes. This land status meant that the public was able to use this land and could camp there for up to 14 days continuously, according to Iron County rules. Paul DeMain (Oneida/Ojibwe) described how they got the idea to get more people onto the land to enjoy its natural beauty and abundance as a way of rallying the public to support preserving it rather than mining it. They set up tents, built wigwams in the woods, created trails, and hosted foraging parties for wild onions and mushrooms. Every person who joined the camp had to sit down first and have a cup of coffee with Ojibwe elder Melvin Gasper, made locally famous when he was arrested for exercising his treaty-ensured rights to gather wild rice. Meals were served every day at the camp, the coffee was always on, and people were encouraged to break bread together. Food was used as a tool to bring people together, as well as to educate people. As DeMain (2019) described it,

> We actually had feasts throughout the year and throughout the time. There was birthday feasts, there was a fish fry, there was all kinds of things where we had traditional cuisine, we had a huge traditional feast at the White Cap Mountain Resort [where they hosted an anti-mining conference]. . . . It has to do with hospitality, camaraderie. The number of things that we learned in regards to Native cuisine up there, especially with people coming and going all the time, we had—there was hunts launched from there. There was fishing expeditions. We had, you know, there was times when you went up there and it could have been a food sovereignty conference type of atmosphere with what was going on there. A lot of people came through there, a lot of indigenous chefs.

The camp was operated for a full year, with the county government then voting in March 2014 to evict the camp from the site (Jablonski, 2014). They moved to nearby private property to continue traditional harvesting and educational activities (Shackleford, 2014). This included tapping maple trees and making and bottling maple syrup, accompanied by a press release that referenced "gold found in the Penokees"—a play on the long history in this region of prospecting for gold and an effort to recast what was considered a valuable resource in those hills. DeMain (2019) recounted that a big goal of the HELP camp was not just to stop a mine to protect Indigenous food resources but also to educate non-Natives about the ways they could be sustainably using the resources in their backyard rather than relying on unsustainable mining revenue:

> You got all these resources—you've got hickory trees, walnut trees. You got maple trees, you got wild onions, you got mushrooms, you got all kinds of sustainable stuff. You got diamond willow that can be carved, you know, because there was people coming up to the camp and just saying wow, you know, what all this stuff that's out in the woods that's just sitting there and being wasted. It's not even being used for sustainability or anything. You know local food stuff. So we were just trying to say you got to get into these cottage industries rather than waiting around for a mining company.

Fortunately, the proposed taconite mine that stood to threaten Anishinaabe foodways was never completed. After two years of bad press and legal challenges, GTAC's company president, Bill Williams, explained that the Hurley, Wisconsin, office was being closed and the project halted, citing the high cost of wetland mitigation as well as a concern that the EPA could intervene to stop the project (Jablonski, 2014). But DeMain warned that, although people may see the closure as a win, as long as there are mineral deposits in the area, there will be people pursuing extraction and thus threatening the local environment and food systems. He plans to continue to educate people about the importance of traditional foods (Jablonski, 2015).

CONCLUSION: YOU CAN'T HAVE FOOD JUSTICE, OR FOOD SOVEREIGNTY, WITHOUT ENVIRONMENTAL JUSTICE

Photo by Elizabeth Hoover.

PHOTO 12.3: Paul DeMain discussing plant foods available in the woods during a plant identification walk at the Red Lake food summit, September 2016.

Eric Holt-Gimenez, former director of Food First, describes how food and hunger efforts tend to split ideologically between those who seek to stabilize the corporate food regime and those who seek to change it. Those who follow a progressive trend are grounded in notions of citizen empowerment and discourse coming from traditions of environmental justice that denounce the ways that people of color and underserved communities are abused by the present food system. This trend invokes the notion of a gradual, grassroots transition to a more equitable and sustainable food system. The progressive food justice trend promotes the local production, processing, and consumption of food and focuses on inventing new business models that better serve economically disadvantaged communities (Holt-Gimenez, 2011, p. 323). By contrast, the concept of food sovereignty is rooted in a more radical trend—seeking to address the root causes of poverty, hunger, and lack of access to culturally appropriate food through radically transforming the food system. The broader food sovereignty movement proposes dismantling the monopoly power of corporations in the food system, redistributing land and the rights to water, seed, and food-producing resources. Holt-Gimenez (2011) describes how food sovereignty proponents engage in direct action against the World Trade Organization, fight to protect peasant land from contamination by genetically modified organisms, and denounce the appropriation of peasant and Indigenous lands for extractive industries (p. 324). For Indigenous people in the United States, food sovereignty means access to healthy, culturally relevant food, land, and information; independence for individuals to make choices about their own consumption and for communities to define their own food systems; keeping food dollars within the community; sustaining land and cultural lifestyles; protecting seeds as living relatives; educating Native youth through food and culture; and sustaining relationships to the environment, food sources, and other people (Hoover, 2018). For any of these factors to be possible, tribes must be able to interact with a clean environment—

clean to the standards of the tribes themselves, far beyond standards assigned by the federal government based on the average U.S. resident. Food sovereignty will be possible for tribes, and food justice ensured, only when subsistence eating from the surrounding environment without the threat of negative contaminant-related health impacts is possible.

True food security—which has been cited as one of the most pressing issues of the 21st century—cannot be achieved through merely increased production controlled by a corporate and unjust system. Rather, social justice is one of the necessary starting points for analyses of, and solutions to, food insecurity (Cadieux & Slocum, 2015). Food justice entails transforming the current food system to eliminate disparities and inequities (Gottlieb & Joshi, 2010). Because of the unique history and political relationships between Native communities and the settler government, achieving environmental justice and food justice for tribes necessitates going beyond "equal protection," under a settler-controlled system. As Akwesasne Mohawk scholars Arquette et al. (2002) assert, "Environmental justice encompasses more than equal protection under environmental laws (environmental equity). It upholds those cultural norms, values, rules, regulations and policies or decisions to support sustainable communities," which might look different for Indigenous and non-Indigenous communities (p. 262). In order to achieve this, to work toward true food sovereignty for Native communities, tribes need to be able to define their own standards for health and well-being, and they need to demand environmental conditions that support the Indigenous food systems necessary to achieve that health and well-being.

DEEPENING OUR UNDERSTANDING

1. What is the most important source of food in your region? You could think about this as either an eco-feature (bay, forest, farm, river) or a species/community (salmon, huckleberries, oak trees). What is potentially threatening that source of food?
2. What is an example of a food local to your current area that is no longer safe to eat? Which communities in the area have been most impacted by this change?
3. Find an example of a tribe local to you whose food system was impacted by environmental contamination.
    a. What is the source of that contamination? Who, if anyone, has been held responsible?
    b. What has been attempted to address this situation—to prevent further contamination, to clean up existing contamination, and to restore the environment to its former condition?
    c. What are the communities impacted by the contamination? Consider the fish, bird, animal, plant, and other-than-human communities, as well as human communities.

d. What would it take to remedy this situation? Which agencies would need to be involved, and what methods would need to be employed?

4. What would "just sustainability" look like for your community?
5. The concept of environmental reproductive justice addresses issues where environmental situations affect not only physical reproduction but also cultural reproduction. What are some environmental conditions in your community that are potentially affecting these multiple levels of reproduction for people?
6. What would food sovereignty look like for your community?

Suggested Film Resources

Homeland: Four Portraits of Native Action (2005)

Tells the inspiring story of four battles in which Native American activists are fighting to preserve their land, sovereignty, and culture.

Directed by Roberta Grossman

Produced by The Katahdin Foundation

Standing on Sacred Ground series (2014)

In this four-part series, Indigenous people from eight different cultures stand up for their traditional sacred lands in defense of cultural survival, human rights, and the environment.

Directed by Christopher McLeod

A Production of the Sacred Land Film Project of Earth Island Institute

References

Adamson, J. (2011). Medicine food: Critical environmental justice studies, Native North American literature, and the movement for food sovereignty. *Environmental Justice, 4*(4), 213–219.

Arquette, M., Cole, M., Cook, K., LaFrance, B., Peters, M., Ransom, J., ... , & Stairs, A. (2002). Holistic risk-based environmental decision making: A native perspective. *Environmental Health Perspectives, 110*(S2), 259–264. doi:10.1289/ehp.02110s2259

Bad River Band of Lake Superior Chippewa Tribe. (2012, April 5). Kakagon and Bad River Sloughs recognized as a wetland of international importance (press release). https://bad-river-tribe.wnzgax0-liquidwebsites.com/kakagon-and-bad-river-sloughs-recognized-as-a-wetland-of-international-importance/

Boswell-Penc, M. (2006). *Tainted milk: Breastmilk, feminisms, and the politics of environmental degradation.* Albany: State University of New York Press.

Broad, G. M. (2016). *More than just food: Food justice and community change.* Oakland: University of California Press.

Cadieux, K. V., & Slocum, R. (2015). What does it mean to *do* food justice? *Journal of Political Ecology, 22.* https://journals.uair.arizona.edu/index.php/JPE/article/view/21076/20664

Cook, K. (2007). Environmental justice: Woman is the first environment. In S. Song (Ed.), *Reproductive justice briefing book: A primer on reproductive justice and social change* (pp. 32–33). https://www.law.berkeley.edu/php-programs/courses/fileDL.php?fID=4051

Demain, P. (2019, March 2). Interview by author, Ponsford, MN.

Donatuto, J. L., Satterfield, T. A., & Gregory, R. (2011). Poisoning the body to nourish the soul: Prioritizing health risks and impacts in a Native American community. *Health, Risk and Society, 13*(2), 103–127. doi:10.1080/13698575.2011.556186

Fitzgerald, E. F., Hwang, S.-A., Langguth, K., Cayo, M., Yang, B.-Z., Bush, B., ..., & Lauzon, T. (2004). Fish consumption and other environmental exposures and their associations with serum PCB concentrations among Mohawk women at Akwesasne. *Environmental Research,94*, 160–170. doi:10.1016/S0013-9351(03)00133-6

Glennie, C., & Alkon, A. H. (2018). Food justice: Cultivating the field. *Environmental Research Letters,13*, 073003. https://iopscience.iop.org/article/10.1088/1748-9326/aac4b2/meta

Gottlieb, R., & Fisher, A. (2000). Community food security and environmental justice: Converging paths towards social justice and sustainable communities. *Race, Poverty & the Environment, 7*(2), 18–20.

Gottlieb, R., Joshi, A. (2010). *Food justice*. Cambridge MA: MIT Press.

Grossman, Z. (2017). *Unlikely alliances*. Seattle: University of Washington Press.

Harper, B. L., Flett, B., Harris, S., Abeyta, C., & Kirschner, F. (2002). The Spokane Tribe's multipathway subsistence exposure scenario and screening level RME. *Risk Analysis, 22*(3), 513–526.

Harper, B., & Ranco, D. (2009). *Wabanaki traditional cultural lifeways exposure scenario*. Prepared for the EPA in collaboration with the Maine tribes. https://www.epa.gov/sites/production/files/2015-08/documents/ditca.pdf

Harris, S. G., & Harper, B. L. (1997). A Native American exposure scenario. *Risk Analysis,17*(6), 789–795.

Harris, S. G., & Harper, B. L. (2000). Using eco-cultural dependency webs in risk assessment and characterization of risks to tribal health and cultures. *Environmental Science and Pollution Research,2*, 91–100.

Harris, S. G., & Harper, B. L. (2011). A method for tribal environmental justice analysis. *Environmental Justice, 4*(4), 231–237.

Holt-Gimenez, E. (2011). Food security, food justice, or food sovereignty? In A. H. Alkon & J. Agyeman (Eds.), *Cultivating food justice: Race, class, and sustainability* (pp. 309–330). Boston, MA: MIT Press.

Hoover, E. (2013). Cultural and health implications of fish advisories in a Native American community. *Ecological Processes,2*, 4. doi:10.1186/2192-1709-2-4

Hoover, Elizabeth. (2017a). Environmental reproductive justice: Intersections in an American Indian community impacted by environmental contamination. *Environmental Sociology, 4*(1), 8-21. doi:10.1080/23251042.2017.1381898

Hoover, E. (2017b). *The river is in us: Fighting toxics in a Mohawk community*. Minneapolis: University of Minnesota Press.

Hoover, E. (2018). "You can't say you're sovereign if you can't feed yourself": Defining and enacting food sovereignty in American Indian community gardening. *American Indian Culture and Research Journal,41*(3). doi:10.17953/aicrj.41.3.hoover

Hoover, E., Cook, K., Plain, R., Sanchez, K.,Waghiyi, V.,Miller, P., ... , & Carpenter, D. O.(2012). Indigenous peoples of North America: Environmental exposures and reproductive justice. *Environmental Health Perspectives, 120*, 1645–1649.

Jablonski, N. (2014, March 25). LCO harvest camp says it will comply with county rules, but won't pack up. WXPR Public Radio. http://www.wxpr.org/post/lco-harvest-camp-says-it-will-comply-county-rules-wont-pack#stream/0

Jablonski, N. (2015, March 2). GTac office closure prompts range of emotions. WXPR Public Radio. http://www.wxpr.org/post/gtac-office-closure-prompts-range-emotions

Johnson, J., & Ranco, D. J. (2011). Risk assessment and Native Americans at the cultural crossroads: Making better science or redefining health?

In G. Ottinger & B. R. Cohen (Eds.), *Technoscience and environmental justice: Transforming expert cultures through grassroots engagement* (pp. 179–199). Cambridge, MA: MIT Press.

LaDuke, W. (1999). *All our relations: Native struggles for land and life.* Cambridge, MA: South End Press.

Langston, N. (2017). The Wisconsin experiment. *Places Journal.* https://doi.org/10.22269/170425

Miller, P. K., Waghiyi, V., Welfinger-Smith, G., Byrne, S. C., Kava, J., Gologergen, J., ... , & Seguinot-Medina, S. (2013). Community-based participatory research projects and policy engagement to protect environmental health on St. Lawrence Island, Alaska. *International Journal of Circumpolar Health, 72*(1). doi: 10.3402/ijch.v72i0.21656

Schell, L. M., Hubicki, L. A., DeCaprio, A. P., Gallo, M. V., Ravenscroft, J., Tarbell, A., ... , & Worswick, P. (2003). Organochlorines, lead, and mercury in Akwesasne Mohawk youth. *Environmental Health Perspectives,111*, 954–961. doi:10.1289lehp.5990

Seeley, R. (2013, July 28). Fight over iron ore mine in North Woods may move to courts. *Wisconsin State Journal.* https://madison.com/wsj/news/local/crime_and_courts/fight-over-iron-ore-mine-in-north-woods-may-move/article_96c69b92-303c-5fd0-9214-4d109fd2b393.html

Shackleford, M. (2014, April 30). Harvest camp turns one. WXPR Public Radio.http://www.wxpr.org/post/harvest-camp-turns-one#stream/0

Stanford, C. (2011, October 11). Food studies: Who's doing the judging in the food justice movement? *Grist.* https://grist.org/article/2011-09-30-food-studies-whos-doing-the-judging-in-the-food-justice-movement/

Tarbell, A., & Arquette, M. (2000). Akwesasne: A Native American community's resistance to cultural and environmental damage. In R. Hofrichter (Ed.), *Reclaiming the environmental debate: The politics of health in a toxic culture* (pp. 92–113). Cambridge, MA: MIT Press.

Whyte, K. P. (2015). Indigenous food systems, environmental justice and settler-industrial states. In M. Rawlinson & C. Ward (Eds.), *Global food, global justice: Essays on eating under globalization* (pp. 143–156). Newcastle upon Tyne: Cambridge Scholars Publishing.

Whyte, K. P. (2016). Food justice and collective food relations. In A. Barnhill, M. Budolfson, & T. Doggett (Eds.), *Food, ethics and society: An introductory text with readings* (pp. 122–134). Oxford, UK: Oxford University Press.

Whyte, K. P. (2018). Food sovereignty, justice, and Indigenous peoples. In A. Barnhill, M. Budolfson, & T. Doggett (Eds.), *The Oxford handbook of food ethics* (pp. 345–366). Oxford, UK: Oxford University Press.

Wildcat, D. R. (2009). *Red alert!* Golden, CO: Fulcrum Press.

Note

1. Definition originally taken from the Just Food website. The website is no longer available, but the definition is cited in Whyte (2016) and Stanford (2011).

13

POVERTY, PRISONS, POLLUTION, AND VALLEY FEVER

Sarah M. Rios

Nico X was incarcerated in three different state prison facilities in California. Serving his long sentence provided him with insider knowledge about how the carceral state delivers health services to prisoners. Nico wrote two letters in 2006 to an environmental justice organization located in the Central Valley of California, where he was incarcerated at the time. In the letters, Nico raises the issue of prisoners' exposure to an environmental disease known as Valley Fever, to pesticide drift from nearby irrigated fields, and to the shoddy and irresponsible medical care provided by prison health authorities.

1ST LETTER

I am writing this letter in hopes that a very serious matter can be brought to light and exposed about Valley Fever and other toxins at (6) six state prisons in the Central Valley. This matter needs to be addressed as prisoners from inner cities are sick and dying in great numbers.... The six (6) prisons in question are highlighted (see the attached page, prisons opened since 1984) as you will see all the prisons in this Valley Fever area are the most over populated and crowded with inner city peoples.... There are lives at stake here, family and children who will be affected by the loss of a father or son, not to mention the future cost of medical expenses, care, and burial as Valley Fever affects the lungs, bones, and brain.... CDCR [California Department of Corrections and Rehabilitation] is trying to cover up all exposures and medical doctors at the prisons are in denial and fail to care for patients exposed in a timely manner until it's full blown Valley Fever and can't be controlled.

2ND LETTER

I'm writing you in the hopes that you might help further my cause in putting a stop to the continued chemical/pesticide drift that the prisoners are being exposed to at numerous state prisons run by the CDCR here in California.... The fields are right next to most of these prisons.... I myself have been at three of the prisons and gassed with chemicals...to the point that medical intervention was needed. I am still very sick from the exposure and we have medical documentation to prove it [b]ut to which the medical department and the CDCR still denies happened. Even with proof of it happening in which we filed a complaint with the [intentionally blank] Agricultural Commissioner, it seems everybody just denies our complaint.... Now they say before they can finish the investigation which started last year they want us to release our medical records to the "worker health & safety branch of the Department of Pesticide Regulation." But on the release the records won't be released to a doctor, no doctors' name is on the release form....

Nico's impassioned and insightful letters reveal that one root of the problem is not just health and healthcare inequities, but also the knowledge regimes that channel the risks created in poisonous places onto powerless people. Prison

---

PHOTO 13.1 Farm labor housing near Wasco, CA, 2015.
Photo by Sarah Rios.

officials and state bureaucrats place prisons in impoverished and polluted places, neglect the healthcare needs of the people they cage, and ignore complaints from inmates whose illnesses they caused. Pressure from recurrent lawsuits from people who contracted Valley Fever while incarcerated, and the court rulings that resulted from those lawsuits, persuaded the California Department of Corrections and Rehabilitation (CDCR) in 2015 to announce a commitment to spend more than $5 million to conduct skin tests to identify whether inmates had developed immunity to Valley Fever. Their intent was not to cure the disease but to lessen the state's responsibilities by preventing inmates classified as vulnerable from serving their sentences in any of the nine state prisons where rates were highly endemic. Since then, the CDCR has updated its exclusion policy to designate only African Americans, Filipinx, and Inuit, along with immune-compromised populations, as barred from admission into the two facilities with the greatest number of reported cases, Avenal State Prison and Pleasant Valley State Prison. The CDCR identified narrow categories of people in danger, tested only some prisoners for immunity, and then sorted a few of them out of infected facilities.

Nico's deeply felt concerns highlighted what the CDCR ignored. Prisoners and rural populations are dangerously exposed to cumulative environmental health burdens from agribusiness, dairy farms, crude oil extraction, and waste management, which dominate the region where prisons stand tall. Between 1986 and 2005, 22 new mega-prisons were built, and more than half were located in the Central Valley's drought-stricken lands near predominantly Mexican American farmworking communities (Braz & Gilmore, 2006; Gilmore, 2007). The Central Valley's pollution not only far exceeds the U.S. Environmental Protection Agency (EPA) standards for acceptable air quality, but it also disproportionately injures the health of the region's largely non-white, poor, immigrant, and rural residents (Harrison, 2011; Office of Environmental Health Hazards Assessment, 2016). Some of the nation's wealthiest food-growing industries depend on the labor of some of the country's poorest families. Between 20 and 25 percent of residents live below the poverty line in Valley Fever–endemic counties, a rate approximately 10 percent greater than the state average (Public Policy Institute of California, 2013).

At the height of the Valley Fever epidemic, no one in a position of authority assessed accurately the combined threats emanating from the fatal nexus of poverty, prisons, and pollution, and from a failing prison healthcare system. All of these factors combine to impose cumulative burdens on people after release from prison. Instead, the immunity and racial identity of a few people who are incarcerated were presumed to be the crux of the problem, even when the origins and complications of Valley Fever were known to stem from environmental and social factors. In this way, the CDCR directed millions of dollars and extensive administrative time toward solutions that evaded the real causes of a major public health problem, refusing to address the cumulative vulnerabilities produced in prisons and the communities that surround them.

THE ENVIRONMENTAL AND SOCIAL IMPLICATIONS OF VALLEY FEVER

Exposure to a soil fungus classified as *Coccidioides immitis* (cocci) causes the illness commonly referred to as Valley Fever. This disease is endemic in the southwestern United States (Centers for Disease Control and Prevention, 2016). The warm climate and low precipitation levels in the Central Valley make Fresno, Kern, Kings, Madera, and Tulare counties especially susceptible to germinating the fungus in the soil, although cases are increasing in San Luis Obispo and Los Angeles counties (Centers for Disease Control and Prevention, 2016; Johnson, 2017). Cocci transform into an infectious disease when the dirt germinating the fungus is dispersed into the air due to a strong gust of wind or excavation of an infected area (Smith, Pappagianis, Devline, & Saito, 1961). Just one breath of the spores can cause Valley Fever.

The disease is not contagious, but it is burdensome. Approximately 40 percent of those who are exposed to the fungi spores will develop a combination of rashes, body aches, rapid weight loss, fatigue, fever, and chest pains that last an average of 42 to 152 days (Tsang et al., 2010). More severe symptoms include infections in the joints, the spinal cord, and the brain. These symptoms require aggressive and timely medical interventions (Centers for Disease Control and Prevention, 2016). Just under 1 percent of patients will require a lifetime of treatment (Tsang et al., 2010). It is estimated that approximately 15,000 people nationwide acquire Valley Fever annually, although many more cases are believed to be unreported (Centers for Disease Control and Prevention, 2017).

Mass incarceration exacerbates the health risks of Valley Fever. Between 2006 and 2010, the rate of infection was 52 times greater at Pleasant Valley State Prison, and 10 times greater at Avenal State Prison, than in the county with the highest rate of Valley Fever in California. The rate of infection at Pleasant Valley State Prison in 2005 was 3,000 cases per 100,000 inmates (Pappagiani, 2007). Over 15 years, more than 200 inmates have died from Valley Fever at the Avenal site alone, and the health of 1,000 others was compromised (*San Quentin News* Staff, 2013).

The symptoms of infection and a general awareness of dust as a potential threat have led experts to believe that individual behavior can minimize, if not prevent, infection. Employers are encouraged to control dust by dampening loose dirt and paving open fields, to provide respiratory protection with HEPA air filters in closed cab equipment, and to instruct employees to use locker rooms to remove and store work clothes so that they do not take spores home (California Department of Public Health, 2015). The National Institute for Occupational Safety and Health (2016) further recommends that employers arrange for prompt medical evaluation and treatment when workers are exposed. Local departments of public health encourage visitors and residents to stay indoors during dust storms, to use an N95 respirator face mask, to clean skin wounds with soap and water, and to take prescribed medication if they become ill (Centers for Disease Control and Prevention, 2016). Such preemptive and time-sensitive strategies purport to reduce the harmful effects of the disease; however, they are ill suited for prisoners who have minimal access to showers and little control over their surrounding environment. They are equally inapplicable to farmworkers, who labor routinely without locker rooms, filtered air, or even running water.

Members of aggrieved racialized groups, especially Mexican, Black, Filipinx, and Indigenous, as well as farmworkers, construction workers, and people incarcerated in prisons, are disproportionately vulnerable to Valley Fever. A careful study by Hector, Rutherford, and Tsang (2011) found that 51 percent of the reported cases of Valley Fever occurred among Hispanics, whereas only 32 percent occurred among whites, even though both groups make up approximately 44 percent of the population of highly endemic counties in California. Black people in highly endemic counties represent 4 percent of the general population, yet they account for 11 percent of the cases during this same period. Blacks and Pacific Islanders are more likely than whites to develop severe symptoms (Cox & Magee, 2004; Hector et al., 2011; Huang, Bristow, Shafir, & Sorvillo, 2012; Mohney, 2013). Native Americans experience the greatest mortality rate from the disease, even though they represent less than 3 percent of the population (Hector et al., 2011; U.S. Census Bureau, 2016). Yet these studies and others do little to explore social factors as possible explanations, focusing instead on biological and genetic causes for disparate outcomes (California Department of Public Health, 2013; Cox & Magee, 2004; Huang et al., 2012; Mohney, 2013). One study that compared hospitalization rates among patients with Valley Fever found a clear class component in that cohort, discovering that low-wage earners and people who rely on subsidized health insurance or charity programs were more likely to use emergency services than were people with private health insurance (California Department of Public Health, 2013). Rather than examining how social structures and harmful environments injure the health of the people with fewer social economic opportunities, biomedical studies focus on genetic information they believe can be linked to disparate outcomes in order to speed development of treatments and hoped-for vaccines. Although effective medical treatments are necessary, and no possible cause of the disease should be off-limits to researchers, an exclusive focus on genetic information ignores what vulnerable groups know about their symptoms and how they forge feasible solutions and remedies.

The following sections highlight the extraordinary bank of knowledge about Valley Fever that emerges from the survival strategies of those suffering from it. The testimonies that follow from an environmental justice advocate, a former prisoner, and a report from a collaboration of community organizations all go a long way toward identifying the specific health injuries that harm both prisoners and rural residents of color in endemic areas, many of whom are farmworkers. They resonate with what epidemiologist Nancy Krieger (2011) identifies as the social ecological approach to illness that grants causal priority to social injustices rather than to biological traits, individual behaviors, and minor genetic differences.

SITUATED KNOWLEDGE: AN ENVIRONMENTAL JUSTICE PERSPECTIVE ON VALLEY FEVER

Valeria is an activist who works with an environmental justice organization in Kern County. She collaborates closely with residents from her neighborhood in the town of Earlimart, in Tulare County. Valeria believes that environmental conditions in the area were responsible for her daughter, Emily, contracting Valley Fever at age six. Valeria is not a farmworker, but she has a family history of relatives migrating from both Mexico and the Philippines to work in the fields. Additionally, her mother, Teresa, engaged in activism advocating restrictions on pesticide use in order to protect the health of Central Valley residents, farmworkers and non-farmworkers alike. Teresa fervently championed the creation of buffer zones free from pesticides near schools and dwellings.[1] Valeria's family history of labor and environmental activism has helped her to become a grassroots expert on the many health risks that farmworkers and other rural residents encounter from exposure to pesticides and dust.

Valeria strongly believes that Emily was exposed to Valley Fever while playing in the front yard of their house in Earlimart. Two vast agricultural fields line the quiet street where the house is located. At the corner intersection near the home is an almond orchard. In the front are seemingly endless rows of vineyards. Between the street and the fields is a narrow five-foot dirt path that farmers use to turn their tractors around into the next row as they fumigate the fields for pests and shake the trees to harvest the almonds. Farmers use every inch of land for production, even when that means that families nearby are exposed to the dangers of dust and chemicals.

There are few barriers protecting residents against the encroaching dust and pesticide drift kicked up by the moving tractors that operate near their homes. Valeria explains that when these fields are doused with fertilizers, fumigants, or other chemicals, residents across the street have to shut their windows in hopes of breathing in less of the poisoned air. The pesticides and dust in the air combine with the high rates of PM 10 (particulate matter, 2.5–10 micrometers) and PM 2.5 (particulate matter, <2.5 micrometers) pollution coming from various industries in Kern County, particularly waste processing sites, chemical production plants, truck and automobile traffic, and oil processing facilities, among others. Particle pollution is measured as the sum of all dry and/or liquid matter suspended in the air, either through direct release from a source or formed out of a chemical reaction in the atmosphere (Airnow, 2016). Pesticides also contribute to "bad" ground ozone, a chemical reaction of volatile organic compounds and nitrogen oxides (U.S. Environmental Protection Agency, 2017). According to the EPA, the health consequences of particle pollution include "irritation of the eyes, nose, and throat, coughing, chest tightness and shortness of breath, reduced lung function, irregular heartbeat, asthma attacks, heart attacks, and premature death in people with heart or lung disease" (Airnow, 2016; U.S. Environmental Protection Agency, 2017). The concentration of particle pollution and ozone in the fields creates serious respiratory problems for residents living near the fields and for farmworkers toiling in them.

When Emily was recovering from Valley Fever, the air quality was a deciding factor in whether her daily activities would take place inside or outdoors. Valeria relied on the School Flag Program to make her decision. This system reports air quality for the day online and through display of a colored flag hanging on the pole below the American flag in front of the school. A green flag indicates good air. A yellow flag reports that the air has moderate quality. Orange means that the air is unhealthy for sensitive groups. Red signals unhealthy air, and purple indicates that the air is very unhealthy. Based on the color of the flag, students at school are forced to adjust their outdoor activities (Airnow, 2015). Restricting Emily's participation outdoors, even during family functions on the weekends, was necessary to protect her health and ensure her recovery. Valeria observed that "it was like she was being punished, while polluters kept doing their business as usual."

Valley Fever made the toxins in dusty fields a matter of immediate concern to Valeria, but it was not the first time that she had come across environmental threats to her family's health. Valeria explains that pesticides used during the past three decades and even earlier still reside in the local drinking water. Traces of nitrate, arsenic, and the pesticide dibromochloropropane (DBCP) are regularly found in groundwater wells in Earlimart and the neighboring towns of Delano and McFarland (Delano-Earlimart Irrigation District Report, 2007). For many years, Valeria complained to her mother about smelling fumigants outside, her eyes burning from pesticides, and feeling dizzy. Valeria drew for me a dramatic but depressing map of her street, illustrating how cancer consumed one resident after another. Many of her neighbors have suffered from similar acute health problems. She pointed to each house that suffered the loss of a loved one to cancer, losses that included her mother and two of her relatives who lived nearby.

Inconclusive but suggestive public health research has prefigured Valeria's observations. In 1989, the California Department of Health Services identified childhood cancer clusters in various parts of Earlimart and in the city of McFarland, 15 miles south (Institute for Rural Studies, 1990). Health researchers sampled the soil and air for contaminants, but the information they gathered at the time was not sufficient to prove, by accepted scientific standards, that the unusually high rates of cancer among children were directly related to the pesticides found in their surrounding environment. The results were inconclusive, due in part to the lack of records kept on the types of pesticides and the amounts applied. In addition, as Julie Sze (2007) argues, studies of cancer clusters are often inconclusive because experts rely on mainstream research methods requiring a large dataset, a high threshold of statistical significance that the scientific community agrees upon, and replicability. The structure of scientific "expertise" is such that it acts as "biased against causation and common-sense notions of causality because scientific standards are held to a higher standard of proof than nonscientific interpretation" (p. 182). This means that the burden of proof is placed on neighbors experiencing environmental pollution. It is left to them to use community-based environmental justice research methods to reveal causation, even if the techniques and strategies may be rendered "meaningless" by credentialed experts because they are outside of the conventions of research knowledge, conventions that are better suited for study in the laboratory than in the plural, diverse, uncontrollable, and unpredictable activities of actual social life.

Valeria's discovery of the patterns of chronic disease on just one residential block displays her advanced knowledge about the causes, consequences, and complications of diseases like Valley Fever. Her framework connects farmworkers' experiences with her own, and identifies specific tools that could be used to prevent, predict, and treat Valley Fever. She urges a reexamination of the power politics that make pesticides and harmful pollution burdens factors in shaping the odds of contracting and recovering from Valley Fever. The framework that Valeria deploys does not emerge out of laboratory tests and sequencing genetic variants, but it does offer tangible goals and solutions that have yet to be explored. In many ways, it resembles social epidemiologist Nancy Krieger's (2011) *ecosocial* theory of health, which posits that the manifestation of illness rests not simply in the physiological deterioration of the body but also in the body's exposure to social and ecological inequality. The body absorbs effects from the social, political, and economic conditions in which everyday life unfolds, which create disparities in "physiology, behavior, and genetic expression" (p. 936). The reaction of the body to social inequality influences the "development, regulation, growth, and death of the biological systems, organs, and cells" (p. 936). Diseases do not discriminate whose health they will injure, but policies, practices, and processes certainly do.

A FORMER PRISONER'S PERSPECTIVE OF VALLEY FEVER

Journalist John Ellis published an interview in the *Fresno Bee* in 2015 with the spokesperson for the California Department of Corrections and Rehabilitation, Luis Patino. Ellis quotes Patino explaining the efforts of the Corrections Department to reduce Valley Fever among inmates:

> "CDCR has been working to mitigate Valley Fever for years," Luis Patino said. "We have put in place numerous measures in our prisons to reduce the amount of dust, and the movement of dust, particularly into buildings. We have also moved inmates deemed at higher risk and who chose to move out of the two prisons in the Valley Fever endemic zone. We have also worked with state and federal public health partners to study further methods of reducing the incidence of Valley Fever in Avenal and Pleasant Valley prisons. To date, more than 2,100 inmates were moved from two prisons and mitigation efforts continue.

Patino would have the general public believe that these "numerous measures" in the prisons have been of long standing, that the facility operators are conscious about the potential threat of mobile dust (both outside and indoors), and that they are considerate of the fears of inmates who desire a quick transfer out of an endemic prison facility into safety. Eager to demonstrate action on and accountability for the well-being of thousands of inmates, Patino's words aggrandize the CDCR as a rational institution in carrying out the exact measures demanded by the federal courts for each patron.

The experiences of Bob, a 60-year-old African American man from Southern California incarcerated at Avenal State Prison, however, contrast with the practices described by Patino. In an interview with me, Bob stated:

> When I first arrived in Receiving and Release [at Avenal State Prison], there nobody ever told us about Valley Fever, how you can catch it, what to do, or stuff like that.... I had went in and I was sitting in line waiting to be housed and I seen the doctor. And you know, police officers and all the staff walking around, and they give us a little orientation. But they never mentioned anything about Valley Fever. So, you know, when I caught it, it was strange to me.... I went to the doctor and the doctor he's sitting back on the door. He didn't tell me nothing about it. He just said, "you probably have pneumonia" or something like that, and that "you'll be okay," right. And all the inmates kept telling me "Bob you got Valley Fever, man. You know that's what's riding this prison here. There's a lot of people who've been here and gotten sick." Some of them had Valley Fever to the point that they were walking like they had polio.... I went from 250 pounds to 190 pounds in a week and a half.... So, like I said my friends and associates there they were getting on me about it. And they said, "man, come on. Go in there man, like you know what you're talking about. That's bull. That's bull crap. You go and take your butt up in there and don't come out of there." So that's what I did.... The next thing I know I fell out inside the doctor's office and they rushed me to the hospital. I stayed in the hospital for like maybe a couple of months trying to be nursed partially back to good health.... I was just, you know, I was stunned by how the doctor, right, he knows the symptoms of Valley Fever because he has dealt with so many people there, the prisoners that had caught the Valley Fever. So, I don't understand still why, how come this guy, the doctor, act like he don't know what's going on.... Things never went right after that...I asked them to ship me out of there because the place was destroying my health.... I think I was 52, 53 years old at the time. You know, I never had high blood pressure. I've never been diabetic. I never had anything wrong with me, period. And I went, I did that there, and I fell sick. But the most sad part about it is that they never told us about Valley Fever in R & R [Receiving and Release] when we first came there. If they would have, I would have taken different precautions when going outside. But since they never told me....

Bob's experience upon arriving at Avenal State Prison contrasts dramatically with the "numerous measures" that Patino trumpets. Exposure to and inadequate treatment for Valley Fever infection caused Bob's temporary sentence for crime to become a lifelong sentence to a slow social death and, potentially, even to physical death. He now relies on medication to keep the infection under control, as well as to regulate other parts of his body that are deteriorating from the disease and from the side effects of the medicine that he must take to stay alive. Bob's unfortunate condition stems from his incarceration in a facility plagued by Valley Fever and from the inadequate features of prison healthcare. It is not a consequence of his personal health habits, physical health, or genetic makeup. Bob discovered that his fellow inmates were more capable of identifying Valley Fever symptoms than the facility's authorized medical provider. Bob's infection and those of countless other inmates, as well as related deaths, are a result of the Department of Corrections administration's practices that minimize its obligations to prisoners behind the scenes while aggrandizing its protective measures in public.

What constitutes adequate healthcare for populations marked with racialized criminality is discussed recurrently but rarely resolved. The CDCR claims to have first become aware in 2005 of the rising rates of Valley Fever and the associated deaths among inmates of color and those medically compromised at Avenal and Pleasant Valley state prisons (U.S. District Court, 2013). In the period between 2005 and 2013, however, the CDCR implemented few measures to protect inmates' health. In 2006, a federal court judge appointed J. Clark Kelso as a monitor of direct operations and instructed him to make improvements in the prison healthcare system to meet constitutional standards. Kelso was authorized specifically to investigate Valley Fever at Pleasant Valley State Prison, where rates were among the highest in the state system. Kelso's study resulted in the posting of laminated signs warning inmates and prison staff about Valley Fever symptoms in state prisons located in endemic zones. Prison administrators also relocated some inmates with pulmonary conditions and heavily immunosuppressed conditions away from endemic prisons under the aegis of the Cocci Exclusion Policy. By 2012, the CDCR and Kelso modified the exclusion policy to include a more elaborate medical classification system and to construct a "Valley Fever transfer list" of inmates who obtained approval for transfer by a physician (see Kelso, 2013). The health hazards assessed in prisons located in Valley Fever endemic zones certainly concerned the CDCR, largely because of potential financial liability and federal court orders that assessed prison conditions as violating constitutionally mandated standards of fair and humane treatment. However, the solutions advanced rarely resolved the problems that the inmates faced.

Inmates were frequently treated as expendable. By 2013, only three inmates were on the transfer list (U.S. District Court, 2013). A recommendation to construct ground coverage throughout the prison property, the same strategy that was used during World War II on military bases in the region to protect troops from infected dust, was never implemented because the CDCR claimed it was not feasible due to an initial cost that "could potentially exceed $750,000 in addition to ongoing maintenance costs" (U.S. District Court, 2013, p. 5). The California Department of Public Health, academic and clinical experts on Valley Fever, and Kelso, the court-appointed receiver, convened in 2006 to examine the health crisis further. Within a year, they issued an additional report with 26 recommendations. The CDCR and the receiver adopted only four of the suggestions, two of which they had already implemented the previous year.

The environmental health analysis ordered by the court could have produced useful recommendations about how to improve the working and living conditions of confinement facilities, but it was delayed, devalued, and, later, discarded. Concern for the health of guards and other personnel employed by the correctional system enjoyed a higher priority for the CDCR. In December 2008, the CDCR's Occupational and Public Health Section requested the first formal health hazard evaluation from the National Institute for Occupational Safety and Health (NIOSH) to be conducted at Pleasant Valley and Avenal state prisons to "'examine Valley Fever cases among prison employees—not inmates"' (*San Quentin News* Staff, 2013, p. 5). Yet even this initiative was cut short. Five months after the section's request and one week prior to the planned site assessment, the state of California "unilaterally cancelled the planned site visit by NIOSH" (U.S. District Court, 2013, p. 7). In court documents, NIOSH explains that the governor of California had decided to create an advisory group within the CDCR to decide "'whether or not pursuing the health hazard evaluation further would be valuable to the State of California'" (p. 7). Shortly afterward, the CDCR disbanded the office that was responsible for overseeing occupational health issues and relocated this responsibility into another undisclosed part of the state bureaucracy. A report by NIOSH explained that "this development along with the lack of support from CDCR management precluded moving forward with the health hazard evaluation" (p. 7). After contacting the leaders of the California Correctional Peace Officers Association, NIOSH was forced to withdraw its request because neither the union nor the CDCR management supported the environmental health analysis. Analyzing and disclosing the source of health hazards for prison staff became trivialized precisely because it would have revealed the vast environmental threats affecting thousands of prisoners' health.

Although the state refused to act, citizen activism started to address the crises of Valley Fever in areas plagued by poverty, prisons, and pollution. Receiving letters from prisoners like Nico and perusing the testimonies from returning offenders like Bob served as a catalyst for forging what George Lipsitz (2006) identifies as "unexpected affiliations" between prison abolitionists and environmental justice activists, which went beyond identifying and protesting against individual acts of discrimination to instead "expose the collective practices and patterns that produce inequality and that keep whole collectivities subordinate to others" (p. 26). Californians United for a Responsible Budget (CURB) and their allies published a report in 2016 that revealed how the rapid growth of the prison population corresponded with the rise in cumulative environmental threats to prisoners' and rural residents' health.

The overpopulation of prisons led to sewage spills into groundwaters that service the prisons and surrounding rural communities, while increased traffic from busing prisoners and automobile travel by officers, attorneys, and visitors contributed to exceeding the safety thresholds for air pollution (CURB, 2016). The siting of prisons in the Central Valley placed inmates in proximity to noxious industries such as agricultural operations, chemical waste evaporation from oil well water ponds, ammonia gases from dairies, and open dust fields that are conducive to the gestation and dissemination of Valley Fever (Braz & Gilmore, 2006; Cole & Foster, 2000; CURB, 2016; Gilmore, 2007; Pardo, 1999). In some cases, the prisoner population is located right across the street from multiple industries operating at maximum capacity that frequently violate environmental safety standards for pollution (G. Aguirre Jr., personal communication, October 2, 2017; CURB, 2016). Unlike other residents of the region, prisoners are denied the right to use water for flushing the toilet. They are restricted to bathing only three times a week, even though more gallons of water are consumed for maintenance than for personal use by inmates (Pemberton, 2015; Rothfield, 2008). The health impacts of pollution in prisons are usually not detected until a significant number of people reveal symptoms of poisoning, despite clear evidence in the hands of state officials about the risks and availability of funds allocated by the state to address them (CURB, 2016).

The protocols adopted by the Department of Corrections in response to Valley Fever focused on genetic predisposition rather than social or environmental causes. They evidence contempt for a population deemed to be "criminals" who are portrayed as threatening to exhaust public resources and, therefore, become seen as undeserving of medical care. Despite the evidence presented by Bob, Nico, and many other current and former inmates, and despite the findings of cumulative vulnerabilities published in the CURB report, the CDCR continued to ignore the conditions that make prisoners sick and that render them unable to secure adequate medical treatment. Thus, expert authorities impeded rather than advanced knowledge about the causes, complications, consequences, and potential cures for Valley Fever.

LESSONS FROM ENVIRONMENTAL JUSTICE

Photo 13.2. ■ A researcher (left) and member of Fresno Environmental Reporting Network (FERN) use a FLIR (Forward Looking Infrared) camera to capture emissions at an oil refinery, 2016.

Photo by Gustavo Aguirre, Jr., coordinator of the Central California Environmental Justice Network.

Nico's letter, Valeria's observations, Bob's testament, and the studies led by community organizers and their allies illuminate the extensive reach and scope of the causes and consequences of Valley Fever. They reveal a refusal to accept health protocols that reduce prisoners to biological and racialized categories of risk while diminishing and denying the social and environmental hazards facing prisoners and rural residents alike. They identify environmental pollutants and their effects on health, suggest strategies to make prison surroundings safer, connect inmate health issues to broader community struggles, and develop and deepen advocacy skills through public participation in acts of contestation. Perhaps most significant, they reveal how the assumptions of the dominant society distort health risks and remedies by portraying illnesses and injuries as singular events rather than as parts of cumulative, collective, and continuous processes.

Side by side, prisoners, prison abolitionists, and environmental justice activists create and cultivate new knowledge through shared experiences with illness that is very different from the privileged knowledge deployed by public health officials and biomedical experts. This knowledge enables better questions to be asked and better remedies to be constructed. Their combined struggle with infectious disease, cumulative vulnerabilities, and administrative indifference complicates the contours of health and environmental justice. What is accomplished when the environmental health wrongs facing prisoners and rural residents are separated into distinct, disaggregated, and unrelated events? How can the knowledge held by both rural and incarcerated residents help to improve scientific expertise about environmental threats to health? How do we privilege the views and experiences of the most vulnerable groups without compromising the health needs of other populations who suffer from Valley Fever?

As we move forward with creating and collecting alternative knowledge bases, selecting a method has to be tempered with the general awareness that every method has its strengths and limitations. Ordinary residents' experiences with and knowledge of health, illness, and the environment need to be central to academic and public policy research. The fields of science and medicine receive considerable and deserved criticism for regularly neglecting to integrate laypersons' knowledge, for generally dismissing evidence that comes from people with limited educational backgrounds and an absence of scientific training (Epstein, 1995). Steven Epstein (1995) warns that considering layperson knowledge as qualitatively inferior to the rigor of scientific research erroneously portrays lone scientists as the sole developers of new knowledge. This top-down approach dichotomizes "experts" from "citizens" along class and racial hierarchies, perpetuating rather than alleviating unequal health outcomes (Sze, 2007; Wynne, 1996). We need a method that honors the perspectives of people in the collective struggle for their right to influence the politics of health and environmental justice in order to illuminate not just new ideas from the ground up but also more complex and pluralistic visions about health and healing.

When medical experts focus exclusively on advancing biomedical research, they default to using racial categories as proxies for genetically related and disease information while occluding the historical ties that connect genomic research to the eugenics framework (Chaufan, 2018). Privileging biomedical approaches over a social ecological framework has historically been used to justify segregation and to distract the public from the unequal and discriminatory social conditions affecting people's health (Ansell, 2017; Roberts, 2011). As David Ansell argues in another context, when it comes to environmental racism, people's ZIP codes are usually more important than their genetic codes. What is needed is a community-based model of environmental health monitoring that is built on the idea that ordinary residents are knowledgeable about their environment and the status of their health.

When laypersons participate in popular epidemiology, which is the process through which ordinary residents gather statistical data, new knowledge is created and new resources are assembled for experts to advance and improve science meaningfully (Sze, 2007). Popular epidemiology in the form of health surveys and alternative environmental research projects can expand the categories that are considered important for inquiry and the methods for interpreting new information (Sze, 2007). Popular participation in research can also promote new research questions that experts would otherwise ignore.

The environmental justice movement often implements a model of grassroots environmental monitoring combined with scientific methods that test for pollution; this is sometimes referred to as "citizen science" or as community-based environmental justice research (CBEJR). CBEJR combines a "multifaceted analytical world view," a social and environmental justice perspective, local insights, and professional scientific techniques (Sze, 2007) to create new knowledge about health and the environment. For example, the "bucket brigades" and health surveys implemented by environmental justice activists in Richmond, California, and in Cancer Alley in Louisiana took samples of air and submitted the samples to a professional laboratory for various tests (Sze, 2007). The "grab samples" and the labs that were privately contracted by residents produced data about toxic releases that were on par with what regulators used to document and monitor air quality. Unlike regulators, citizen scientists draw on local knowledge about local risks to pinpoint the places where pollution can be measured most productively, for example, in rural communities near noxious industries and within neighborhoods rather than in geographic peripheries of a municipality or from aggregate data constructed on a regional basis.

When ordinary residents gather information about environmental and health threats, it opens up the possibility to fuel new scientific studies for local officials and encourages residents to seek unexpected and unanalyzed sources of information that can yield valuable data. As Julie Sze notes, community-centered science illuminates the analytical links that connect race, poverty, health, and the environment in ways that expand scientific knowledge and that can lead to better environmental health policies.

DEEPENING OUR UNDERSTANDING

When learning and thinking about health and environmental justice, consider the strengths and limitations of the tools made available to measure cumulative vulnerabilities, environmental burdens, and health consequences. The EPA offers the EJSCREEN website to map and examine various environmental resources and risks for all U.S. ZIP codes.

Step 1: Spend 20 minutes familiarizing yourself with how to generate an environmental justice map of your ZIP code. Start with this link to help you navigate the website: https://www.epa.gov/ejscreen/learn-use-ejscreen

Then, launch the EJ Screen Mapping Tool website:
https://ejscreen.epa.gov/mapper/

Step 2: Track your own community's ZIP code on the mapping tool. You can also compare what you find to other environmental health data available from the Centers for Disease Control and Prevention: https://ephtracking.cdc.gov/InfoByLocation/

Step 3: Think critically about what the EJSCREEN site reports and how it compares to your own observations. What, if anything, is missing from the environmental justice map? What details did the EPA observe that you missed? How can these differences influence people's health? Try to identify key ways that local knowledge can be combined with "expert knowledge" to better assess environmental health threats.

(Optional) Step 4: When possible, document these differences with film footage or pictures. The documents could be posted on the classroom website for further discussion.

For example, you or your classmates might identify the lack of sidewalks in rural communities that inhibit residents from participating in active lifestyles and that leave dust free to be kicked up and circulate. Others might point out urban food deserts, sometimes referred to as "liquor store diets," as important information about environmental health. The unavailability of hospitals and urgent care centers combined with the lack of primary care providers, expensive for-profit ambulance services, and long waiting lines in community health clinics could make up part of the built-in environment and are injurious to health. Access to open land space such as parks and community gardens, public transportation routes, and healthy food markets could be identified as promoting health.

References

Airnow. (2015). Air quality flag program: Know your air quality to protect your health. https://www.airnow.gov/index.cfm?action=flag_program.index

Airnow. (2016). Particle pollution. https://airnow.gov/index.cfm?action=aqibasics.particle

Ansell, D. A. (2017). *The death gap: How inequality kills.* Chicago, IL: University of Chicago Press.

Braz, R., & Gilmore, C. (2006). Joining forces: Prisons and environmental justice in recent California organizing. *MARHO: The Radical Historian's Organization Inc.*, 95–111.

California Department of Public Health. (2013). Valley fever and African Americans, Filipinos, and Hispanics. https://www.cdph.ca.gov/Programs/CID/DCDC/CDPH%20Document%20Library/VFRaceEthnicity.pdf

California Department of Public Health. (2015). Preventing work related coccidioidomycosis (Valley Fever). https://www.cdph.ca.gov/Programs/CCDPHP/DEODC/OHB/HESIS/CDPH%20Document%20Library/CocciFact.pdf

Centers for Disease Control and Prevention. (2016). Coccidioidomycosis: Risks and prevention. http://www.cdc.gov/fungal/diseases/coccidioidomycosis/risk-prevention.html

Centers for Disease Control and Prevention. (2017). Number of reported Valley Fever cases. https://www.cdc.gov/fungal/diseases/coccidioidomycosis/statistics.html

Chaufan, C. (2018). What can the Slim Initiative in Genomic Medicine for the Americas (SIGMA) contribute to preventing, treating, or decreasing the impact of diabetes among Mexicans and Latin Americans? *Kalfou: A Journal of Comparative and Relational Ethnic Studies, 5*(1), 24–35.

Cole, L. W., & Foster, S.R. (2000). *From the ground up: Environmental racism and the rise of environmental justice movement.* New York: New York University Press.

Cox, R., & Magee, M. (2004). Coccidioidomycosis: Host response and vaccine development. *Clinical Microbiology Reviews, 17*(4), 804–839.

CURB. Californians United for a Responsible Budget. (2016). We are not disposable: The toxic impacts of prisons and jails. http://curbprisonspending.org/wp-content/uploads/2016/10/CURB-WeAreNotDisposableReport.pdf

Delano-Earlimart Irrigation District. (2007). Ground-water management plan. Provost & Pritchard Engineering Group, Inc. sgma.water.ca.gov/basinmod/docs/download/933

Ellis, J. (2015, June 6). Inmates sue saying they got Valley Fever while in state prison. *Fresno Bee*. http://www.fresnobee.com/news/local/article23306760.html

Epstein, S. (1995). The construction of lay expertise: AIDS activism and the forging of credibility in the reform of clinical trials. *Science Technology and Human Values, 20*(4), 408–437.

Gilmore, R. W. (2007). *Golden gulag: Prisons, surplus, crisis, and opposition in globalizing California*. Berkeley, CA: University of California Press.

Harrison, J. L. 2011. *Pesticide drift and the pursuit of environmental justice*. Cambridge, MA: MIT Press.

Hector, R., Rutherford, G. W., Tsang, C.A., Erhart, L. M., McCotter, O., Anderson, S. M., ... & Galgiani, J.N. (2011). The public health impact of coccidioidomycosis in Arizona and California. *International Journal of Environmental Research and Public Health, 8*, 1150–1173.

Huang, J. Y., Bristow, B., Shafir, S., & Sorvillo, F. (2012). Coccidioidomycosis-associated deaths, United States, 1990–2008. *Emerging Infectious Disease, 18*(11), 1723–1729.

Institute for Rural Studies. (1990). Childhood cancer clusters: Department of Health Services studies in the San Joaquin Valley. *Rural California Report, 2*(5), 1–6. https://www.cirsinc.org/component/search/?searchword=cancer%20cluster&searchphrase=all&Itemid=55

Johnson, B. (2017, November 14). How to protect yourself from Valley Fever. ABC News. abc30.com/health/how-to-protect-yourself-from-valley-fever/2648472/

Kelso, C.J. (2013). Notice of filing of report and response of receiver regarding plaintiffs motion re: Valley Fever. Report submitted pursuant to court order, dated March 21, 2013. California Correctional Health Care Services.

Krieger, N. (2011). *Epidemiology and the people's health*. New York: Oxford University Press.

Lipsitz, G. (2006). Unexpected affiliations: environmental justice and the new social movements. *Works and Days, 47/48* (1&2), 25–44.

Mohney, G. (2013, July 5). Valley Fever outbreaks lead California to move inmates. ABC News. https://owl.english.purdue.edu/owl/resource/583/03/

National Institute for Occupational Safety and Health. (2016). Coccidioidomycosis (Valley Fever): Prevention in a work setting. http://www.cdc.gov/niosh/topics/valleyfever/prevention.html

Office of Environmental Health Hazards Assessments. (2016). CalEnviroScreen Mapping Tool. https://oehha.ca.gov/calenviroscreen

Pappagiani, D. (2007). Cocciodioidomycosis in California state correctional institutions. *New York Academics of Sciences, 1111*, 103–111.

Pardo, M. (1999). *Mexican American women activists: Identities and resistance in two Los Angeles communities*. Philadelphia, PA: Temple University Press.

Pemberton, P. (2015, December 22). California men's colony sewage spill caused by inmates throwing trash into toilets. *San Luis Obispo Tribune*. http://www.sanluisobispo.com/news/local/article51150065.html

Public Policy Institute of California. (2013). Just the facts: Poverty in California. http://www.ppic.org/main/publication_show.asp?i=261

Roberts, D. (2011). *Fatal invention: How science, politics, and big business re-create race in the twenty-first century*. New York: The New Press.

Rothfield, M. (2008, December 29). Drink up—Assuming you like arsenic, that is. *Los Angeles*

Times. http://articles.latimes.com/2008/dec/29/local/me-arsenic29

San Quentin News Staff. (2013, August). Valley Fever prompts prison court to order. *San Quentin News*. http://www.columbia.org/pdf_files/sanquentinnews6.pdf

Smith, C., Pappagianis, D., Devline, H. B., & Saito, M. (1961). Human coccidioidomycosis. *Microbiology and Molecular Biology Reviews, 25*, 310–320.

Sze, J. (2007). *Noxious New York: The racial politics of urban health and environmental justice*. Cambridge, MA: The MIT Press.

Tsang, C. A., Anderson, S. M., Imholte, S. B., Erhart, L. M., Chen, S., Park, B. J., ... & Sunenshine, R. H. (2010). Enhanced surveillance of coccidioidomycosis, Arizona, USA, 2007–2008. *Emerging Infectious Diseases, 16*(11), 1738–1744. doi:10.3201/eid1611.100475

U.S. Census Bureau. (2016). 2016 Census of the United States of America. Washington, DC: U.S. Government Printing Office.

U.S. District Court. *Order granting plaintiffs motion for relief re: Valley Fever at Pleasant Valley and Avenal state prison*. Case NO. C01-1351 TEH: xii. (June 24, 2013). https://www.clearinghouse.net/chDocs/public/PC-CA-0018-0112.pdf

U.S. Environmental Protection Agency. (2017). Federal Insecticide, Fungicide, and Rodenticide Act of 1996. Retrieved from https://www.epa.gov/laws-regulations/summary-federal-insecticide-fungicide-and-rodenticide-act

Wynne, B. (1996). May the sheep safely graze? A reflexive review of the expert-lay knowledge divide. In S. Lash, B. Szerzynski, & B. Wynne (Eds.), *Risk environment and modernity: Toward a new ecology*. Los Angeles, CA: SAGE.

Note

1 See Tracey Perkins's telling interview with Teresa DeAnda, available online at https://rememberingteresa.org/

14

BECOMING STORMS

Indigenous Water Protectors Fight for the Future

Kaitlin Reed
Beth Rose Middleton Manning
Deniss Josefina Martinez

WATER

gets what it wants. Water finds
a way. It seeks out the possible

crack, the promising path
of least resistance. Water refuses

to believe in walls, stone, root,
isn't afraid of flood-force.

Water makes a way:
pushes over, under,

around, through. Water rams
sand and rocks down

the ribbed black throat
of a culvert,

busts through terracotta stones
and thick mortar, tosses

rubble like redbud petals, swarms
over the broken remnants of restraint

and rushes the road like a glorious army,
exulting with gravelly April song. Water

gets what it wants, one way or another.
Water dreams of storms
like you.

—Deborah Miranda (2008)

Indigenous peoples worldwide have developed understanding and knowledge with water—for life is impossible without this relationship. As a structure of land dispossession, settler colonialism also understands the power of water; thus, the control and codification of water rights is central to the settler colonial project. Today Indigenous peoples across the globe are standing up to protect water relatives and future generations. Processes of settler colonialism harm both water and Indigenous peoples, and we fight for decolonization together. As the impacts of climate change continue to unfold, it is easy to lose optimism, but Deborah Miranda reminds us that "water finds a way…[and] refuses to believe in walls."

PHOTO 14.1
Robert Mutch / Shutterstock

In a special double issue of *Decolonization* entitled "Indigenous Peoples and the Politics of Water," Melanie K. Yazzie and Cutcha Risling Baldy (2018) demonstrate that "struggles over water figure centrally in concerns about self-determination, sovereignty, nationhood, autonomy, resistance, survival, and futurity." In other words, Indigenous politics of water cannot be separated from struggles for decolonization. This is because settler colonialism specifically has been inseparable from violence committed on land, water, and people. In fact, settler colonialism is fundamentally concerned with land and resources. Voyles (2015) argues that "settler colonialism is so deeply about resources that environmental injustices, whether on Native lands or lands of others, must always be viewed through the lens of settler colonialism" (p. 23).

For Native communities, notions of environmental justice are far older than the environmental justice born in the 1960s. For many Indigenous communities, environmental justice struggles began with the invasion and colonization of our lands. As such, within Indigenous communities, environmental justice must be framed through the issue of colonization. "As Indian lands are assaulted, so are Indian peoples.... Environmental destruction is simply one manifestation of the colonialism and racism that have marked Indian/White relations since the arrival of Columbus in 1492" (Weaver, 1996, p. 3). As such, environmental injustices are prevalent in Indian Country. Environmental abuses range from industrial pollution to toxic dumping to nuclear testing and mining (LaDuke, 1999; LaDuke, & Cruz, 2012; Vickery & Hunter, 2016; Weaver, 1996; Whyte, 2016). Felix Cohen—considered to be the father of modern federal Indian law—likened Native Americans to miners' canaries: "[T]he multiple ills that are visited upon them are only a prelude and a harbinger of what is to be expected for society as a whole" (Weaver, 1996, p. 17). Thus, environmental justice issues in Indian Country must be addressed with more historical nuance than claims of environmental racism born in the 1960s. Similarly, understandings of political ecology—an approach to understanding the multilayered political-economic-ecological drivers of environmental issues—are transformed by Indigenous studies' foregrounding of Indigenous epistemologies and self-determination, and ethnic studies' engagement with coloniality or ongoing colonial relations. A resulting Indigenous political ecology allows specific attention to Indigenous perspectives on the intertwined causes and consequences of, and responses to, environmental crisis (Carroll, 2015; Middleton, 2015, 2019). An Indigenous political ecology seeks to (re)right relationships to land and water by re-centering the social and political relationships with nonhuman relatives.

MNI WICONI (WATER IS LIFE): INDIGENOUS RELATIONS WITH WATER

In this section we will focus on decolonial Indigenous thought and practice surrounding water. Indigenous water protectors—from Standing Rock to Mauna Kea to Oak Flat—demonstrate the centrality of settler colonial processes and orientations to environmental injustice for Indigenous communities. Moreover, Indigenous ways of relating to and caring for water provide a critical framework for environmental justice scholarship and activism to reframe settler orientations to water. This framing of "living well" to mean "interdependency and respect among all living things" is articulated by Yazzie and Risling Baldy (2018) as "radical relationality." Radical relationality "brings together the multiple strands of materiality, kinship, corporeality, affect, land/body connection, and multidimensional connectivity...[that] provides a vision of relationality and collective political organization that is deeply intersectional and premised on values of interdependency, reciprocity, equality and responsibility" (p. 2). Drawing from several Indigenous feminists, Yazzie and Risling Baldy describe radical relationality as the ability to see water as self and to see self as water. Water is ancestor, relative, teacher, narrator, theory, and an agent deserving of the same respect and care we give to other members of the community. Water runs throughout human veins and connects us to everything. "The water that we drink is the water the salmon breathes, is the water the trees need, is the water where Bear bathes, is the water where the rocks settle" (Yazzie & Risling Baldy, 2018, p. 1).

Courtesy of the Yurok Tribe, 2018

In the aftermath of Standing Rock, numerous Indigenous scholars have focused on Indigenous relationships to water and Indigenous water stewardship practices. As Gilio-Whitaker (2019) observes, "[T]here are few issues more sensitive than water: its availability, quality, accessibility, and even its power to destroy life and permanently alter the existence of communities" (p. 83). Nick Estes (Lakota), in his book *Our History Is Our Future: Standing Rock versus the Dakota Access Pipeline, and the Long Tradition of Indigenous Resistance*, argues that

> Mni Wiconi and these Indigenous ways of relating to human and other-than-human life exist in opposition to capitalism, which transforms both humans and nonhumans into labor and commodities to be bought and sold. These ways of relating also exist in opposition to capitalism's twin, settler colonialism, which calls for the annihilation of Indigenous peoples and their other-than-human kin. (Estes, 2019, p. 16)

Indeed, indigenous ways of relating to and caretaking for water are inherently resistant to settler colonial logics and offer strength to indigenous communities. Fox et al. (2017) connect the processes of river restoration and community restoration and demonstrate that "river restoration is embedded in human-water relationships" (p. 531). Through indigenous knowledge and cultural and spiritual practices, river "restoration has the potential to not only restore ecosystem processes and services, but to repair and transform human relationships with rivers and create space politically for decolonizing river governance" (p. 521). In a similar vein, Zoe Todd (Métis) examines human-fish relations in the context of settler colonialism in Canada. She argues that if we understand water as its own agent, we must understand contamination of waters as a violation—one that human beings are responsible for and must rectify (Todd, 2017). As Pawnee legal scholar Walter Echo-Hawk (2013) argues, the United States is in desperate need of a new land (and water) ethic to effectively address the environmental consequences of capitalism and climate change. Indigenous relationships with water can provide insights to environmental justice activism and scholarship and, ultimately, rework to restructure environmental destructive structures.

Institutions can be structured to reflect the concept of relationality and the commitments it entails. On May 9, 2019, for example, the Yurok Tribe passed Resolution 19–40, which recognizes the rights of the Klamath River:

> Therefore be it resolved: That the Yurok Tribal Council now establishes the Rights of the Klamath River to exist, flourish, and naturally evolve; to have a clean and healthy environment free from pollutants; to have a stable climate free from human-caused climate change impacts; and to be free from contamination by genetically engineered organisms.
>
> Therefore be it resolved: that this resolution serves as written notice to the United States of America, the State of California, and all other entities which threaten and endanger the freshwaters, ecosystem, and species of the Klamath River, that it has become necessary to provide a legal basis to protect the Klamath River, its ecosystem, and species for the continuation of the Yurok people and the Tribe for future generations.

Through policy and practice, Indigenous communities around the world have found unique ways to honor, protect, and be in community with water. For example, Hupa scholar Cutcha Risling Baldy describes a Hupa "waterview," instead of worldview, in which Hupa directions are based on one's location on the river. Downriver and upriver are common directions instead of north, south, east, and west. As Risling Baldy explains, "Hupa understand our place in the world based on how water views the world, or even how water views us" (Yazzie & Risling Baldy, p. 2). Water is also deep in the fabric of Nium stories, serving as a source of oral cartography that both physically connects the landscape in the San Joaquin River watershed and narratively connects stories as a common character. Water appears as an agent of transformation and is venerated as sacred, immortal, and healing (Aldern & Goode, 2014). Aldern and Goode (2014) say that Nium stories "hold water" in several senses: They describe real and lived experiences that people have with water; they feature water as a powerful and instructive character; and they hold knowledge that literally raised water tables, reduced interception of rainfall by trees, and eliminated competition for water among desired plants. Re-storying the Sierra Nevada is therefore an issue not only of revitalizing culture but also of restoring relationships and, therefore, entire landscapes.

The Yurok resolution, Hupa orientations to space, and Nium water stories demonstrate radical relationality. Radical relationality is not simply different from capitalist and colonial understandings of natural systems; seeing water and land as alive and as members of the community is a direct threat to colonial and capitalist power structures. "Within this framework of relationality, water is not seen as a resource to be weaponized for the interests of capital by corporations that harness, obstruct, pollute, and discipline water through infrastructure projects like dams and pipelines to boost the capitalist economies of settler nation-states. No, within an Indigenous feminist framework, water is a relative with whom we engage in social (and political) relations premised on interdependency and respect" (Yazzie & Risling Baldy, pp. 2–3). Radical relationality is essential to the environmental justice movement because it decentralizes colonial orientations to land and water and acknowledges the agency and sacredness of our environment.

WATER AND SETTLER COLONIALISM

Settler colonialism is a form of colonialism wherein settlers create a new home for themselves on land apart from their homeland. Anthropologist Patrick Wolfe has argued that settler colonialism is not an event that occurred in the past and is now complete; rather, settler colonialism is a structure that must be continually perpetuated and reproduced. This differs from extractive forms of colonialism wherein the colonial power seeks to extract natural resources and human bodies for wealth accumulation and labor. In the context of settler colonialism, settlers must destroy Indigenous populations to replace them (Wolfe, 2006)—or to reproduce settler colonial structures and populations (Arvin, 2013). Settler colonialism also differs from other forms of extractive colonialism because there is an insistence on "settler sovereignty over all things in their new domain" (Tuck & Yang, 2012, p. 5). When we ask our students the difference between a settler and an immigrant, initially we receive a few confused looks, but then a brave volunteer may say that "immigrants must follow the laws of the place they are coming to." However, in the context of American settler colonialism, settlers ignored the many diverse and complex Indigenous systems of jurisprudence across Turtle Island. Settler-invented justifications or "legal fictions," such as Manifest Destiny and the Doctrine of Discovery, thereby were seen as entitling settlers to Indigenous lands and, by extension, to Indigenous bodies (living or dead), cultures, and spiritualities (Echo-Hawk, 2010). Yale history professor Ned Blackhawk (Te-Moak Western Shoshone) explains:

> From the use of the U.S. Army to combat and confine Indian peoples, to the state-sanctioned theft of Indian lands and resources, violence both predated and became intrinsic to American expansion. Violence enabled the rapid accumulation of new resources, territories, and subject peoples.... From the initial moments of American exploration and conquest, through statehood, and into the stages of territorial formation, violence organized the region's nascent economies, settlements, and polities. Violence and American nationhood, in short, progressed hand in hand. (Blackhawk, 2006)

Similarly, in *An Indigenous Peoples' History of the United States*, Roxanne Dunbar-Ortiz (2014) argues that settler colonialism requires violence because "people do not hand over their land, resources, children, and futures without a fight, and that fight is met with violence" (p. 8). In short, settler colonialism is a genocidal policy.

Settler colonialism holds particular orientations toward land. Settler colonial ecology understands the Earth as a natural resource—a nonliving object to be utilized for human consumptive purposes. Humans within this socioecological context orient their relationship to land and water as owner to property or real estate. Ideological conceptions of *terra nullius*, or "empty land," marked the landscape of North America as open for business. Notions of Manifest Destiny engendered a moral justification to colonize and develop territories. The justification for land dispossession was the Doctrine of Discovery,[1] as articulated by Chief Justice John Marshall in 1823. The Doctrine of Discovery differentiated the settler state as landowner, and Indigenous peoples as land users. Settler colonial orientations, informed by European perspectives, that view land as a material resource demarcated Indigenous subsistence practices as wasteful, as they did not generate profit. In contrast, within Indigenous worldviews, Earth is universally understood as a living entity and all creation is related. Tuck and Yang (2012) argue that

> [w]ithin settler colonialism, the most important concern is land/water/air/subterranean earth (land, for shorthand, in this article). Land is what is most valuable, contested, required. This is both because the disruption of Indigenous relationships to land represents a profound epistemic, ontological, cosmological violence. This violence is not temporally contained in the arrival of the settler but is reasserted each day of occupation. This is why Patrick Wolfe (1999) emphasizes that settler colonialism is a structure and not an event. In the process of settler colonialism, land is remade into property and human relationships to land are restricted to the relationship of the owner to his property. Epistemological, ontological, and cosmological relationships to land are interred, indeed made pre-modern and backward. Made savage. (p. 5)

Thus, Indigenous relationships to and knowledges of water were not respected by incoming settlers. Rather, from the perspective of the U.S. federal government, Native peoples throughout Turtle Island were viewed as "wasting" water and other natural resources because they were not using them for wealth accumulation (Voyles, 2015). This justified the taking of natural resources—specifically, water resources.

The codification of water rights was (and continues to be) an integral part of the settler colonial project (Matsui, 2009). Dependent on Lockean theories of private property, settler colonial water rights in the western United States were established on the basis of the Prior Appropriation Doctrine. "Developed as a response to the unique needs of western water users," the Prior Appropriation Doctrine "holds that the first water user to file a water claim with the state has priority over all subsequent water users" (McCool, 2002, p. 11). The central principle is "first-in-time is first-in-right," a glorified "finders' keepers, losers' weepers" arrangement. Contemporary senior water rights stem from this process, which systematically excluded Indigenous peoples. For example, in California under the Land Claims Act of March 1851, all persons who claimed California lands were required to submit their claims for review and approval. Given the legalized, state-funded genocide committed against California Indians during this time (see Johnston-Dodds, 2002; Lindsay, 2012; Madley, 2016; Reséndez, 2016), the fact that California Indians were not citizens, and the lack of communication to California Indians of this law, no California Indians were able to file their land claims under this statute. Along with these land claims came water rights—rights still upheld as *senior* rights for settlers who filed in this time period. Even if California Indians were eligible to file for water rights, the legal immunity of white settlers committing a genocide served as an effective deterrent to tribal members to venture to county or state offices to assert their Aboriginal water right. Indeed, Native American tribes did not have a legal mechanism to reserve water rights within the settler state until 1908 (see *Winters v. United States*, 1908).

The second broad principle of western water rights is beneficial use, which requires that water be put to one or more established uses. A water rights holder who does not put their water to a designated beneficial use may lose their right—hence, the water law mantra "use it or lose it." Which uses of water are recognized as "beneficial" directly "reflects social values and scientific understandings" (Halvorsen, 2018, p. 941). Indeed, as Robison et al. (2018) recognize,

> colonial entities (governments, corporations, communities, etc.) have been primary recipients of the institutions' material benefits, while Indigenous Peoples have often been subject to inequitable allocation rules, distorted funding and resource arrangements, and non-representation within water governance bodies and processes. (p. 901)

California currently has 23 beneficial uses that may be applied on water bodies statewide. These include activities such as mining, irrigation, and hydroelectric power, as well as habitat preservation, endangered species protection, and groundwater recharge. While environmental protection is now an accepted beneficial use, initial state-recognized beneficial uses subscribed to settler colonial orientations to land as a source of wealth accumulation. From a settler colonial perspective, beneficial uses were initially uses that produced income and reduced waste.

While the dispossession, manipulation, and pollution of waters represents a profound settler colonial violence, we argue that settler colonialism itself must be conceptualized as an environmental injustice. Inherently tied to the accumulation of natural resources, settler colonialism erases the socioecological contexts in which Indigenous people live, and works to eliminate the possibility of collective continuance (Voyles, 2015; Whyte, 2017). Indigenous environmental justice perspectives come from "the original instructions" that have been given to us by our Creator (Black, 2011, 2016; Kassi, 1996; Lyons, 2008; Nelson, 2008; Nelson & Shilling, 2018). For example, Yurok people have a reciprocal relationship with salmon established through our creation story: We steward the river and fish for salmon, and salmon sustain our generations. Damming the river, therefore, is an environmental injustice for Yurok as downriver people, because they deal with a disproportionate amount of environmental impacts, primarily water quality issues (Middleton Manning et al., 2019; Reed, 2019). Damming the river prevents salmon from returning home and creates a toxic river that salmon cannot survive in; therefore, Yurok people cannot fulfill their physical, social, and spiritual purposes. Settler colonial orientations to land result in its destruction, thereby erasing the socioecological context of Yurok people (Reed, 2019). Therefore, to attain environmental justice in Indian Country, one must necessarily engage with ongoing processes of settler colonialism and questions of decolonization.

In *As Long as Grass Grows: The Indigenous Fight for Environmental Justice, From Colonization to Standing Rock,* Colville tribal member Dina Gilio-Whitaker (2019) asks what an indigenized environmental justice looks like. Starting from "the assumption that colonization was not just a process of invasion and eventual domination of Indigenous populations by European settlers but also that the eliminatory impulse and structure it created in actuality began as environmental injustice" (p. 12), Gilio-Whitaker argues that the capitalist framework and white supremacist origins of the environmental movement must be reexamined. An indigenized environmental justice framework "must acknowledge the political existence of Native nations and be capable of explicitly respecting principles of Indigenous nationhood and self-determination" (p. 12). Historically, environmental justice activists have appealed to their rights as citizens of the United States (Mutz, Bryner, & Kenney, 2002). Tribes, however, are sovereign nations, and any articulation of environmental justice must incorporate tribal sovereignty and self-determination (Krakoff, 2002, 2011; Ranco et al., 2011; Tsosie, 1996; Vickery & Hunter, 2016; Weaver, 1996; Whyte, 2011, 2016). Incorporating tribal sovereignty as a tenet of environmental justice will require tribal environmental policies (Rechtschaffen & Guana, 2002). Such policies must be responsive to traditional ecological knowledge, western science, economics, and tribal systems of ethics (Tsosie, 1996, 2007).

RESISTANCE: STRATEGIC PROTECTION OF WATER

While Standing Rock and other, similar struggles "reflect how Indigenous people are (re)activating water as an agent of decolonization" (Yazzie & Risling Baldy, 2018, p. 1), sometimes Indigenous resistance to settler colonialism is more subtle, yet no less impactful. As Diver (2018) reminds us in "Native Water Protection Flows Through Self-Determination," fighting for Indigenous health and well-being in relation to water often involves working both within and against settler institutions. We briefly discuss two legal ways Indigenous peoples are engaged in water protection: through advocating for an expanded definition of beneficial uses, and applying "treatment as a state" status to exercise jurisdiction over tribal water quality.

Beneficial Uses

In the past decade, California Indians have proposed additional beneficial uses to be added to the list of 23 statewide beneficial uses: specifically, tribal traditional and cultural uses (CUL) and tribal subsistence fishing (T-SUB). Recognizing a particular water use as "beneficial" and applicable to a specific water body requires the management and treatment of water to ensure that that use can continue (Halvorsen, 2018). T-SUB was differentiated from another newly adopted beneficial use, subsistence fishing (SUB), which was largely supported by environmental justice advocates to protect subsistence fishers throughout the state. CUL, SUB, and T-SUB constitute "additional" beneficial use categories that can be adopted by individual regional water boards and applied to specific waterways (State Water Resources Control Board, 2016). Once they are applied, water quality objectives may be altered to ensure that the beneficial use is protected.

Currently, only one region of the state has adopted CUL and T-SUB beneficial uses on water bodies under its jurisdiction, and advocacy continues to get other regional water quality control boards to adopt the beneficial uses (i.e., Rios, 2018). From an Indigenous environmental justice perspective, the CUL and T-SUB beneficial use designations are long overdue. Tribal members' relationship to fisheries throughout California, and tribal members' level of consumption of fish from California waterways, are not accounted for in current beneficial use designations and associated water quality standards. Aside from CUL and T-SUB, the beneficial uses and associated water quality objectives assume a qualitatively and quantitatively lower level of use and subsistence than that practiced by many California tribal members. As such, generally assumed beneficial uses are effectively *sickening* tribal members by disregarding their extensive and varied use of fish and other aquatic species (i.e., Shilling et al., 2014). Beneficial use designations must protect all Californians, most significantly first Californians, who have been stewarding these resources and waterways since time immemorial.

The 2014 California Tribes Fish-Use study by Shilling et al. (2014) looks at tribal members' current suppressed rate of fish consumption, which is approximately half of the traditional rate of consumption. The study included surveys with 23 percent of tribes in the state, who use 25 percent of waterways in the state. This indicates that, if all tribes participated in the study, their uses of fisheries would involve all waterways throughout the state. As such, all waterways should be subject to CUL and T-SUB designations. Polluters discharging wastewater into waterways should have to prove that they are not negatively impacting the CUL and T-SUB designations, rather than tribes having to prove that these beneficial uses apply to the waterways in question. Tribal traditional stewardship of and interdependence with the species that live in local waterways has been constant in California throughout history. In a context of working for Indigenous environmental justice, it is time that regulations recognize, respect, and protect the oldest beneficial uses of water in the state.

ASSERTING JURISDICTION

Generally, the U.S. Environmental Protection Agency (EPA) delegates authority to states to develop and enforce regulations that meet or exceed the federal standards. In recognition of tribal self-governance, and of the gaps in environmental regulation and enforcement that occur on tribal lands in the absence of federal enforcement or state jurisdiction, beginning in 1986 the EPA began to approve delegation of environmental regulatory authority to federally recognized tribes (i.e., Diver, 2018; Larned 2018). Tribes that have applied for and received "treatment as a state" status can establish and enforce regulatory standards for water quality within tribal lands (including water quality flowing on to tribal lands). Under the *Montana* (1980) exception, tribes have jurisdiction over nonmembers only if the nonmembers' activities are impacting the "political integrity, economic security, and health and welfare of the Tribe." Impairing water quality certainly qualifies as impacting health and welfare as well as political integrity and possibly economic security. Tribal rights to promulgate and enforce water quality standards and tribally specific beneficial uses have been affirmed by the Supreme Court (*Albuquerque v. Browner*, 1996). Because enforcement may "disrupt economic norms" (Larned 2018, p. 81), it becomes controversial. Tribal entities may struggle with enforcing water quality standards against polluters that are also significant regional employees or otherwise engaged in a relationship with the tribe.

Achieving Indigenous environmental justice requires a reorientation of institutions (settler and Indigenous) to one another. Rather than engaging as adversaries or participating in the colonial ruse of settler state authority, settler states respect Indigenous institutions as senior institutions, and as institutions that are enacting and upholding beneficial uses that retain water quality and quantity for human as well as nonhuman beings. Legal scholars Robison et al. (2018) define the realization of Indigenous water justice as

> the exercise of Indigenous Peoples' right to self-determination vis-a-vis water...[including] "construction of a new relationship between Indigenous peoples and the State under terms of mutual respect, encouraging peace, development, coexistence, and common values." (p. 920)

Indigenous environmental justice and, indeed, survival for all humans in a context of climate change calls for a legal and institutional reframing of relationships between humans and nonhumans as mutually constitutive and relational. This reading of "radical relationality" calls for legal and institutional reform that de-centers profit and re-centers responsibility for reciprocal care of one another, humans and water, as articulated in Indigenous origin narratives that establish covenants between humans and other beings.

NEXT STEPS / A CALL TO ACTION

The truth is that the world is changing. As the climate warms and we see what seems like an endless string of environmental catastrophe, chaos, and injustice, it is important to remain firm on the idea that we are powerful. We are powerful, and we have many responsibilities that we must meet. As Yazzie and Risling Baldy (2018) remind us,

> Our mountain, human, animal, and water relatives with whom we have reignited a promise of mutual care and protection await our next move. We are not starting from scratch; the seeds have already been planted, the cracks in empire already made. The tide of history is with us. So long as we remain committed to building a successful liberation movement and embracing a far-reaching relational politics of life, the web of radical relationality will only grow until it blankets the world in stunning beauty and restores the balance that our stories and prophecies have always foretold. (p. 10)

A hope like this is not trivial to maintain. Hope takes work, it takes a community, and it takes flexibility and conviction. Our work is not to impose order. It is to return to be in relationship with land and water and be accountable to that relationship moving forward. This can be expressed in individual expressions of love, ceremony, and caretaking but also in building collective movements that overthrow the strangleholds of settler colonialism, white supremacy, unchecked corporate capitalism, patriarchy, and war. This will be no small feat. Indigenous peoples will never quit fighting for water or future generations. Remain hopeful because we have the most powerful ally—water itself. As Deborah Miranda reminds us, water seeks out the cracks in the empire; rest assured, "Water gets what it wants, one way or another. Water dreams of storms like you."

DEEPENING OUR UNDERSTANDING

1. Find out where your drinking water comes from and conduct research on the following questions:

    a. What Indigenous nation(s) steward that watershed?

    b. What is the Indigenous name for that water body (in their language)? What does it mean?

    c. Are there any stories about that water body?

    d. Are there any current threats to the well-being of that water body? If so, how are Indigenous communities fighting to protect that water? What can you do to help?

2. Tribes can assert their jurisdiction over traditional lands and nontribal members through the use of environmental legal mechanisms such as treatment as a state (TAS) under various environmental laws (i.e., the Clean Water Act and the Clean Air Act). Find an example of a tribe with TAS status and respond to the following questions:

    a. Do the tribe's beneficial uses and water or air quality objectives differ from those of the surrounding state? In what ways?

    b. Find an instance in which the tribe asserted jurisdiction over lands off the reservation or over non-Native people or companies. How did these entities respond to being under tribal jurisdiction? Why?

3. Consider an environmental disaster affecting Indigenous lands, such as the Gold King Mine spill that impacted the San Juan River, which flows through the Navajo Nation.

    a. What questions and methods did Indigenous scientists or Indigenous community members apply to address the impacts of the disaster?

    b. Were their questions and methods different from those of non-Indigenous scientists? In what ways? Who do you think they see themselves as responsible to?

References

Aldern, J. D., & Goode, R. W. (2014) The stories hold water: Learning and burning in North Fork Mono homelands. *Decolonization: Indigeneity, Education & Society, 3*(3).

Arvin, M. (2013). *Pacifically possessed: Scientific production and Native Hawaiian critique of the "almost white" Polynesian race.* Ph.D. dissertation, Department of Ethnic Studies, University of California, San Diego.

Black, C. F. (2011). *The land is the source of the law.* New York: Routledge Press.

Black, C. F. (2016). *A mosaic of Indigenous legal thought: Legendary tales and other writings.* New York: Routledge Press.

Blackhawk, N. (2006). *Violence over the land: Indians and empires in the early American West.* Cambridge, MA: Harvard University Press.

Carroll, Clint. (2015). *Roots of our renewal.* Minneapolis: University of Minnesota Press.

City of Albuquerque v. Browner, 97 F.3d 415 (10th Cir. 1996).

Diver, S. (2018). Native water: Protection flows through self-determination: understanding tribal water quality standards and "treatment as a state." *Journal of Contemporary Water Research and Education, 153*, 6–30.

Dunbar-Ortiz, R. (2014). *An Indigenous peoples' history of the United States.* Boston, MA: Beacon Press.

Echo-Hawk W. R. (2010). *In the courts of the conqueror: The 10 worst Indian law cases ever decided.* Golden, CO: Fulcrum Publishing.

Echo-Hawk W. R. (2013). *In the light of justice: The rise of human rights in Native America and the UN Declaration on the Rights of Indigenous Peoples.* Golden, CO: Fulcrum Publishing.

Estes, N. (2019). *Our history is our future: Standing Rock versus the Dakota Access Pipeline, and the long tradition of Indigenous resistance.* London, UK: Verso Books.

Fox, C. A., Reo, N. J., Turner, D. A., Cook, J., Dituri, F., Fessell, B., ... , & Riley, C. (2017). "The river is us; the river is in our veins": Re-defining river restoration in three Indigenous communities. *Sustainability Science*, 1–13.

Gilio-Whitaker, D. (2019). *As long as grass grows: The Indigenous fight for environmental justice, from colonization to Standing Rock.* Boston, MA: Beacon Press.

Halvorsen, E. (2018). Compact compliance as a beneficial use: Increasing the viability of an interstate water bank program in the Colorado. *University of Colorado Law Review, 89*, 937–966.

Johnston-Dodds, K. (2002). *Early California laws and policies related to California Indians.* Sacramento: California Research Bureau.

Kassi, N. (1996). A legacy of maldevelopment. In J. Weaver (Ed.), *Defending Mother Earth: Native American perspectives on environmental justice.* (ed. Jace Weaver). Maryknoll, NY: Orbis Books.

Krakoff, S. (2002). Tribal sovereignty and environmental justice. In K. M. Mutz, G. C. Bryner, & D. S. Kenney (Eds.), *Justice and natural resources: Concepts, strategies, and applications.* Washington, DC: Island Press.

Krakoff, S. (2011). Radical adaptation, justice, and American Indian nations. *Environmental Justice,* 4(4), 2017–212.

LaDuke, W. (1999). *All our relations: Native struggles for land and life.* Boston, MA: South End Press.

LaDuke, W., & Cruz, S. A. (2012). *The militarization of Indian Country.* Ann Arbor: Michigan State University Press.

Larned, S. M. (2018). Water is life: The Native American tribal role in protecting natural resources. *Environmental and Earth Law Journal,* 52–94.

Lindsay, B. C. (2012). *Murder state: California's Native American genocide, 1846–1873.* Kearney: University of Nebraska Press.

Lyons, O. (2008). Listening to natural law. In M. K. Nelson (Ed.), *Original instructions: Indigenous teachings for a sustainable future*. Rochester, VT: Bear & Company.

Madley, B. (2016). *An American genocide: The United States and the California Indian catastrophe, 1846–1873*. New Haven, CT: Yale University Press.

Matsui, K. (2009). *Native people and water rights: Irrigation, dams, and the law in Western Canada*. Montreal, Quebec, Canada: McGill-Queen's University Press.

McCool, D. (2002). *Native waters: Contemporary Indian water settlements and the second treaty era*. Tucson: University of Arizona Press.

Middleton, B. R. (2015). Jahát Jatítotòdom*: Toward an Indigenous political ecology. *International Handbook of Political Ecology*.

Middleton Manning, B. R. (2019). Geographies of Hope in Cultural Resources Protection. *Environment and Planning E: Nature and Space*.

Middleton Manning, B. R., Talaugon, K., Young, T. M., Wong, L., Fluharty, S., Reed, K., ... , & Myers, R., II. (2019). Bi-directional learning: Identifying contaminants on the Yurok Indian Reservation. *International Journal of Environmental Research and Public Health*.

Miranda, D. (2008). Water. In M. Dubin & K. Hogeland (Eds.), *Spring salmon hurry to me: The seasons of Native California* (pp. 49–50). Berkeley, CA: Heyday Press.

Montana v. US, 450 U.S. 544 (1981).

Mutz, K. M., Bryner, G. C., Kenney, D. S. (2002). *Justice and natural resources: Concepts, strategies, and applications*. Washington, DC: Island Press.

Nelson, M. (2008). *Original instructions: Indigenous teachings for a sustainable future*. Rochester, VT: Bear & Company.

Nelson, M., & Shilling, D. (2018). *Traditional ecological knowledge: Learning from indigenous practices for environmental sustainability*.Cambridge, UK: Cambridge University Press.

Ranco, D. J., O'Neill, C. A., Donatuto, J., & Harper, B. L. (2011). Environmental justice, American Indians and the cultural dilemma: Developing environmental management for tribal health and wellbeing. *Environmental Justice, 4*(4), 221–230.

Rechtschaffen, C., & Guana, E. (2002). Native American issues. In *Environmental justice: Law, policy, and regulation* (pp. 421–460). Durham, NC: Carolina Academic Press.

Reed, K. (2019). *The environmental & cultural impacts of cannabis cultivation on Yurok tribal lands*. Ph.D. dissertation, Department of Native American Studies, University of California, Davis.

Reséndez, A. (2016). *The other slavery: The uncovered story of Indian enslavement in America*. New York: Houghton Mifflin Harcourt.

Rios, G. (2018, September 5). Now is the time, California tribes! Make sure tribal cultural beneficial uses are included in your region's 2018 Triennial Basin Plan Update. *Procipio Law Blog*. https://bloggingcircle.wordpress.com/2018/09/05/now-is-the-time-california-tribes-make-sure-tribal-cultural-beneficial-uses-are-included-in-your-regions-2018-triennial-basin-plan-update/

Robison, J., Cosens, B., Jackson, S., Leonard, K., & McCool, D. (2018). Indigenous water justice. *Lewis & Clark Law Review, 22*, 841–921.

Shilling, F., Negrette, A., Biondini, L., & Cardenas, S. (2014). California tribes fish-use: Final report. State Water Resources Control Board, U.S. Environmental Protection Agency (Agreement # 11-146-250 between SWRCB and UC Davis). https://www.waterboards.ca.gov/water_issues/programs/mercury/docs/tribes_%20fish_use.pdf.

State Water Resources Control Board. (n.d.). Beneficial uses: Definitions. https://www.waterboards.ca.gov/about_us/performance_report_1314/plan_assess/docs/bu_definitions_012114.pdf

State Water Resources Control Board. (2016). State Water Resources Control Board Resolution No. 2016-0011.

Todd, Z. (2017). Fish, kin and hope: Tending to water violations in *amiskwaciwâskahikan* and Treaty Six Territory. *Afterall: A Journal of Art, Context, and Enquiry, 43*, 102–107.

Tsosie, R. (1996). Tribal environmental policy in an era of self-determination: The role of ethics, economics, and traditional ecological knowledge. *Vermont Law Review, 21*, 225–333.

Tsosie, R. (2007). Indigenous people and environmental justice: The impact of climate change. *University of Colorado Law Review, 78*, 1625–1677.

Tuck, E., & Yang, K. W. (2012). Decolonization is not a metaphor. *Decolonization: Indigeneity, Education & Society, 1*(1), 1–40.

Vickery, J., & Hunter, L. M. (2016). Native Americans: Where in environmental justice. *Society & Natural Resources, 29*(1), 36–52.

Voyles, T. B. (2015). *Wastelanding: Legacies of uranium mining in Navajo Country.* Minneapolis: University of Minnesota Press.

Weaver, J. (1996). *Defending Mother Earth: Native American perspectives on environmental Justice.* Maryknoll, NY: Orbis Books.

Whyte, K. P. (2011) Environmental justice in Native America. *Environmental Justice, 4*(4), 185–186.

Whyte, K. P. (2016). Indigenous experience, environmental justice and settler colonialism. In B. E. Bannon (Ed.), *Nature and experience: Phenomenology and the environment.* Lanham, MD: Rowman & Littlefield.

Winters v. United States, 207 U.S. 564 (1908).

Wolfe, P. (2006). Settler colonialism and the elimination of the Native. *Journal of Genocide Research, 8*(4): 387–409.

Yazzie, M. K., & Risling Baldy, C. (2018). Introduction: Indigenous peoples and the politics of water. *Decolonization: Indigeneity, Education & Society, 7* (1).

Yurok Tribe. (2019). Resolution 19-40: Resolution Establishing the Rights of the Klamath River. http://files.harmonywithnatureun.org/uploads/upload833.pdf

Note

1. The Doctrine of Discovery, originally referred to as the Papal Bulls in 1493 by the Pope, established a spiritual, political, and legal justification for the colonization and seizure of land not inhabited by Christians. It was used by the Supreme Court in the 1820s justify the taking of the United States.

NEW FRONTIERS AND OLD QUESTIONS

PART V

15

NARRATIVES OF STRUGGLE AND RESISTANCE IN THE FIGHT AGAINST ENVIRONMENTAL RACISM IN AFRICAN NOVA SCOTIAN COMMUNITIES

Ingrid Waldron

THE CASE OF AFRICVILLE

No other black Canadian community has served as a more classic example of environmental racism than Africville—a tight-knit African Nova Scotian community that was located just north of Halifax. Like other African Nova Scotians, Africville residents were descendants of African slaves and freedmen, black Loyalists from the United States, the Nova Scotian colonists of Sierra Leone, the Maroons from Jamaica, and the refugees of the War of 1812.

The community of Africville experienced several injustices related to the provision of basic municipal services, infrastructure, exposure to environmental hazards, and, eventually, displacement. Although the City of Halifax collected taxes in Africville, it did not provide the community with basic utilities and infrastructure offered to other parts of the city, such as paved roads, sanitary water, sewage, public transportation, garbage collection, recreational facilities, fire protection, street lights, and adequate police protection (Fryzuk, 1996; Halifax Regional Municipality, n.d.; Historica Canada, n.d.a; Nelson, 2001; Tavlin, 2013; Waldron, 2018). In addition to the lack of basic amenities, marginalized communities often reside in areas with poor quality or unwanted infrastructures, such as landfills (Campbell et al., 2013; MacNeil et al., 2018). Such entities can affect drinking water, as 46 percent of Nova Scotians rely on private wells for their supply of water (Government of Nova Scotia, 2019).

In 1947, Halifax rezoned Africville for industrial development. In the 1950s, the city built an open-pit dump in Africville, which many considered to be a health menace (Historica Canada, n.d.a). A number of industries later became environmental hazards in the wake of Africville's rezoning, including a fertilizer plant, a slaughterhouse, a tar factory, a stone- and coal-crushing plant, a cotton factory, a prison, two infectious disease hospitals, and three systems of railway tracks (Fryzuk, 1996; Nelson, 2001). By 1965 the city had embarked on an urban renewal campaign that involved the expropriation and bulldozing of homes and the forcible displacement of Africville residents to derelict or rented public housing. The last residents left in 1969.

After Africville residents were forced to relocate (to the North End of Halifax and other parts of the province), they formed the Africville Action Committee in 1969 and, later, the Africville Genealogy Society in 1983 to seek redress (Historica Canada, n.d.b). Throughout the 1990s, the Africville Genealogy Society put pressure on city hall to renegotiate with residents for compensation and, in 1996, resorted to launching a lawsuit against the City of Halifax, requesting compensation for the current value of their lands.

Fourteen years later, in 2010, then–Halifax mayor Peter Kelly apologized to the former residents of Africville. This apology was supported by the allocation of land and $3 million for the construction of the Seaview African United Baptist Church, a replica of the one that stood in Africville. The community was split over the final settlement, however. The Africville Genealogy Society was in favor of it, while those opposed perceived the political decision-making process that led to the settlement as discriminatory because it failed to promote public

Photo 15.1. Gone but never forgotten: Photographic portrait of Africville in the 1960s (Nova Scotia Archives, 1965).

participation of those directly affected by the final decision (Halifax Regional Municipality, n.d.; Nelson, 2001; Tavlin, 2013). Consequently, the suit was revived in 2016 (Borden Colley, 2016).

ENVIRONMENTAL RACISM IN AFRICAN NOVA SCOTIAN COMMUNITIES

The story of Africville provides an analytical entry point for examining other cases of environmental racism in African Nova Scotian communities because it illustrates some of the main aspects of environmental racism, including environmental policies, practices, or directives that disproportionately disadvantage individuals, groups, or communities (intentionally or unintentionally) based on race (Bullard, 1993, 2002).

Environmental racism was not an issue I came to organically. I had long been turned off by the environmental movement because of my own perceptions of environmental activists as overzealous and judgmental zealots. However, my perceptions changed in spring 2012 when an activist asked to meet with me to discuss a new project on environmental racism in Indigenous and black communities in Nova Scotia, which was later named the Environmental Noxiousness, Racial Inequities and Community Health Project (ENRICH Project). While I knew little about environmental racism, I agreed to get involved because I recognized that I could contribute to research on environmental health inequities, the area of my ongoing work. And, although I would still not call myself an environmentalist, I now recognize the urgency of environmental issues in a way that I hadn't before, particularly for people of color and Indigenous peoples.

From its inception, the mission of the ENRICH Project has been to support ongoing and new efforts by Mi'kmaw and African Nova Scotian peoples to address the social, economic, political, and health effects of disproportionate pollution and contamination in their communities. The ENRICH Project seeks to bridge the boundaries between academic scholarship, theory, and analysis and grassroots activism. This chapter is inspired by the work my team and I have been carrying out since 2012 with African Nova Scotian communities to address environmental contamination, which except for the historical case of Africville, receives little attention in Nova Scotia or the rest of North America.

African Nova Scotian Communities and Settler Colonialism

Many people of African descent landed in Nova Scotia between the late 1700s and the early 1900s. Between 1783 and 1785, more than 3,000 people arrived, as part of the black Loyalist migration at the close of the American Revolution. As mentioned earlier, many of those who came during the War of 1812 were freed or escaped slaves. During the 1920s, hundreds of Caribbean immigrants like the Maroons from Jamaica, referred to as the "later arrivals," flocked to Cape Breton (the northeastern tip of Nova Scotia) to work in coal mines and the steel factory. The majority of African Nova Scotians continue to reside in rural and isolated communities as a result of institutionalized racism during the province's early settlement (Black Cultural Centre for Nova Scotia, n.d.; Maddalena et al., 2010; Nova Scotia Museum, n.d.; Waldron, 2010, 2015).

I locate my analysis of environmental racism within Canada's settler-colonial existence, particularly as it relates to ongoing institutionalized racism that began during Nova Scotia's early settlement of African Nova Scotians. Environmental racism in African Nova Scotian communities is a manifestation of spatial arrangements that emerged out of a legacy of colonialism, racism, and segregation in Nova Scotia and Canada. Settler colonialism encompasses contemporary forms of colonialism in British-descended "settler colonies" such as Canada, the United States, and Australia (Shoemaker, 2015; Veracini, 2011; Wolfe, 1999). The goal of settler colonialism is the transformation of the new colony into "home"—a process that lasts indefinitely. Its central features include profit seeking through land acquisition, resource extraction, and exploitation (Coulthard, 2014; Shoemaker 2015; Veracini, 2011; Wolfe, 1999). The location of environmentally hazardous facilities in African Nova Scotian communities implicates African Nova Scotian peoples in the landscape of a white settler nation.

Shelburne

Shelburne, in rural southwestern Nova Scotia (see Map 15.1), remains a prominent African Nova Scotian community today (Nova Scotia Museum, n.d.; Waldron, 2016). Shelburne's South End, which is home to the majority of African Nova Scotian residents, had been near the Morvan Road landfill (locally referred to as the Shelburne Town Dump) for more than 60 years and is also the only area in Shelburne without access to the town's drinking water supply (Waldron, 2016). Thus, many of the households near the landfill rely on wells.

Although factual and accurate information is not currently available about the type and quantity of waste deposits made at the landfill or about the background to its establishment, anecdotal evidence provided by community members suggests that it accepted municipal and medical waste from the adjacent hospital, the community college, the former naval base, and the municipality's industrial park, with little record keeping or documentation (Waldron, 2016). The landfill is suspected to be leaching toxic contaminants into the residents' drinking water (Ore, 2018). Residents are also concerned about the years of smoke inhalation due to burning of garbage that occurred during the peak of the landfill's operation (Ore, 2018).

Photo 15.2. The now decommissioned Shelburne dump. Photo by Robert Devet.

In recent decades, the landfill had "officially" been more restricted but with little supervision of day-to-day administration and control by the Town of Shelburne. In recent years, the town had "reopened" the landfill, although it had failed to consult the community in any meaningful way about this. Visual investigation conducted by community members, as well as photos taken by one community member in 2016, showed evidence of continued use of the landfill—for at least the past 20 years—as a poorly controlled repository for assorted materials not intended for the facility.

Lincolnville

Like Shelburne, Lincolnville is a small, rural African Nova Scotian community. Driven from land promised to them, the residents were forced to move inland away from the white population and to become squatters on a barren, rocky

MAP 15.1 ■ African Nova Scotian Communities (Canadian History Workshop, n.d)

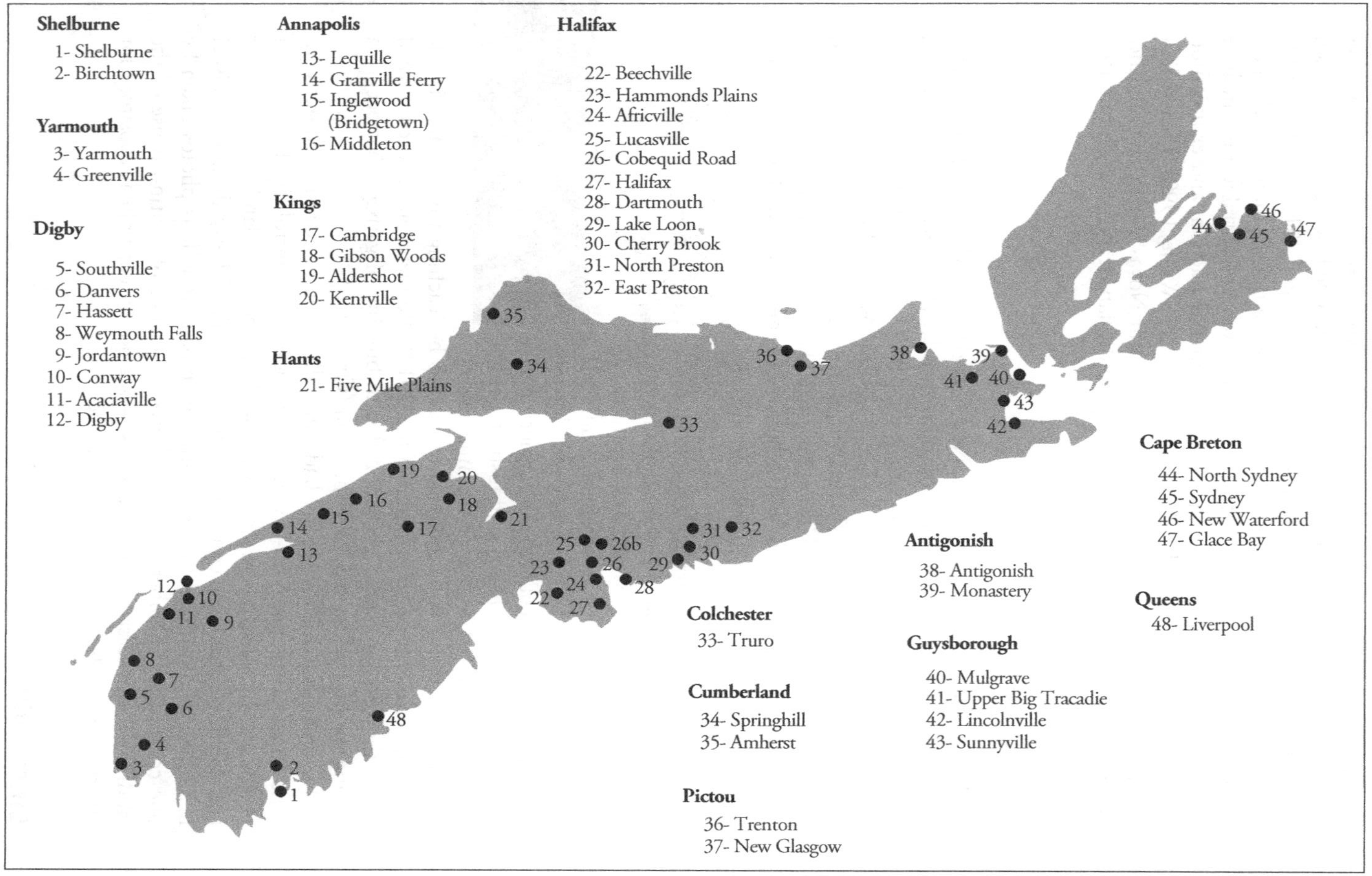

Shelburne
1- Shelburne
2- Birchtown

Yarmouth
3- Yarmouth
4- Greenville

Digby
5- Southville
6- Danvers
7- Hassett
8- Weymouth Falls
9- Jordantown
10- Conway
11- Acaciaville
12- Digby

Annapolis
13- Lequille
14- Granville Ferry
15- Inglewood (Bridgetown)
16- Middleton

Kings
17- Cambridge
18- Gibson Woods
19- Aldershot
20- Kentville

Hants
21- Five Mile Plains

Halifax
22- Beechville
23- Hammonds Plains
24- Africville
25- Lucasville
26- Cobequid Road
27- Halifax
28- Dartmouth
29- Lake Loon
30- Cherry Brook
31- North Preston
32- East Preston

Colchester
33- Truro

Cumberland
34- Springhill
35- Amherst

Pictou
36- Trenton
37- New Glasgow

Antigonish
38- Antigonish
39- Monastery

Guysborough
40- Mulgrave
41- Upper Big Tracadie
42- Lincolnville
43- Sunnyville

Cape Breton
44- North Sydney
45- Sydney
46- New Waterford
47- Glace Bay

Queens
48- Liverpool

Source: Black Loyalist Museum. Reprinted with permission.

landscape (NSPIRG, n.d.). In 1974, a first-generation landfill was opened about a half-mile away from the community. In 2006, the District of Guysborough (in which Lincolnville resides) closed the first landfill and opened a second-generation landfill that accepts waste from across northern Nova Scotia and Cape Breton. The decision to develop this landfill was based on the provincial government's need to decrease spending and generate needed tax revenues. Although Guysborough County determined the suitability of the landfill site and conducted extensive testing of surface water and groundwater, they did not consider the negative social impacts. The council proceeded with approvals and the construction of the landfill because they claimed that the public did not substantively oppose the project through the official environmental assessment process.

In the second landfill's first month of operation, about 55,780 metric tonnes (61,500 tons) of solid waste was received (Save Lincolnville Campaign, n.d.). According to regional environmental organizations, hazardous items such as transformers and refuse from offshore oil spills have been deposited at the landfill. In light of these issues, residents in Lincolnville have been concerned for some time about traces of carcinogens—including cadmium, phenol, and toluene—being above acceptable limits in the community's surface water and groundwater, from which the residents drink (Benjamin, 2008).

Despite the community's efforts to address the harms from the landfill over the past several decades, government has been largely unresponsive to their requests to meet or to hold mandatory community consultations. Deacon and Baxter (2013) contend that Lincolnville was unfairly chosen as a site for the landfill and that the proponents used intimidation tactics throughout the siting process. Residents believe that the mandatory community consultation period for sharing their concerns about the environmental assessment process was neither accessible, nor inclusive; that they were not adequately notified about the process or consulted about how the landfill project would affect residents; that they were denied the opportunity to reject the development; that they did not have a clear understanding of the terms used during the siting process; and that the process reinforced residents' sense of exclusion from decision making. Residents also perceived the consultation process as mere "lip service," because the decision to site the landfill in Lincolnville had already been made by the minister of environment. Although residents were provided with only one public drop-in session, government officials from the municipality alleged that residents failed to participate in the process or voice their concerns.

Photo 15.3. The community of Lincolnville living with landfills (East Coast Environmental Law Association, n.d.) Photo by ENRICH Project.

THE GEOGRAPHIC PATTERNING OF DISEASE: UNDERSTANDING THE ROLE OF PLACE IN THE HEALTH OF AFRICAN NOVA SCOTIANS

Black communities in Nova Scotia (and Canada more broadly) are considerably vulnerable to illness and disease associated with their greater exposure than white communities to environmental risks and other structural determinants of health, such as unemployment and underemployment, income insecurity and poverty, food insecurity, neighborhoods with poor infrastructures, poor-quality housing, criminalization, and a legacy of oppression (Waldron, 2018). According to Kisely, Terashima, and Langille (2008), the incidences of circulatory disease, diabetes, and psychiatric disorders are higher in African Nova Scotians than in other communities. They attribute these disparities to unequal access to disease prevention opportunities and long-term psychosocial stressors that elevate cortisol levels, leading to negative health effects.

A Structural-Determinants-of-Health Framework

In this section, I use the term "*structural* determinants of health" because

> the concept of social determinants of health, by definition, tends to exclude or marginalize other types of determinants not typically considered to fall under the category of the "social"—for example, spirituality, relationship to the land, geography, history, culture, language, and knowledge systems. (Reading, 2015, p. xii)

A structural-determinants-of-health framework supports the notion that an analysis of the social context of inequality is important for understanding why environmental health inequities, resulting from greater environmental exposures, can't be understood separate from the complex web of inequalities that are shaped by how society is organized (in part a remnant of a colonial legacy). Public policies and institutional practices are structured in ways that create and sustain social, economic, and political inequalities by preventing peoples from accessing valued goods and resources. In other words, understanding how structural inequalities impact health allows us to appreciate the many structurally rooted factors that create health inequalities beyond simply the social.

Structural determinants are deeply embedded, representing historical, political, ideological, economic, and social foundations from which all other determinants evolve (Reading, 2015; Waldron, 2018). They include the distribution of resources, income, and goods and services, as well as the circumstances of people's lives, such as their access to quality health care, schools, education, work, leisure, housing, and the environment. The more unequal the distribution of these resources among different populations, the lower the overall level of health and well-being and the greater the health inequities. A conceptualization of health that focuses on place and the geographic patterning of disease provides a robust framework that recognizes the need to understand health as an outcome of both long-standing structural, social, political, and economic determinants and inequalities (income insecurity, poverty, educational underachievement, poor housing) and environmental health inequities resulting from disproportionate exposures to contaminants and pollutants from nearby industry.

In an ENRICH Project study on health risk perceptions in African Nova Scotian communities near environmental contamination (Waldron, 2016), participants observed that good health is dependent on a number of structural determinants. These include access to health care (e.g., medical centers, home care), services (e.g., stores), jobs, healthy and affordable food, high-quality and low-rent housing, exercise, and reliable and low-cost transportation; accessibility of sidewalks, crosswalks, and parks; community safety; and community cohesiveness.

Scholarship provides strong evidence of environmental health inequities across racial dimensions, showing that Indigenous and other majority non-white communities in Canada are exposed to greater health risks compared to white communities because they are more likely to be spatially clustered around waste disposal sites and other environmental hazards (Atari et al., 2012; Maantay, 2002; Mascarenhas, 2007; Masuda et al., 2008; Sharp, 2009; Teelucksingh, 2007). Several studies found higher rates of certain illnesses in non-white communities, including cancer, upper respiratory disease, congenital anomalies, cardiovascular disease, skin diseases, and allergies (Bharadwaj et al., 2006; Crouse et al 2015; Cryderman et al., 2016; Rowat, 1999).

The potential short-term health effects that are linked to the presence of a landfill in one's community include respiratory issues; skin irritations; headaches and fatigue; allergies; physiological stressors; and in some cases, decreased birth weight, reproductive issues, and birth defects (Rushton, 2003). Several scholars have found that waste disposal sites are associated with increased incidences of adult cancers, such as leukemia and brain, stomach, liver, lung, rectum, prostate, and bladder cancers (Goldberg et al., 1999; Moy et al., 2008, Vrijheid, 2000). If you compare the map of African Nova Scotian communities above with the map of toxic facilities below, you will note that that most toxic facilities are located in or near African Nova Scotian communities.

The probability of groundwater contamination is greater in areas surrounding landfills due to the presence of leachate (MacNeil et al., 2018; Mor et al., 2006). Leachate is a liquid containing organic and inorganic compounds that gather at the bottom of a landfill and can percolate into groundwater. In rural areas, the possibility of biological and chemical contaminants from septic tanks is great because of the likelihood of coliform bacteria, nitrates, pharmaceuticals, hormones, per- and polyfluoroalkyl substances (PFASs), and other compounds contaminating groundwater supply (MacNeil et al., 2018; Schaider et al., 2016). In rural areas where agriculture or forestry is more common, pesticide runoff is another possible source of contamination (MacNeil et al., 2018; Soutter & Pannatier, 1996).

Toxic metalloid arsenic and other naturally occurring mineral deposits also affect water quality. Arsenic is carcinogenic to humans (American Cancer Society, 2016). Even with low levels of contamination (10 μg/L to 100 μg/L), it is linked to a range of diseases, including urinary tract and skin cancers, cardiovascular disease, and diabetes (Dummer et al., 2015; MacNeil et al., 2018). In an ENRICH Project study, a Shelburne resident attributed high rates of cancer and liver and kidney disorders in her community to the landfill nearby:

> Everyone knows that in all surrounding communities the dump can be seen in the south end of the community. A significant amount of people in close proximity to the dump have died from cancer and was or is suffering from an array of other health problems such as various forms of cancer, increased blood pressure, changes in nerve reflexes, brain, liver, and kidney disorders. Immune system [and]...digestive system. (Waldron, 2016)

MAP 15.2 ■ The ENRICH Project's map shows the proximity of African Nova Scotian and First Nations communities to degraded environments.

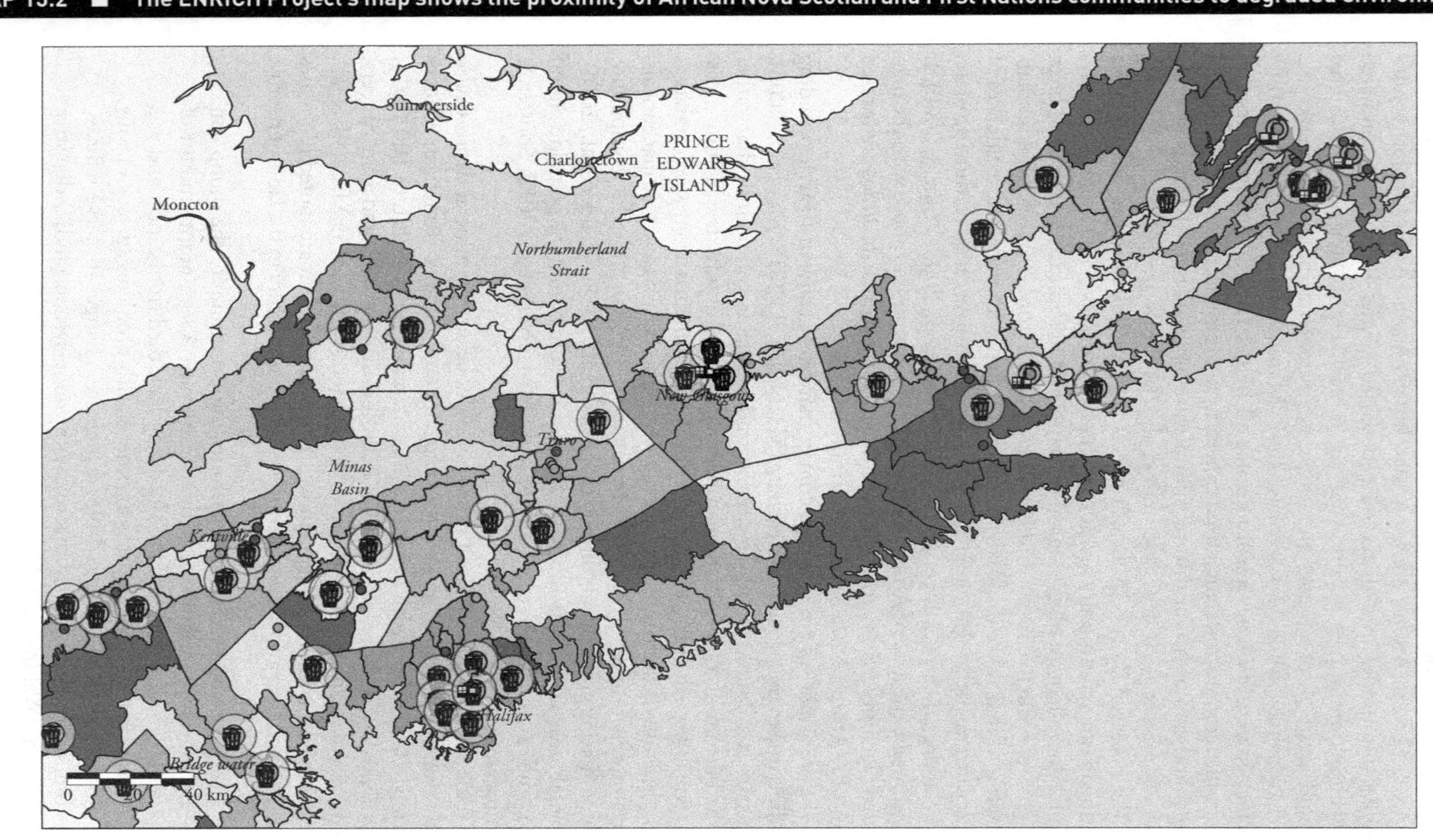

Source: ENRICH Project and ArcGIS. Content is the intellectual property of Esri and is used herein with permission. Copyright © 2020 Esri and its licensors. All rights reserved. See https://enrichproject.org/map/ for an interactive version.

The link between environmental exposures and psychological stress has also been documented by several scholars, who observe that industry and other environmental hazards are often appraised as stressful by residents (Dawson & Madsen, 2011; Downey & Van Willigen, 2005; Kondo et al., 2014). Downey and Van Willigen (2005) assert that those who live closer to industrial activity perceive greater neighborhood disorder, personal powerlessness, and depression, leading to greater symptoms of psychological distress. A resident in Lucasville who participated in the ENRICH study, describes the emotional toll of living near an equestrian farm:

> And, then talking about these physical industries, there's also a mental cost to that as well—to our mental health and well-being. When we know that all this is happening around us and nobody is concerned enough to do anything about it, that takes a toll on your mental health. So, it's not only your physical health that's being affected here, it's also our mental well-being. (Waldron, 2016)

This resident expresses that the psychological toll on the community is compounded by a sense of abandonment the community experiences due to government inaction.

Another resident of Lucasville expressed her concerns about the link between ill health and the equestrian farm in her community:

Photo 15.4. Memento Horse Farm in Lucasville. Photo by Robert Devet.

> It does probably have more of an effect on our physical health. But like I said, we don't know what we're breathing in. We don't know the moment when that reaches our well water, if there's E. coli. We don't know when we breathe that air in or that smell, which smells, like it's a smell you shouldn't be smelling.... But we don't know when we breathe it in how harmful it is in our lungs.... And you keep remembering all these uncles and cousins and grandfathers and people that died from cancer over the years. And when you compare it to another sector down the road or over there in the next province that didn't have this problem, you see, well, they lived to be old and they never had any health problems. So yes, it has a health impact. All these things have a health impact on us. It's stress, it's mental, it's mental harassment. (Waldron, 2016)

As this resident observes, the knowledge of physiological effects contributes to stress, which has its own physiological effects. We know that illness can result from environmentally hazardous activities that transmit pollutants in several ways, such as through water and air. The cancer risks associated with landfill-based disposal methods are approximately five times higher than the risks associated with waste-to-energy incineration (Moy, Krishnan, Ulloa, Cohen, & Brandt-Rauf, 2008). It is for this reason that community members in Lincolnville attribute high rates of prostate, stomach, lung, and skin cancers in their community to the landfill. According to a Lincolnville resident who participated in the ENRICH Project study, rates of cancer in her community have risen since the first-generation landfill was placed in the community:

> If you look at the health of the community prior to 1974 before the landfill site was located in our community, our community seemed to be healthier. From 1974 on until the present day, we noticed our people's health seems to be going downhill. Our people seem to be passing on at a younger age. They are contracting different types of cancer that we never heard of prior to 1974. Our stomach cancer seems to be on the rise. Diabetes is on the rise. Our people end up with tumours in their body. And, we're at a loss of, you know, of what's causing it. The municipality says that there's no way that the landfill site is affecting us. But, if the landfill site located in other areas is having an impact on people's health, then shouldn't the landfill site located next to our community be having an impact on our health too? (Waldron, 2016)

Increasing rates of illness and disease, as well as a lack of response by government, have prompted residents in Shelburne and Lincolnville to take action by organizing and mobilizing other residents to participate in protests, petition signings, letter-writing campaigns, legal complaints, and other activities, which I discuss in the next section.

MOBILIZING AND ACTIVISM IN AFRICAN NOVA SCOTIAN COMMUNITIES

The most significant challenge African Nova Scotian communities face in addressing environmental racism is the absence of legal tools that acknowledge, respond to, and address environmental concerns in their communities.

South End Environmental Injustice Society: The Morvan Road Landfill

Shelburne provides an excellent case study of the role that grassroots activism can play in addressing environmental racism. The Morvan Road landfill was located on the southeastern outskirts of Shelburne, where the largest concentration of African Nova Scotian residents has lived for decades (Waldron, 2018). At the end of 2015, I hired a local activist to facilitate focus groups with members of her community to discuss the landfill's impact on the community's health and well-being over the years. Following these focus groups and a meeting the ENRICH Project held in April 2016, several members of the community banded together to form the South End Environmental Injustice Society (SEED). Since forming, SEED has collaborated with concerned citizens; the ENRICH Project; social, environmental, and health organizations; and federal, provincial, and municipal governments to ensure cleanup and closure of the landfill.

In June 2016, after their inspection found several concerns, including an oil spill, Nova Scotia Environment issued a directive to the Shelburne Town Council and hired private-sector engineers to follow up. The landfill was finally closed in late 2016 (Waldron, 2018). Since then, SEED has been pursuing a collaboration with the town to (a) establish a reliable database on the relationship between the landfill and its socioeconomic, health, and environmental effects; (b) implement a plan for environmental, health, and community infrastructure and associated remedial issues; and (c) identify potential federal and provincial funding for the remedial work.

In September 2018, SEED conducted a water-testing project in collaboration with faculty at Nova Scotia Community College and Rural Water Watch, a new nongovernmental organization that was formed by members of the ENRICH Project team. The purpose of this project was to identify the pollutants to which community members have been exposed. In addition, the collaboration intends to document the chronological development of the landfill, its use, and its administration; document results from any previous testing; conduct soil testing; assess the ongoing health impacts of the landfill, including cancer rates; seek compensation for residents who have contracted cancer over the past several decades; address mold in homes; and investigate the impact of the landfill on property values in the past, present, and future.

Lincolnville Reserve Land Voice Council: The Second-Generation Landfill

In 2006, the Concerned Citizens of Lincolnville (renamed the Lincolnville Reserve Land Voice Council) was formed to oppose the opening of the second-generation landfill. The community alleges that the Municipality of the District of Guysborough improperly consulted them about this development. When the Concerned Citizens met with the municipality, they discussed their desire to be compensated for the economic fallout from the landfill, as well as its impact on the community's health. They also insisted that they receive a portion of the revenue if the landfill remained in their community. After persistent lobbying and a letter to the minister of environment, community members received a survey from the municipality gauging the community's willingness to pay for a well water treatment and storage facility in Lincolnville. Not surprisingly, community members responded that they should not have to front the capital costs for cleanup efforts associated with someone else's garbage from a landfill they never wanted (Benjamin, 2008). As one Lincolnville community member explained, they have been pressuring the government to remove the landfill since the early 1970s:

> We've had marches. We had the standoff at the dump and a march in Halifax in 2006. We had all kinds of protests about this. And we protested in 1974 when they started just dumping the stuff in a hole. We had a big protest. The police were called on us and all that. I don't know how many times the police were called and how many demonstrations I've been to where the cops have been called on us just because we were walking, saying, you know, "take your garbage somewhere else and take the rats somewhere else." (Waldron, 2016)

The protests in Lincolnville over the years have been supported by organizations from multiple sectors in the province. The objective of this provincewide campaign was to expose and challenge institutionalized oppression and send a message to municipal and provincial levels of government (including the Municipality of the District of Guysborough and Natural Resources Canada) that they opposed the landfill and that it was a clear example of environmental racism (Save Lincolnville Campaign, n.d.).

The Save Lincolnville Campaign demanded the following: that the Guysborough municipality review alternative locations and commence the closure and relocation of the landfill site; that the land be redeveloped and recovered; that all municipal resource management programs and planned waste management infrastructure in Nova Scotia be preceded by an inclusive and transparent consultative process with all sectors of the community; that full reparations and compensation be given for land displacement, health costs, and environmental contamination; that public policy be grounded in mutual respect and justice for all people; and that all people be guaranteed universal protection against nuclear testing and extraction, production, and disposal of toxic/hazardous wastes and poisons, regardless of race or class (Save Lincolnville Campaign, n.d.). Despite the considerable mobilizing the community has done over the past 40 years, their requests continue to be ignored. In 2016, some community members organized to support Mary Desmond (a member of the Lincolnville Reserve Land Voice Council) in her unsuccessful bid for a councillor seat in District Two, which includes Lincolnville. Their hope was that a win by Desmond would have advanced their cause in addressing the landfill.

That same year, members of the ENRICH Project's Water Monitoring Working Group, in collaboration with the Lincolnville Reserve Land Voice Council, launched a community-based water-monitoring project, which enabled residents, academia, industry, government agencies, and local institutions to collaborate to monitor, track, and address water quality and associated environmental concerns. Residents are trained on the ground to collect data alongside scientists, which helps to build trust among all parties and empowers residents (Whitelaw, Vaughan, & Craig, 2003, p. 410). ENRICH's water-monitoring project had three objectives: (a) to determine if there was contaminated water flowing in the direction of the community from the landfill site, (b) to build the community's capacity to test their water, and (c) to provide community members with basic knowledge about contaminants and groundwater sampling. This work involved sampling wells for bacteria and major ions and elements that are typically included in a water analysis. Results show that one of the sites tested positive for both coliforms and *E. coli*, while another site tested positive for coliforms. Surface water is most likely entering these wells, which is consistent with coliform contamination (Bonner et al., 2016). The project was completed in August 2016, when members of the working group returned to the community to share findings, answer questions, and address concerns.

In the summer of 2016, I was working to forge a new relationship between the Lincolnville Reserve Land Voice Council and Ecojustice—Canada's largest environmental law charity. They are currently exploring possible legal remedies to address the landfill, including examining overlap between human rights law and environmental law to determine if government has violated residents' human rights and the Canadian Charter of Rights and Freedoms. Human rights violations may include the exposure of racialized communities to harm, as well as under-provision of clean drinking water and environmental services. It is important to note that Canada does not officially recognize access to clean drinking water as a constitutional right. In July 2010, the right to water and sanitation was recognized by the United Nations General Assembly through Resolution 64/292. The United Nations states that clean drinking water is "essential for the full enjoyment of the right to life" (Council of Canadians, n.d.). Of the 193 countries that make up the United Nations, 181 of them signed on to recognize the right to clean drinking water. Canada has yet to sign on (MacNeil et al., 2018; Mitchell, 2015).

CONCLUSION

Addressing environmental racism in Nova Scotia and Canada must include an environmental justice lens that acknowledges and addresses structural as well as environmental determinants of health and involve the community in consultations and monitoring. Policymakers, environmental organizations, activists, and others involved in environmental justice struggles in Nova Scotia must acknowledge the central role that racism plays through the enduring impacts of colonialism and capitalism in African Nova Scotian communities. Reducing inequitable siting of polluting industries can't be achieved unless deliberate attention is paid to educating environmentalists and others about the systemic ways in which racist ideologies get written into environmental decision-making and policy.

Environmental policy in Nova Scotia must also begin to address, in a more forthright way, the cumulative health, including mental health, impacts of environmental racism. In doing so, it must acknowledge the complex web of inequalities that create greater biological susceptibility and social vulnerability and that, consequently, drive and sustain health disparities between white, Indigenous, and black communities. In addition, given that clean drinking water has yet to be acknowledged as a right in Canada, changes to policy are warranted, as well as the establishment of a government-subsidized program that will help cover the costs associated with proper water testing of privately owned wells in rural Nova Scotia. This is particularly crucial in cases where high rates of illness and disease in specific communities suggest that contaminated groundwater may be to blame (MacNeil et al., 2018). In Nova Scotia, private well owners are responsible for testing drinking water from wells. The province recommends testing for bacteria every six months and chemicals every one to two years (Nova Scotia Environment, 2018). Despite these recommendations, only 4 percent of private well owners in Nova Scotia test their well water every year because there is a general lack of concern about water quality and because testing is viewed as inconvenient and costly (Chappells et al., 2015; MacNeil et al., 2018). And, for African Nova Scotian communities that are among the lowest income groups in the province, testing is particularly costly. Consequently, they will have ongoing anxieties about the safety of their water (Ore, 2018).

Culturally relevant participatory democracy is also an important approach for generatively engaging African Nova Scotian communities in environmental issues. An approach that draws on the experiences of and listens to marginalized communities with more opportunities for public participation and consultation should improve decision-making processes related to environmental policy. In other words, such an approach can ensure that the people most affected by policy decisions are involved early on and throughout the process. It is important to note, however, that although Section 35 of the Constitution Act (Government of Canada, 1982) sets out a legal requirement to consult with and accommodate Indigenous peoples where treaty rights or title interests are engaged, there is no such legal requirement for African Nova Scotians or any other racial minority group in Canada.

Community-based monitoring (CBM) is increasingly being used to address community concerns of various kinds around the world (Garg & Laskar, 2010; MacNeil et al., 2018). Whitelaw, Vaughan, and Craig (2003) define it as "a process where concerned citizens, government agencies, industry, academia, community groups and local institutions collaborate to monitor, track and respond to issues of common community concern" (p. 410). CBM is increasingly being relied on in Nova Scotia and other parts of Canada because of a general mistrust of how the government is addressing environmental concerns (Conrad, 2006; MacNeil et al., 2018). CBM is a collaborative approach that involves the different stakeholders in planning and management processes, capacity building, community assessment, interface meetings, and final evaluations (Garg & Laskar, 2010; MacNeil et al., 2018). Concerned citizens and activists are provided with various opportunities to become educated, contribute to environmental protection, and participate in building morale and raising awareness of potential environmental hazards (Killcreas, 2012; MacNeil et al., 2018). CBM is also a useful tool for groups that wish to mobilize residents and conduct research that will validate health concerns brought forward by communities that are near environmental hazards (Campbell et al., 2013).

DEEPENING OUR UNDERSTANDING

Work in a group to research a local case of environmental racism and environmental justice:

1. Choose a community of color in North America (each group should choose a different community) that has endured similar experiences of environmental racism and that has mobilized to achieve environmental justice.

2. Document how the social, economic, political, and health effects of a historical or contemporary case of environmental racism has been addressed through research; policy; legal remedies; and community mobilization, activism, and advocacy.

3. Share what you learn through an approved platform (e.g., Facebook page, website, presentation). Useful resources would include text that provides information (discussed in #4) on the case; narratives or direct quotes, taken from newspaper articles and other sources, from researchers, community members, politicians or government officials and other policymakers, health professionals and agencies, or lawyers; online videos; journal articles; and links to social media pages.

4. The text/written information should outline the following:
   a. A description of how this case reflects the concept/definition of environmental racism.
   b. A community profile: city; historical profile of the community, including income, socioeconomic and political challenges this community has faced over the years, and the like.
   c. A description of the environmental hazards this community has faced or is facing, and information on the location of these hazards in relation to the community.
   d. A discussion of the social, economic, political, and health impacts of the environmental hazard on the community.
   e. A discussion of how this case reflects an environmental justice approach by providing a definition and the main components of an environmental justice approach—and by outlining how the issues have been addressed through research; policy; legislation; legal remedies; and community mobilization, activism, and advocacy (e.g., civil disobedience, protests, marches).

References

American Cancer Society. (2016). Known and probable human carcinogens. https://www.cancer.org/cancer/cancercauses/general-info/known-and-probable-human-carcinogens.html

Atari, D. O., Luginaah, I., Gorey, K., Xu, X., & Fung, K. (2012). Associations between self-reported odour annoyance and volatile organic compounds in "Chemical Valley", Sarnia, Ontario. *Environmental Monitoring and Assessment, 185* (6), 4537–4549.

Benjamin, C. (2008, August 7). Lincolnville dumped on again. *The Coast.* thecoast.ca/halifax/lincolnville-dumped-on-again/Content?oid=993619.

Bharadwaj, L., Nilson, S., Judd-Henrey, I., Ouellette, G., Parenteau, L., Tournier, C., ... , & Bear, A. (2006). Waste disposal in First-Nations communities: The issues and steps toward the future. *Journal of Environmental Health, 68* (7), 35–39.

Black Cultural Centre for Nova Scotia. (n.d.). Black migration in Nova Scotia. bccnsweb.com/web/our-history/

Bonner, F., Menendez Sanchez, W., Bonner, C., Clarke, A., & Beckett, R. (2016). *Lincolnville water monitoring report.* Halifax, Nova Scotia, Canada: Dalhousie University.

Borden Colley, S. (2016, November 30). Hundreds of former Africville residents could join class-action lawsuit. CBC News Nova Scotia. cbc.ca/news/canada/nova-scotia/africville-proposed-class-action-lawsuit-in-court-1.3874538

Bullard, R. D. (1993). *Confronting environmental racism: Voices from the grassroots.* Boston, MA: South End Press.

Bullard, R. D. (2002). Confronting environmental racism in the 21st century. *Global Dialogue: The Dialogue of Civilization, 4,* 34–48.

Campbell, R. L., Caldwell, D., Hopkins B., Heaney, C. D., Wing S., Wilson, S. M., O'Shea, S., & Yeatts, K. (2013). Integrating research and community organizing to address water and sanitation concerns in a community bordering a landfill. *Journal of Environmental Health, 75* (10), 48–50.

Canadian History Workshop. (n.d.). Black Loyalists in Nova Scotia. https://canadianhistoryworkshop.wordpress.com/2013/03/15/black-loyalists-in-nova-scotia/

Chappells, H., Campbell, N., Drage, J., Fernandez, C. V, Parker, L., & Dummer, T. J. B. (2015). Understanding the translation of scientific knowledge about arsenic risk exposure among private well water users in Nova Scotia. *Science of the Total Environment, 505*, 1259–1273.

Conrad, C. (2006). Towards meaningful community-based ecological monitoring in Nova Scotia: Where are we versus where we would like to be. *Environments, 34* (1), 25–36.

Coulthard, G. S. (2014). *Red skin, white masks: Rejecting the colonial politics of recognition*. Minneapolis: University of Minnesota Press.

Council of Canadians. (n.d.). *Our right to water: A people's guide to implementing the United Nations' recognition of the right to water and sanitation*. https://canadians.org/sites/default/files/publications/RTW-intl-web.pdf

Crouse, D. L., Peters, P. A., Villeneuve, P. J., Olivier Proux, M., Shin, H. H., Goldberg, M. S., … , & Burnett, R. T. (2015). Within- and between-city contrasts in nitrogen dioxide and mortality in 10 Canadian cities: A subset of the Canadian Census Health and Environment Cohort (Canchec). *Journal of Exposure Science and Environmental Epidemiology*, *5*, 482–89.

Cryderman, D., Letourneau, L. Miller, F., & Basu, N. (2016). An ecological and human biomonitoring investigation of mercury contamination at the Aamjiwnaang First Nation. *EcoHealth, 13* (4), 784–95.

Dawson, S. E., & Madsen, G. E. (2011). Psychosocial and health impacts of uranium mining and milling on Navajo lands. *Health Physics, 101* (5), 618–25.

Deacon, L., & Baxter, J. (2013). No opportunity to say no: A case study of procedural environmental injustice in Canada. *Journal of Environmental Planning and Management, 56* (5), 607–23.

Devet, R. (2017a). The now decommissioned Shelburne dump. *The Nova Scotia Advocate*. https://nsadvocate.org/2017/01/02/a-community-of-widows-the-shelburne-dump-and-environmental-racism/

Devet, R. (2017b). Memento Farm in Lucasville. *The Nova Scotia Advocate*. https://nsadvocate.org/2017/07/11/historic-black-community-of-lucasville-continues-to-fight-horse-farm-feels-abandoned-by-hrm/

Downey, L., & Van Willigen, M. (2005). Environmental stressors: The mental health impacts of living near industrial activity. *Journal of Health and Social Behaviour*, *46* (3), 289–305.

Dummer, T. J. B., Yu, Z. M., Nauta, L., Murimboh, J. D., & Parker, L. (2015). Geostatistical modeling of arsenic in drinking water wells and related toenail arsenic concentrations across Nova Scotia, Canada. *Science of the Total Environment, 505*, 1248–1258.

East Coast Environmental Law Association. (n.d.). The community of Lincolnville living with landfills. https://www.ecelaw.ca/images/PDFs/ER_event/Lincolnville_final.pdf

Fryzuk, L. A. (1996). Environmental justice in Canada: An empirical study and analysis of the demographics of dumping in Nova Scotia. Unpublished master's dissertation. Dalhousie University, Canada.

Garg, S., & Laskar, A. R. (2010). Community-based monitoring: key to success of national health programs. *Indian Journal of Community Medicine, 35* (2), 214–216.

Goldberg, M. S., Siemiatyck, J., DeWar, R., Desy, M., & Riberdy, H. (1999). Risks of developing cancer relative to living near a municipal solid waste landfill site in Montreal, Quebec, Canada. *Archives of Environmental Health, 54* (4), 291–296.

Government of Canada. (1982). *Constitution Act*. laws-lois.justice.gc.ca/eng/const/page-15.html#h-38

Government of Nova Scotia. (2019). Private wells. https://novascotia.ca/nse/water/privatewells.asp

Halifax Regional Municipality. (n.d.). *Remembering Africville: A source guide*. https://www.halifax.ca/

about-halifax/diversity-inclusion/african-nova-scotian-affairs/africville/remembering-africville-a

Historica Canada. (n.d.b). Africville. thecanadianencyclopedia.ca/en/article/africville/

Killcreas, A. H. (2012). The power of community action: Environmental injustice and participatory democracy in Mississippi. *Mississippi Law Journal, 81* (4), 769–812.

Kisely, S., Terashima, M., & Langille, D. (2008). A population-based analysis of the health experience of African Nova Scotians. *Canadian Medical Association Journal, 179* (7), 653–658.

Kondo, M. C., Gross-Davis, C. A., May, K., Davis, L. O., Johnson, T., Mallard, M., ... , & Branas, C.C. (2014). Place-based stressors associated with industry and air pollution. *Health & Place, 28*, 31–37.

Maantay, J. (2002). Mapping environmental injustice: Pitfalls and potential of geographic information systems in assessing environmental health equity. *Environmental Health Perspectives, 110* (Suppl. 2), 161–71.

MacNeil, R., Poirier, M., Solaimani Baghainia, S., & Tsehtik, M. (2018). *Case B–Environmental racism: Living next to a landfill in Shelburne, NS.* Halifax, Nova Scotia, Canada: Dalhousie University.

Maddalena, V., Thomas Bernard, W., Etowa, J. B., Davis-Murdoch, S., Smith, D., & Marsh-Jarvis, P. (2010). Cancer care experiences and the use of complementary and alternative medicine at end of life in Nova Scotia's Black communities. *Journal of Transcultural Nursing, 21*, 114–22.

Mascarenhas, M. (2007). Where the waters divide: First Nations, tainted water and environmental justice in Canada. *Local Environment, 12* (6), 565–77.

Masuda, J. R., Zupancic, T., Poland, B., & Cole, D. C. (2008). Environmental health and vulnerable populations in Canada: Mapping an integrated equity-focused research agenda. *The Canadian Geographer, 52* (4), 427–50.

Mitchell, K. (2015, April 22). Why it's time for Canada to recognize our right to water. Ecojustice. https://www.ecojustice.ca/why-its-time-for-canada-to-recognize-our-right-to-water/

Mor, S., Ravindra, K., Dahiya, R. P., & Chandra, A. (2006). Leachate characterization and assessment of groundwater pollution near municipal solid waste landfill sites. *Environmental Monitoring and Assessment, 118*(3), 435–456.

Moy, P., Krishnan, N., Ulloa, P., Cohen, S., & Brandt-Rauf, P. W. (2008). Options for management of municipal solid waste in New York City: A preliminary comparison of health risks and policy implications. *Journal of Environmental Management, 87* (1), 73–79.

Nelson, J. J. (2001). The operation of whiteness and forgetting in Africville: A geography of racism. Unpublished doctoral dissertation. University of Toronto, Canada.

Nova Scotia Archives. (1965). Gone but never forgotten: Bob Brooks' photographic portrait of Africville in the 1960s. https://novascotia.ca/archives/africville/archives.asp?ID=13

Nova Scotia Environment. (2018). About private drinking water supplies. https://novascotia.ca/nse/water/privatewatersupplies.asp

Nova Scotia Museum. (n.d.). Remembering Black Loyalists, Black communities in Nova Scotia. novascotia.ca/museum/blackloyalists/communities.htm

NSPIRG (Nova Scotia Public Interest Research Group). (n.d.). Save Lincolnville campaign. nspirg.ca/projects/past-projects/save-lincolnville-campaign/#sthash.9YV4BKxl.dpuf

Ore, J. (2018, April 4). A community of widows: How African-Nova Scotians are confronting a history of environmental racism. CBC Radio. http://www.cbc.ca/radio/thecurrent/features/facing-race/a-community-of-widows-how-african-nova-scotians-are-confronting-a-history-of-environmental-racism-1.4497952

Reading, C. (2015). Structural determinants of Aboriginal peoples' health. In M. Greenwood, S. de Leeuw, N. M.Lindsay, & C. Reading (Eds.), *Determinants of Indigenous peoples' health in Canada: Beyond the social* (pp. xi–xxix). Toronto: Canadian Scholars Press.

Rowat, S. C. (1999). Incinerator toxic emissions: A brief summary of human health effects with a note on regulatory control. *Medical Hypothesis, 52* (5), 389–96.

Rushton, L. (2003). Health hazards and waste management. *British Medical Bulletin, 68* (1), 183–97.

Save Lincolnville Campaign. (n.d.). Our demands. savelincolnville.h-a-z.org/demands.php

Schaider, L. A., Ackerman, J. M., & Rudel, R. A. (2016). Septic systems as sources of organic wastewater compounds in domestic drinking water wells in a shallow sand and gravel aquifer. *Science of the Total Environment, 547*, 470–481.

Sharp, D. (2009). Environmental toxins, a potential risk factor for diabetes among Canadian Aboriginals. *International Journal of Circumpolar Health, 68* (4), 316–26.

Shoemaker, N. (2015). A typology of colonialism. historians.org/publications-and-directories/perspectives-on-history/october-2015/a-typology-of-colonialism

Soutter, M., & Pannatier, Y. (1996). Groundwater vulnerability to pesticide contamination on a regional scale. *Journal of Environmental Quality, 25* (3), 439–444.

Tavlin, N. (2013, February 4). Africville: Canada's secret racist history. *Vice*. vice.com/en_ca/read/africville-canadas-secret-racist-history

Teelucksingh, C. (2007). Environmental racialization: Linking racialization to the environment in Canada. *Local Environment, 12* (6), 645–61.

United Nations General Assembly. (2010). Resolution adopted by the General Assembly on 28 July 2010 (pp. 1–3). http://www.un.org/es/comun/docs/?symbol=A/RES/64/292&lang=E

Veracini, L. (2011). Introducing settler colonial studies. *Settler Colonial Studies, 1* (1), 1–12.

Vrijheid, M. (2000). Health effects of residence near hazardous waste landfill sites: A review of epidemiological literature. *Environmental Health Perspectives, 108* (1), 101–112.

Waldron, I. R. G. (2010). *Challenges and opportunities: Identifying meaningful occupations in low-income, racialized communities in the North End*. Halifax, Nova Scotia, Canada: Dalhousie University.

Waldron, I. R. G. (2015). Findings from a series of workshops entitled "In Whose Backyard?—Exploring toxic legacies in Mi'kmaw and African Nova Scotian communities." *Environmental Justice, 8* (2), 1–5.

Waldron, I. R. G. (2016). *Experiences of environmental health inequities in African Nova Scotian communities*. Halifax, Nova Scotia, Canada: Dalhousie University.

Waldron, I. R. G. (2018). *There's something in the water: Environmental racism in Indigenous and black communities*. Halifax, Nova Scotia, Canada: Fernwood Publishing.

Whitelaw, G., Vaughan, H., & Craig, B. (2003). Establishing the Canadian Community Network. *Environmental Monitoring and Assessment, 88*, 409–418.

Wolfe, P. (1999). *Settler colonialism and the transformation of anthropology: The politics and poetics of an ethnographic event*. London: Cassell.

16

FANTASTIC PRAGMATIC

The Enduring Effects of the 1993 Encounter between Black Panthers and Black Brazilian Activists

João Costa Vargas

Whereas the other chapters in this text focus more explicitly on environmental justice, this chapter will delve deeper into an enduring pattern of imposed and global marginalization on black communities that precipitates and normalizes environmental injustices. My entry point is the 1992 Los Angeles rebellion. Contrary to how it was portrayed by the worldwide news media monopoly (and various "specialized" research and academic accounts; see Vargas, 2010, chap. 4), the revolt could not have happened without a collective agreement on the necessity of expressing a sense of injustice and betrayal outside of the formal mechanisms of grievance redressal. This was not a spontaneous "riot," carried out by "criminal elements," "hoodlums," and opportunists. Rather, it was the result of a recognition of the marginalization of black communities. As it had happened in Harlem in 1964, Watts in 1965, Detroit in 1967, and more recently in Cincinnati in 2011 and Ferguson in 2014, the rebellion in L.A. was sparked by an incident of police brutality against one or more black persons, most notably Rodney King. Yet, whereas police abuse featured prominently in the local residents' grievances, it was seen as only a part of a constellation of antiblack forces. Such forces generated and were furthered by the intersecting and mutually compounding effects of unemployment and underemployment; residential hypersegregation; lack of day care facilities and supervised youth activities; homelessness; exposure to environmental hazards in the air, water, and land; food insecurity; punitive schooling; police brutality; targeted mass incarceration; and early death by a litany of preventable and treatable ailments including HIV/AIDS, sickle cell anemia, asthma, diabetes, hypertension, and clinical depression.[1]

Wanting to better understand the 1992 rebellion, in 1995, I found a relatively expensive,[2] roach- and mice-infested apartment as close as possible to the infamous intersection of Florence and Normandie Avenues, one of the epicenters of the 1992 revolt. I went to the office of the Coalition Against Police Abuse (CAPA) on Western Avenue, following a comment I heard about a Black Panther who had close ties to black Brazilian activists. I introduced myself to CAPA's coordinator, Michael Zinzun.

Zinzun was receptive of my offer to work at the coalition. A member of the Black Panther Party from the late 1960s until the early 1970s, and a co-founder of CAPA in 1975, he had been in Brazil in 1993, when he attended a conference in Rio de Janeiro on the black diaspora convened by a local nongovernmental organization (NGO), the Institute for the Study of Black Culture (*Instituto de Pesquisas da Cultura Negra*, IPCN). IPCN is a black activist congregation that, due to censorship during the military dictatorship (1964–1985), strategically presented itself as a research and cultural organization. The Afro-Brazilian activists who participated in the event on the diaspora belonged to a variety of organizations that were not always politically synchronized yet identified with and were supportive of the Unified Black Movement (*Movimento Negro Unificado*, MNU) and its broad umbrella. They were members of various black collectives based in universities, women's forums, *Candomblé* houses, study and publication groups, samba schools, and traditional black

---

PHOTO 16.1
John Gomez / Shutterstock

neighborhoods, known as *favelas*. One of these favelas was Jacarezinho, an incubator of a rich tradition of union organizing, music, poetry, and auto-construction (Vargas, 2005). Zinzun and his delegation—which included activists, teachers, artists, professors, formerly incarcerated people, and gang members working on establishing peace among warring factions—arrived in Rio a couple of days after the Candelária massacre, when armed men, including off-duty police officers, killed eight children who were sleeping on the steps of a downtown church.

CONCEPTUALIZING ANTIBLACK GENOCIDE

This chapter examines the ways in which a critical analysis of a black transnational collaboration that took place more than 25 years ago between U.S. and Brazilian activists helps explain the contemporary challenges of conceptualizing and combating gendered antiblack genocide. Drawing on recent ethnographic data, as well archival research,[3] I analyze (a) the 1993 encounter's political motivations, its theoretical bases, and the practical consequences that followed, and (b) the ways in which contemporary black political formations in Brazil and the United States utilize similar analytical and political frameworks to grapple with the multiple and enduring manifestations of black people's structural disposability.

How does the utilization of this diasporic conceptual universe, the fantastic pragmatic, both enable and hinder the practice of *autonomy* and the theoretical and practical coming to terms with *gendered antiblack genocide*? What can be gleaned from that transnational encounter between black organizers more than quarter-century ago that explains both the successes and the shortcomings of contemporary black organized efforts, including those linked to the transnational Black Lives Matter movement?

In Brazil, past and current collective efforts, such as police monitoring programs in black neighborhoods and the yearly course on the black diaspora offered to black activists by Criola, a black women's NGO in Rio, trace at least part of their genealogies to the 1993 encounter. Whereas the encounter, per se, may not have been the direct trigger of these initiatives, the event manifested a series of concepts, strategies, and views of the world that were central to their forming. This chapter focuses on such concepts, strategies, and views of the world. If black lives are to be rendered viable and thriving, what societal conditions of possibility need to be in place? What theoretical concepts and political strategies are necessary to achieve such conditions?

This chapter is organized into five sections:

1. The diasporic encounter (the fantastic), where the theoretical and political context is explained
2. Confronting antiblack genocide (the pragmatic), which focuses on some of the practical challenges the black collectives faced
3. Crafting autonomy, where the employment of self-determination is explained

4. The Popular Movement of the Favelas and *Operation Ghetto Storm,* which focuses on two examples of black initiatives whose concepts gain meaning when juxtaposed to the encounter
5. The challenges of the fantastic pragmatic, which concludes the piece and suggests a research exercise based on some of the concepts developed here

The Diasporic Encounter (The Fantastic)

Black Power, as a political philosophy, oriented much of Zinzun and his group's political worldview and, indeed, led them to Rio. Black Power argued that black people in the United States were part of a planetary community united by experiences of oppression and marginalization.

> **"*Black Power means that black people see themselves as part of a new force, sometimes called the "Third World"; that we see our struggle as closely related to liberation struggles around the world. We must hook up with the struggles. (Ture & Hamilton, 1992, p. xix)*"**

Black people's exclusive affiliation to the U.S. empire-state was thus suspended, at least theoretically.[4] In this attempt at reconnecting with a planetary community of struggle, one's assumed citizenship was relativized, and, more generally, there emerged an alternative concept and practice of humanity. It was the very project of Western civilization that was called into question.

The encounter between black Brazilian and U.S. activists sought recognition, validation, and strength in the diaspora. It was also a calculated attempt at crafting a new social world in whose emergence the black subject would have a central role. If, in the perspective of Cedric Robinson (2000), blacks of the diaspora consistently and transhistorically assert their ontological totality over individual perspectives, and emphasize the metaphysical over the pragmatically given, then the transnational encounter, from the beginning, was all about the fantastic.[5]

Fantastic: the etymology of the term suggests that, in its late 14th-century usage, it meant "existing only in imagination," from Middle French *fantastique*, which in turn derived from *fantasticus* in Medieval Latin, *phantasticus* in Late Latin, meaning "imaginary," and from Greek *phantastikos,* meaning "able to imagine."[6] The transnational encounter sought and realized the fantastic in that, prior to the actual meeting, the diaspora, as a yet-to-be but nevertheless present in its past-ness and thus graspable, was already in the activists' imagination.

The diaspora was not only an imagined transhistorical community of struggle but also the imagination of recognition, shared affect, and a world to be constructed. It did not matter that the U.S.-based persons knew little Portuguese; it did not matter that the Brazilians could communicate in English only with considerable difficulty. It was more about recognizing a shared being-in-the world, which included both the dystopian experience of genocide and a graspable utopian energy. The grammar of urgency and dreams took precedence over the words spoken. The fantastic gives the event one of its defining traits: its refusal to settle for the past and the present; its unwavering undercurrent that pushes, relentlessly (but not always obviously), against the most daunting forms of unfreedoms, suffering, and death; and its utopian determination, announcing the full realization of black people's full lives, in times and spaces both graspable and yet-to-be conceived.

Confronting Antiblack Genocide (The Pragmatic)

To be sure, there were urgent pragmatic issues to be discussed and addressed during the encounter. Central among them were survival and black autonomy. To achieve them, what theories, practices, and common understandings were required? From the moment of its foundation, the Brazilian Black Movement had been explicit about the overarching challenge of interrupting antiblack genocide. For the movement, it was understood that Brazilian society was fundamentally structured on the hatred of the black. Although this hatred was veiled by a national narrative of racial harmony and mixing, it surfaced in a variety of discourses rooted in and disseminated though commonsensical understandings, pseudoscientific knowledge, research and teaching institutions, and the empire-state. Such discourses became more prominent at the end of the 18th century, as the legal abolition of slavery demanded a restructuring of social and racial hierarchies—a phenomenon not unlike that in the United States.[7]

One discourse was of the whitening of the republic's genetic stock, to be achieved by immigration policies that favored European whites, and the ensuing multigenerational progressive diluting and eventual physical disappearance of black persons due to the assumed ultimate dominance of whiteness. Another discourse was the avoidance of racial mixing with blacks. The latter is more obviously associated with eugenic philosophies that became prevalent in Brazil in the 1930s, but the former was just as effective in calling into question the black presence in the empire-state. Both were unmistakably antiblack (Nascimento, 1989).

Antiblack hatred surfaced in the patterns of everyday and institutional forms of discrimination and violence that resulted in the measurable premature and preventable death of black people. From residential segregation, to police brutality, to blocked access to well-being and quality medical care, and to exposure to environmental toxins, black people in Brazil were never meant to survive (Vargas, 2010). Symptomatically, the mobilization that sparked the April 1978 formation in São Paulo of the *Movimento Unificado Contra a Discriminação Racial* (Unified Movement Against Racial Discrimination), which in 1979 would give rise to the MNU, followed the death of Robson Luz, a black cab driver, at the hands of the police (see, for example, Covin, 2006).

For black Brazilians to survive, it was necessary to craft worldviews and concrete programs that provided the means of ontological and social reaffirmation. The Unified Black Movement/MNU drew from a vast assemblage of black formations of social, spiritual, and material support. Congregating spaces of worship, study, research, and political mobilization, MNU employed organizing experience accumulated in openly black political and cultural collectives such as the 1930s *Frente Negra Unificada* (Unified Black Front) and the 1940s and 1950s *Teatro Experimental do Negro* (Black Experimental Theater). Various groups intersected with these formations, such as the *Associação das Empregadas Domésticas* (Domestic Workers Association) and the *Conselho Nacional de Mulheres Negras* (Black Women's National Council).[8] MNU also relied on long-established forms of association less visible to the non-black public. Operating mostly according to underground and undetected social frequencies, such collective forms of autonomous association made possible the creation, circulation, and sharing of material and spiritual resources; the dissemination of information; land stewardship; and strategies of political mobilization. While unmistakably hierarchical, women, gays, lesbians, and transgender people traditionally played key roles in such black bureaucracies that are often described as vertical and authoritarian yet consultative and transparent.[9]

Among the U.S.-based activists, the claim of genocide was not as centrally and openly articulated, although it was an important conceptual undercurrent. William Patterson and the Civil Rights Congress's public campaign, based on the 1951 publication *We Charge Genocide: The Historic Petition to the United Nations for Relief from a Crime of the United States Government against the Negro People,* sensitized a wide spectrum of black social environments, congregations, and political clusters.[10] Genocide was presented as multifaceted because it included, most obviously, physical and symbolic violence inflicted by antiblack organized groups, public bureaucracies such as the police, and elected officials who defended the use of force against blacks to prevent them from voting. Yet it also included social processes that, if not immediately related to black early death, were nevertheless contributing factors to suffering and diminished life chances. Unemployment, low wages, imposed residential segregation, and political disenfranchisement deepened black people's vulnerability to death by preventable causes (Patterson, 1951, pp. 5–19). This formulation also suggested that genocide is part of a continuum: For its multiple facets to occur consistently and without widespread and sustained contestation, they necessitate a cultural agreement about the disposability of black people.

This understanding of genocide as multifaceted and part of a continuum provides an explanation for the Black Panther Party's 10-point program as well as its community initiatives, purposefully called *survival programs*. Each of the 10 points suggests a facet of oppression to which black people are subjected. For example, point one stated, "We want freedom. We want to determine the destiny of our Black Community." Here it was suggested that black social, economic, and political spaces were controlled by non-blacks, who not only profited from this state of subordination but also enforced dehumanizing social conditions defined by vulnerability to disease and violence, either by state negligence or by state-sanctioned apparatuses such as law enforcement. Taken as a whole, the 10-point program suggests an understanding of U.S. society as structurally arranged against the psychological, legal, political, and physical survival of black people. The structural argument suggests that, besides the actual forms of antiblack oppression, a cultural consensus about black disposability undergirds the empire-state's mundane sociability as well as its institutional dynamics.[11]

Even though the grammar of genocide was not articulated as explicitly among the U.S. black activists, it was very much present in the ways in which their survival programs were elaborated and put into practice. At one point, the Black Panther Party had more than 60 survival programs, including food, breakfast for children, shoes and clothing, job training, testing for sickle cell anemia, transportation for the elderly and incapacitated (and to and from prisons, for relatives and loved ones of those incarcerated), ambulance, cooperative housing, music, dance, child care, and pest control. They were all free. The idea was that, with enough free programs, freedom would be achieved (Bloom & Martin, 2016) and antiblack genocide averted.

For example, CAPA, drawing on the Black Power and Panther diasporic legacies, ran the Off the Roach Program. It resulted from cutting-edge research on entomology, epidemiology, and toxicology. In a 1973 article, *Jet* magazine featured the initiative:

> Zinzun declared his "war on roaches" when the Pasadena Freeway was built and many houses were torn down in his neighborhood, sending the insects scurrying for new habitats. Zinzun and his brother, Juan, learned that the roaches carry infectious disease organisms on their bodies, particularly salmonella—a bacteria that can cause food poisoning or diseases of the genital track. After consulting with a professor of entomology at UCLA, Zinzun discovered that boric acid was an effective killer of roaches, so he set up the Pasadena Community Information Center. To date, his group has treated 700 homes... Zinzun said that he hopes to "off" the roaches in 1,000 local homes by July. ("California Man," 1973, p. 32)

This initiative—one among many others developed over four decades of community activism—led to Michael Zinzun's invention of the boric acid gun, a safe and effective technology to eliminate pests, which was then used by the young people operating the pest control program. CAPA also initiated a recycled plastic dome program for the homeless. The concept was to know and control the entire process: from the collection of discarded plastic, to its transformation into plastic sheets, and to the assemblage of the sheets into the livable domes. In CAPA's backyard, a prototype dome was used for silk-screening and photography classes, community meetings, and as a concrete reminder of the possibilities that arise once the imagination of the fantastic meshes with the pragmatic attempts at solving actual social problems (Vargas, 2006).

Crafting Autonomy

Black Brazilian and U.S. activists came to the encounter motivated by shared expectations about the fantastic nature and concrete demands and possibilities of the diaspora. Utopian, yet pragmatically oriented, and all too aware of the foundational antiblackness laid bare in recurring manifestations of genocide—they sought to establish effective collaborative initiatives based on the concept of autonomy. For Brazilians and U.S. activists, autonomy did not have the same meaning, nor was it manifested in similar forms. Black autonomy did not seem as central a concern for black Brazilian organizers. Four factors may explain, at least partially, this seemingly more reserved approach to autonomy: (a) In contrast to U.S. blacks, black Brazilians have always been a significant demographic group, if not the majority (53.6 percent of the total population in 2015, according to the official census); (b) black collective formations, as explained earlier, were mostly underground and, over generations, had been quite effective in providing material, spiritual, and political support; (c) the Brazilian ideology of racial democracy, stressing multiracial social harmony, arguably constituted a significant obstacle to assert one's autonomous political blackness; and (d) the Brazilian dictatorial state (1964–1985) had successfully repressed openly oppositional black organizations and individuals and, thus, created added hesitation even in times of formal democracy.[12]

An important exception to this relatively reticent approach to autonomy is the emergence, in the late 1980s and early 1990s, of black women's collectives that sought independence from both black nationalist formations and feminist groups. For example, the black women's groups Geledés, founded in 1988, and Criola, founded in 1992, maintained that nationalist formations operated as if "all Blacks were men," and feminist congregations revealed a perspective that "all women are white."[13] At the same time that Geledés and Criola sought self-determination, they drew from black women's networks already in place (such as domestic workers' formal and informal associations, artist collectives, and Candomblé houses). They pursued support from and collaborations with national and international like-minded organizations, funding agencies, and various domestic and international official bureaucracies.

As practitioners of Black Power, the U.S. activists considered black self-determination a foundational and guiding principle. More explicitly than their Brazilian counterparts, they recognized that processes of exclusion and violence affected black people disproportionately and distinctively and, thus, fundamentally. A result of this recognition was the assertion of unapologetically black autonomous analytical and political agendas. Exercised for generations in a variety of communities in the United States and the black diaspora, the concept of black autonomy was systematized in Kwame Ture and Charles Hamilton's *Black Power* as

> a call for black people in this country to unite, to recognize their heritage, to build a sense of community. It is a call for black people to begin to define their own goals, to lead their own organizations and to support those organizations. It is a call to reject the racist institutions and values of this society. (Ture & Hamilton, 1992, p. 44)

Developed in political action, which included study groups that Zinzun coordinated at CAPA, this critical analysis of institutional racism acknowledged that, while operating independently of personal values and actions, it "relies on the active and pervasive operation of anti-black attitudes and practices" (Ture & Hamilton, 1992, p. 5). Zinzun's community study groups focused on the works of activist intellectuals such as Karl Marx, W. E. B. Du Bois, Mao Zedong, Frantz Fanon, Robert Williams, and Assata Shakur. As well, the seminars engaged topics with which the activists were involved in their daily practice such as surveillance policies and technologies; transnational capitalism and organized resistance against its forms of underpaid labor, especially performed by women; AIDS/HIV, cancer, birth-related problems, and health disparities more generally; and residential segregation, homelessness, and alternative housing materials.

As was the case with the political orientations of *Black Power,* the U.S. activists were not separatists. Nor were they revolutionaries in the sense that they sought to overthrow established forms of power. To the contrary, Ture bluntly reaffirmed *Black Power*'s political perspective: "It preaches reform" (Ture & Hamilton, 1992, p. 187). Black Power had a clear anticapitalist vision and wanted a society free from exploitation. Yet it argued for political programs and actions within the law, programs that sought to capture power through two related and sequential steps: first, the constitution of autonomous black local organizing; and second, the establishment of multiracial alliances. Ture reminded his readers that "all action proposed in the book is legal" (Ture & Hamilton, 1992, p. 188). CAPA's programs were squarely within the parameters of legality, as were the related black Brazilian initiatives that I analyze in the next section.

More to the point, Ture and Hamilton defined the concept of Black Power as part of a tense and productive dichotomy between, on the one hand, the assertion of autonomy, and on the other hand, the requirement of alliances with non-blacks. Regarding the first imperative of autonomy, they affirmed the following:

> Before a group can enter the open society, it must first close ranks.... By this we mean that group solidarity [acquired through the cultivation and actualization of autonomy] is necessary before a group can operate effectively from a bargaining position of strength in a pluralistic society. (Ture & Hamilton, 1992, p. 44)[14]

Thus, while there was an unmistakable emphasis on black self-determination, the ultimate goal was to operate efficiently in the legal-political world of alliances, elections, and occupation of relevant decision-making spaces in the empire-state apparatus.

The diasporic encounter made apparent the varied takes on antiblack genocide and autonomy each group endorsed. But it also marked interesting diasporic thematic convergences since the 1990s. On the Brazilian side, there have been new developments that suggest a greater attention to forms of black autonomous organizing. Some of them, like the 2001 favela movement in Rio, were direct products of the diasporic encounter; others, such as the various national black mobilizations, including the 2015 black women's march, suggest a political kinship based on the diasporically shared recognition of the imperative of black self-determination. On the U.S. side, there has been a greater engagement with and utilization of the concept of genocide. For example, in 1997, inspired by the 1951 effort anchored in *We Charge Genocide,* the National Black United Front charged the United States of Genocide before the United Nations (Vargas, 2018). Although this common genocide vocabulary cannot be credited to the encounter, it does indicate that the encounter happened at a moment when political vocabularies in the diaspora were becoming more obviously aligned. This alignment culminates with the transnational influences and reverberations of the Black Lives Matter (BLM) movement. The encounter thus constitutes a window into a moment just before new common concepts and organizational strategies begin to emerge transnationally.

One is tempted to say that Brazilians have become more attuned to autonomy while U.S. black activists have made significant inroads in the analytical and strategic development of the concept of genocide. And the fantastic pragmatic, in practice, seems to often emphasize the pragmatic to the detriment of the fantastic. Yet the ensuing two decades have been far more complex. This becomes evident in the next section's analysis of two black initiatives: one in Brazil, the other in the United States. Particularly in some of the BLM programs, the fantastic reemerges as a fundamental tool in the collective efforts to avert antiblackness.

The Popular Movement of Favelas and *Operation Ghetto Storm*

Some of the black political initiatives that emerged in Brazil and the United States in the past 25 years can be explained, at least partially, by the possibilities and challenges announced in the transnational encounter whose contours were analyzed earlier in this chapter. I will focus on two of them. First, the Popular Movement of Favelas (*Movimento Popular de Favelas*, MPF), which emerged in 2000 and disbanded a year later, not only drew deeply from diasporic knowledge but also provided inspiration and political lessons, particularly about the strategic use of autonomy, that are central to a number of black activist projects today. Second, I will focus on the 2012 *Operation Ghetto Storm* publication by the Malcolm X Grassroots Movement (MXGM).

Both initiatives helped (re)establish an epistemological and tactical terrain for subsequent political projects that embraced autonomy and recognized antiblack genocide. They made compelling cases for a type of unapologetic black worldview that rejected accepted theoretical, institutional, and practical wisdom. They configured an incisive critique and project of society that recognized the centrality of antiblackness in formations of multiracial democracy. And as authoritative as the critique presented itself, it was informed by a pulsating utopian drive according to which a new society, built on the unfettered and transhistorical black diasporic imagination, was as much of a collective dream as it was a practical imperative.[15]

The Popular Movement of the Favelas

Originated in 2000, the Popular Movement of Favelas congregated 76 of Rio's historic black communities. Characterized by a radical emphasis on collective decision making, autonomy from the traditional black movements, and a conscious attempt at experimenting with forms of radical egalitarian organizing, MPF made a considerable impact during the short time of its existence. Openly criticizing previous attempts at bringing Rio's favelas together, MPF modeled its horizontal organizational structure on the Landless Workers' Movement (*Movimento dos Trabalhadores Rurais Sem Terra*, MST),[16] inspired by liberation theology and the pedagogy of Paulo Freire. Although MPF was not as meticulous about the mechanics of representation as MST, the organization did understand the need to avoid concentrating power in the hands of a few individuals. Given MPF's oppositional politics and its accumulated knowledge about the ways in which progressive groups were surveilled, the favela activists accurately predicted that they would soon become the target of the empire-state's illegal attempts at demobilizing them. Intimidation, imprisonment on bogus charges, and assassinations were part of the Brazilian empire-state's repertoire for curbing dissident movements. Not unlike the FBI counterintelligence programs, known as COINTELPRO, that were successful at demobilizing the Black Panther Party in the United States by means of infiltration, the dissemination of false information, fabricated criminal charges, imprisonment, and assassinations (Churchill & Vander Wall, 1990), Rio de Janeiro state's military and civil police forces, and their counterinsurgency apparatuses, were known for their relentless, illegal, and brutal, often lethal tactics (Vargas, 2006). As soon as

MPF gained local, state, and national attention, its more visible activists were harassed, charged with collaborating with drug dealers, and some of their and their relatives' houses were searched and shot at. Others were constantly harassed by phone and followed by agents in unmarked cars.[17]

Even though MPF collaborated with a few NGOs in Rio,[18] it sought independence from traditional black associations, progressive research institutions, the government, and leftist formations. MPF members engaged demands formulated in and by the favelas, and not in universities, middle-class study groups, and other organizations that demanded social, cultural, and material capital. This explicit emphasis on favela self-determination, while effective in establishing an autonomous agenda of research and organizing, also meant that MPF found itself often isolated and excluded from decision-making forums dominated by traditional black and progressive blocs. For example, during the preparatory meetings for the 2001 World Conference Against Racism (WCAR), MPF members were instructed to negotiate favela-specific demands, including their legal recognition as urban *quilombos* and their right to self-determination through elected neighborhood associations; the end of police occupation; and their constitutional and human right to housing, education, and health. Yet they were systematically rejected by members of the traditional MNU, established NGOs, and leftist parties. While MPF members expressed their frustration at what they considered the elitism of such organizations, those organizations, in turn, looked at MPF with suspicion. Much of that suspicion, I found out by talking with those involved, derived from the assumption that MFP was a product of—or at least was involved with—the favela drug cartels and their desire to influence the political scene. Such erroneous assumptions not only were widespread among non-favela activists, including black activists, but they also were frequently repeated on television, in newspapers, and in academic publications.[19]

Although MPF's emphasis on self-determination resulted in exclusion from established forums of black politics, it generated a series of favela-centered programs that became sources of inspiration for activists in other cities and time periods. The Zinzun Center for monitoring police abuse was one such program. Established in the Jacarezinho favela, it followed many of the strategies used at CAPA. Using cameras and a network of local researchers who recorded incidents and interviewed victims, the Zinzun Center aimed not only to record and systematize data on police activity in the neighborhood but also to transform cases of police abuse into opportunities to create public discussions on the multiple social problems affecting the favela.[20] By initially focusing on police abuse, favela activists were able to relate it to residential segregation, unemployment, and deficient public education, transportation, sanitation, water supply, and health care systems. Such public forums reinaugurated a concerted attempt at debating, from a critical perspective, the historical and contemporary challenges faced by black people in the alleged racial democracy.

Today, there are several examples in Brazil of black autonomous initiatives that target police abuse and attempt to implement community policing, in major urban centers and in smaller cities. These programs are usually run by local youths affiliated with an expanding network of black activists who share information and strategies, even though they are not always in agreement about how to approach the state machine and the political establishment.

Some consider it legitimate and effective to pressure elected representatives and participate in elections; others claim the electoral system, and indeed the empire-state apparatus that sustains and depends on it, is irremediable and antiblack and, thus, should be avoided; and still others adapt different strategies as the context change.[21]

Using the framework of antiblack genocide, drawing from the increasingly available government and independently generated data on police brutality and lethality and, above all, producing their own data and means of publicization via newspapers, magazines, and websites, these police monitoring programs display an obvious ideological kinship with MPF and, by extension, with the 1993 diasporic encounter. As activists told me, MPF's methods and performance were fundamental references for imagining and enacting their black political projects. MPF emphasized autonomy, on public displays of black political agency, on favela-based concepts and programs, and on an unprecedented fearlessness with which the empire-state machine was confronted and the public sphere was occupied: Such actions provided blueprints to be tested and modified in the activist practice. Examples of MPF's impactful innovations included its protests in majority-white shopping malls, where they featured black children from favelas, and transformed the discomfort their presence caused, as platforms from which to address patterns of residential segregation and exclusion; its rotating weekly meetings, always in favela communities, rather than at the traditional downtown, university, NGO, and friendly government spaces; its ability to use the media monopoly to its favor by creating public visibility as a shield against police harassment and other government forms of abuse; its use of daring (and often dangerous) methods of denouncing the state of war in favelas by placing their activists in the recurring crossfire between drug dealers and the police, and filming the scenes; its ability to attract and maintain collaborations with international foundations, political parties, and activist groups; and its insistence on conceptually relating myriad antiblack social phenomena to each other.

Operation Ghetto Storm

The 2012 publication of *Operation Ghetto Storm* marked an important moment in U.S. black politics. *Operation Ghetto Storm* places itself squarely in the tradition of engaged research as conducted by Ida B. Wells, which resulted in, among others, the 1882 publication *On Lynching,* and as exemplified by the above-mentioned *We Charge Genocide*, published in 1951. *Operation Ghetto Storm* followed the well-publicized 2012 assassination of an unarmed black youth, Trayvon Martin, by a self-appointed neighborhood crime watch captain,[22] claiming,

> the practice of executing Black people without the pretense of a trial, jury, or judge is an integral part of the government's current overall strategy of containing the Black community in a state of perpetual colonial subjugation and exploitation. (Akuno, 2012, p. 3)

This assertion of the antiblack nature of the U.S. empire-state resonated with conceptual foundations of the 1993 diasporic encounter, particularly, the critical reevaluation of U.S. black people's belonging and citizenship. Although MXGM's activists were not part of that encounter, the critical appraisal of the notion of the United States as a viable republic for black people goes hand in hand with a diasporic consciousness. In other words, the concept of genocide facilitates and reveals a transnational perspective because it defines the U.S. empire-state as the primary perpetrator of systematic antiblack oppression and death and, thus, requires legal and political forums outside of the United States where claims can be evaluated and judged. The concept of genocide also invites and draws from analyses across geopolitical boundaries and, thus, establishes potential bridges of recognition and collaboration. When Zinzun and Brazilian activists told each other that "you are me, and I am you," they were asserting the fundamental realizations that their respective empire-states of origin systematically excluded and disregarded them. The cultural and everyday mechanisms by which black people's lives are rendered disposable may vary across the diaspora, but it is indisputable that black people consistently and disproportionately die early of preventable causes—either directly produced by the empire-state, such as police lethality, or by negligence, such as deficient or inexistent medical care and exposure to environmental toxins (Bullard, 2000; Santos, 2008; Vargas, 2008; Waiselfizs, 2012). Thus, MXGM's report can be linked to a diasporic tradition of critical analysis and action that was represented in, but exceeds, the Rio encounter between U.S. and Brazilian activists.

By registering the killing of at least 313 black people in 2012—"or one *Every 28 Hours!*"—*Operation Ghetto Storm* makes the case that, were it not for MXGM's investigation, such patterns would be ignored both by officials and by the larger public. Implied is the idea that the systematic extralegal assassination of black people by the police has become normative and culturally acceptable. In an attempt to denaturalize both black U.S. citizenship and the assumption that the U.S. democracy is compatible with black well-being, *Operation Ghetto Storm* asks,

> How can the supposedly "most democratic" country on Earth be the largest jailer on the planet? What types of "legitimate" democratic processes result in nearly half of the country's prison population being Black, while Black people only comprise 13% of the total population of the United States? (Akuno, 2012, p. 3)

The conclusion can only be that the democratic regime of the United States (and, for that matter, of Brazil) not only tolerates, but expects, produces, and normalizes the systematic suffering and premature death of its Afro-descended population (Vargas & James, 2013).

In the black diaspora, particularly in the United States and in Brazil, the moment at which *Operation Ghetto Storm* emerged was critical for black mobilization. The year 2013 marked the beginning of the BLM movement; there appeared several local and state black-led collective efforts for discussing and opposing police brutality; and public protest against police abuse not only became widespread but also gained the attention of elected officials, politicians running for office, and cultural producers.[23] Although the language of genocide did not feature centrally in BLM and related programs, the emerging collectives offered comprehensive analyses that connected police brutality to a host of antiblack social facts such as residential hypersegregation, chronic unemployment, punitive and deficient schooling, vulnerability to preventable disease and blocked access to treatment, and exposure to environmental toxins.

Embracing the fantastic—"to dream, to unhinge the chains that bind us"[24]—the BLM's Black Futures Month featured creative and analytical works. Stressing the need to engage unbounded creativity, Black Futures Month "imagines a joyful, free and liberated future for black folks." Expressed through various artistic forms, such imaginings encompass autonomous farming projects that provide alternatives to food insecurity and land injustice; a reflection on the intersections of blackness and Islam; a reckoning of the interconnections between environmental racism and health disparities affecting black people disproportionately; and Barbara Ransby's "Revolutionary Musings," considering, once again, the challenge Martin Luther King Jr. posed in his speech "A Radical Revolution of Values," where he indicated the need of conquering the "giant triplets of racism, extreme materialism and militarism" (Black Lives Matter, 2018).

CONCLUSION: CHALLENGES OF THE FANTASTIC PRAGMATIC

The 1993 encounter between black U.S. and Brazilian activists serves as a conceptual matrix with which we can identify and make sense of diasporic practices in disparate times and spaces.[25] Neither a beginning nor an ending, the encounter is like an in-between space and time, connected to past, present, and future collective initiatives constituting a diasporic black radical tradition (Robinson, 2000). Thus, the analytical challenge is to evaluate whether, how, when, and why the tradition's reemergence occurs. Such creative, theoretical, and practical events take multiple and sometimes unexpected forms. Yet they all suggest an unrelenting collective determination against the profoundly dehumanizing effects of antiblackness.

When we focus on two concepts of this political imagination—autonomy and genocide—it becomes evident that the encounter drew from previous attempts at engaging these concepts in different places and generated offshoots not immediately recognizable as related to previous diasporic efforts. Following the encounter, whereas the Brazilians became apparently more intentional about forging modalities of autonomy, the U.S. activists renewed their critical practices related to the recognition of antiblack genocide.

Autonomy can be linked more easily to the elements of the fantastic; the pragmatic is associated more directly with the necessity of coming to terms with, and deterring, the multiple facets of genocide. Yet, as evidenced in the diasporic forms of agency described earlier, there are pragmatic arguments about the need for autonomy. For example, autonomy allows for the elaboration of one's own conceptual and political vocabulary, based on specific, and often untranslatable, black experiences. CAPA, MPF, MXGM, and BLM activists recognized the imperative of grappling with the incommensurable manifestations of antiblack police lethality, residential segregation, and early death by preventable causes. "Incommensurable" means that, although the activists acknowledge parallels between their experiences and those of non-blacks who are oppressed, they stress the need to come to terms with the defining and singular characteristics of black experiences.[26] Furthermore, black autonomy, in line with Black Power, constitutes a necessary step before entering an arena of political alliances for electoral ends (Ture & Hamilton, 1992).

Besides the need for survival programs, as those elaborated by the Black Panthers Party, there is a related necessity of creativity and utopian musings. Such a necessity is a product of the realization that it is only via radically new perspectives on human sociability that the ubiquitous, structural, and transgenerational facts of antiblackness can be grasped and opposed. It is a product of understanding the very bases of social cognition, and their corresponding institutional apparatuses and cultural assumptions, as rooted in, dependent on, and reproducing antiblack genocide. The pragmatic fantastic needs to be not only transcendental but also adaptable; at times confrontational; underground at other times. Conditions permitting, it travels (literally and/or figuratively); and depending on the political context, it stays dormant, operating at frequencies not immediately detectable, only to then reemerge, transfigured and transfiguring. The fantastic pragmatic thus lives at times (and places) stressing more the pragmatic, and at times (and places) stressing more the fantastic. This chapter has shown that the elements of the fantastic pragmatic shared, elaborated, and put into practice through mostly the unspoken lingua franca of the diaspora, are as evident and diffused today as they were prior to the 1993 encounter.

An intriguing challenge is this: How are black U.S. and Brazilian initiatives able to reconcile the two seemingly contradictory orientations of autonomy and alliances with non-black collectives? On the one hand, diasporic activists assert their practice on notions of black autonomy. A potential effect of reconciling these two orientations is that, as is usually required in multiracial political alliances, the emphasis is on common denominators of oppression. When that is the case, the singular—and, some would say, incommensurable (Wilderson, 2005; Yancy, 2004)—aspects of antiblackness may have to be bracketed in favor of a political platform that stresses shared and relatable experiences. This may be one reason for the difficulty in calling out antiblack genocide in a context of multiracial political alliances.

DEEPENING OUR UNDERSTANDING

Analyze a black political formation that emphasizes autonomy and multiracial political alliances. This political formation can be based in Brazil, the United States, or anywhere else in the diaspora. To do so, take into account the following steps and questions:

1. Gather and analyze information on the political formation's program and actions. You can do so by consulting the Internet, newspapers, magazines, documentaries, films, books, and various archives.
2. Which of these would you consider to be about or connected to environmental justice? Explain.
3. How does this political formation affirm and practice autonomy, and how does it go about seeking and establishing collaborations with non-black groups?

4. How does or could the formation's emphasis on autonomy influence the ways in which it conceptualizes and tries to achieve its environmental justice goals or demands?
5. How does the formation's emphasis on alliances with non-blacks influence the ways in which it conceptualizes and tries to achieve its environmental justice goals or demands?
6. Does this formation stress antiblack genocide? Why or why not? How is this concept connected to environmental justice?
7. How does the notion of the fantastic pragmatic enable the analysis and evaluation of the political formation on which you focused?

References

Akuno, K. (2012). *Operation Ghetto Storm.* http://www.operationghettostorm.org

Alston, L. J., Libecap, G. D., & Mueller, B. (1999). *Titles, conflict, and land use: The development of property rights and land reform on the Brazilian Amazon frontier.* Ann Arbor: University of Michigan Press.

Anderson, C. (2003). *Eyes off the prize: The United Nations and the African American struggle for human rights.* New York: Cambridge University Press.

Black Lives Matter. (2018, June 2). About [Black Lives Matter]. https://blacklivesmatter.com/about/

Bloom, J., & Martin, W. E., Jr. (2016). *Black against empire: The history and politics of the Black Panther Party.* Berkeley: University of California Press.

Bullard, R. (2000). *Dumping in Dixie: Race, class, and environmental quality.* Boulder, CO: Westview Press.

California man leads battle to "off" roaches.(1973, June 14). *Jet,* 32.

CETIM (Centre Europe-Tiers Monde / Europe-Third World Center). (1999). *Land concentration in Brazil: A politics of poverty.* Genève, Switzerland: CETIM https://www.cetim.ch/land-concentration-in-brazil-a-politics-of-poverty/

Churchill, W., & Vander Wall, J. (1990). *The Cointelpro Papers: Documents from the FBI's secret wars against dissent in the United States.* Cambridge, MA: South End Press.

Covin, D. (2006). *The Unified Black Movement in Brazil.* Jefferson, NC: McFarland & Co.

Desmond, M. (2016). *Evicted: Poverty and profit in the American city.* New York: Broadway Books.

Gooding-Williams, R. (1993). *Reading Rodney King/reading urban uprising.* New York: Routledge.

Hartman, S. (1997). *Scenes of subjection: Terror, slavery, and self-making in nineteenth-century America.* New York: Oxford University Press.

Horne, G. (2013). *Black revolutionary: William Patterson and the globalization of the African American freedom struggle.* Urbana: University of Illinois Press.

Hull, G., Bell Scott, P. , & Smith, B. (1982). *All the women are white, all the blacks are men, but some of us are brave.* Old Westbury, NY: Feminist Press.

James, J. (1999). *Shadowboxing: Representations of black feminist politics.* New York: Palgrave Macmillan.

Jung, M.-K. (2015). *Beneath the surface of white supremacy: Denaturalizing U.S. racisms past and present.* Stanford, CA: Stanford University Press.

Nascimento, A. D. (1989). *Brazil, mixture or massacre?: Essays on the genocide of a black people.* Dover, MA: Majority Press.

Paixão, M. R. (2010). *Relatório das desigualdades raciais no Brasilç 2009–2010.* Rio de Janeiro, Brazil: Garamond Universitária.

Patterson, W. (1951). *We charge genocide: The historic petition to the United Nations for relief from a crime of the United States government against the Negro people.* New York: Civil Rights Congress.

Patterson, W. (1971). *The man who cried genocide: An autobiography.* New York: International Publisher.

Pinto, A. F. (2010). *Imprensa negra no Brasil do século XIX.* São Paulo, Brazil: Selo Negro.

Robinson, C. (2000). *Black Marxism: The making of the black radical tradition.* Chapel Hill: University of North Carolina Press.

Santos, S. B. (2008). Brazilian black women's NGOs and their struggles in the areas of sexual and reproductive health. Ph.D. dissertation. University of Texas, Austin.

Ture, K., & Hamilton, C. (1992). *Black Power: The politics of liberation.* New York: Vintage Books.

Vargas, J. H. (2005, January/June). Apartheid Brasileiro: Raça e segregação residencial no Rio de Janeiro. *Revista de Antropologia, 48*(1), 75–131.

Vargas, J. H. (2006). *Catching hell in the City of Angels: Life and meanings of blackness in South Central Los Angeles.* Minneapolis: University of Minnesota Press.

Vargas, J. H. (2008). *Never meant to survive: Genocide and utopias in black diaspora communities.* Lanham, MD: Rowan & Littlefield.

Vargas, J. H. (2010a). Geographies of death: An intersectional analysis of police lethality and the racialized regime of citizenship in São Paulo. *Ethnic and Racial Studies, 33*(4), 611–636.

Vargas, J. H., & James, J. A. (2013). Refusing blackness-as-victimization: Trayvon Martin and the black cyborgs. In G. A. Yancy (Ed.), *Pursuing Trayvon Martin: Historical contexts and contemporary manifestations of racial dynamics* (pp. 193–204). Lanham, MD: Lexington Books.

Vargas, J. H. (2014, December 24). Black disidentification: The 2013 protests, Rolezinhos, and racial antagonism in post-Lula Brazil. *Critical Sociology, 42*, 1–15.

Vargas, J. H. (2017). Por uma mudança de paradigma: Antinegritude e entagonismo estrutural. *Revista de Ciências Sociais*, 48 (2), 83–105.

Vargas, J. H. (2018). *The denial of antiblackness: Multiracial redemption and black suffering.* Minneapolis: University of Minnesota Press.

Waiselfizs, J. J. (2012). *Mapa da violência 2012: A cor dos homicídios no Brasil* (CEBELA, FLACSO). Brasília, Brazil: SEPPIR.

Wilderson, F. (2005, January). Gramsci's black Marx: Whither the slave in civil society? *We Write, 2*(1), 1–17.

Yancy, G. (2004). *Who is white? Latinos, Asians, and the new black/nonblack divide.* Boulder, CO: Lynne Rienner.

Yancy, G., & Jones, J. (2013). *Pursuing Trayvon Martin: Historical contexts and contemporary manifestations of racial dynamics.* Lanham, MD: Lexington Books.

Notes

1. See Vargas (2010b), chap. 4.
2. For a collection of essays on the 1992 revolt, see Gooding-Williams (1993).

3. On the persistence of relatively high-priced residential rentals in segregated areas, see, for example, Desmond (2016).

4. Although my original research on the Coalition Against Police Abuse was conducted while collaborating with the organization between 1996 and 2006, since 2012 I have been immersed in the CAPA Collection at the Southern California Library. I thank Michele Welsing and Yusef Omowale at the library for their support, orientation in navigating the archives, and critical insights. At the library, Dylan Rodriguez, Damien Sojoyner, and Shana Redmond also provided encouragement and shared their always generative thoughts.

5. On the ways in which antiblackness and social policy impacted each other in the first half of the 20th century in Brazil, see, for example, Davila (2003). For how post-abolitionist policies, legal practices, and social norms reformulated black subjugation in the United States, see, for example, Hartman (1997).

6. See, for example, Covin (2006).

7. The black press of the 19th century is another example of black autonomy. See Pinto (2010).

8. *We Charge Genocide* was an editorial success, selling more than 45,000 copies, 5,000 of which were sold in its first week (Vargas, 2010b, p. 3). For an analysis of the research and political genealogy of *We Charge Genocide*, see Anderson (2003). For a detailed account of the ways in which William Patterson's political trajectory intersected with the publication of *We Charge Genocide*, see Patterson's autobiography (Patterson, 1971), and Horne (2013).

9. On the conceptual and practical aspects of the Black Panther Party, see, for example, Bloom and Martin (2016).

10. This, of course, is a reference to the subtitle of Hull, Bell Scott, and Smith (1982), *But Some of Us Are Brave*.

11. This quotation exemplifies the only reference to antiblack processes explicitly named as such in *Black Power*, which, given the many examples about black people's singular experiences that appear in the book, indicates a vocabulary choice rather than a conceptual awareness.

12. For an account and analysis of the black diasporic fantastic pragmatic informing contemporary efforts at securing cooperative zones and collective land stewardship, see, for example, Akuno and Nangwaya (2017).

13. Founded in 1984, MST mobilized the people most affected by Brazil's profoundly unequal and unjust pattern of land ownership. At that time, it was estimated that almost 80 percent of all farm land belonged to latifundia (Alston, Gary, & Mueller,1999, pp. 67, 68). Brazil is an extreme case of land concentration in the hands of major landowners: 2.8 percent of landowners own more than 56 percent of arable land, 45 percent of the total area is occupied by only 1 percent of agricultural holdings. Moreover, 50 percent of small holdings have access to only 2.5 percent of the total area and employ about two-thirds of the rural population (CETIM, 1999). The Workers' Party administrations were not effective in changing this pattern of land concentration. See Gini for land distribution.

14. For example, the house of Rumba Gabriel's mother was shot at in 2001. Since Gabriel did not have any declared local enemies, it was deduced that the police carried out the shooting. The president of the Jacarezinho neighborhood association, elected after an unprecedented transparent process, Gabriel was one of the most visible members of MPF. He was later arrested on bogus charges of collaborating with drug dealers. To this day, more than 17 years later, he is still attempting to have his name cleared (Vargas, 2006). It is also telling that, similar to what happened when I began working at CAPA, when I became involved with MPF I started receiving threatening anonymous phone calls. I was also surveilled and followed by someone in an unmarked car. The person doing the surveillance was a civil police agent, who tried to bar my trip back to the United States at the Rio international airport (Vargas, 2010a).

15. For example, MPF held strategic meetings and used some of the research conducted by the Brazilian Institute for Social and Economic

Analysis (*Instituto Brasileiro de Análises Sociais e Economicas*), a well-known, progressive organization based in downtown Rio.

16. For CAPA's manual of community organizing, see the appendix in *Catching Hell in the City of Angels* (Vargas, 2006). Like MPF, the Zinzun Center was short lived. Although it was able to begin systematizing police abuse cases in computers donated by the U.S. activists, and provided training for young people in collecting and recording the data, the office was shut down when MPF activists began to be harassed, charged on bogus accusations, followed, imprisoned, and shot at.

17. For an analysis of the ways in which some black Brazilian organizations decided to reject the electoral system, see Vargas (2017). For an analysis of Brazil and the United States as empire-states, see Jung (2015) and Vargas (2018).

18. There are various publications on this topic. See, for example, the collection of essays in *Pursuing Trayvon Martin* (Yancy & Jones, 2013).

19. That 2016 Democratic presidential candidates Bernie Sanders and Hillary Clinton addressed police lethality and mass incarceration in their campaigns is a reflection of the political moment on which BLM had a considerable influence. See Vargas (2018).

20. Phrases in quotations marks appear in Black Lives Matter (2018), under the headings "What We Believe" and "Black Futures Month."

17

FROM DUMPING TO DISPLACEMENT

New Frontiers for Just Sustainabilities

Julian Agyeman

Stephen Zavestoski

This chapter is concerned with the following urban sustainability paradox: Cities are simultaneously incubators of innovation and experimentation for sustainable societies *and* propagators of new forms of environmental injustice. We draw on our backgrounds as a geographer/urban planner and an environmental sociologist to offer insight to those endeavoring to understand and resolve this paradox. At the foundation of our analysis is the question of how neoliberal rollbacks of state and public-sector functions concede to the private sector more and more of the processes of planning and shaping urban spaces and places. We believe that sustainable neighborhoods should be for everyone, especially those who have had to endure the legacies of 20th-century urban planning failures: concentrated air pollution, contaminated brownfields, lack of greenspace, and automobile dependency linked to inadequate public transit. Yet as 21st-century urban planning—which is often motivated by buzzwords like livable cities, smart cities, carbon neutral cities, and resilient cities—reshapes the urban landscape, it appears that not everyone is benefitting equally.

Sustainability has become a commodity that sells. And if left to market forces, it is a commodity that will be distributed unevenly. Take, for example, the concept of "Complete Streets," which reimagines streets as places for multiple forms of transportation and street life—a vision that is often translated into bike lanes, wide sidewalks amenable to sidewalk cafes and more green space, and traffic-calming measures that create safer and more inviting streetscapes. Complete Streets represent a dream for developers and realtors who envision live-work lofts and high-end condos above the newly transformed streeetscape. Some might contend that Complete Streets are intuitive and even compelling. Who could be opposed to making streets safer for pedestrians and cyclists or to creating more vibrant urban spaces? But underneath lies a darker tale of exclusion.

Imagine, for instance, a city that uses sustainability infrastructure, like the placement of bike racks, to stop homeless people from sleeping under a bridge. Imagine a tree-lined, walkable, cyclable, quiet neighborhood. Then take a look in a realtor's window. You're likely to see listings that describe bike lanes and farmers markets as desirable amenities that justify the outrageous rents and home prices. It is in precisely such neighborhoods that bike lanes, a key strategy in urban health and sustainability, are being called "gentrification superhighways." The WalkScore app, which measures neighborhood walkability, and the BikeScore app, which measures neighborhood cyclability, are Big Data tools owned by the residential real estate company Redfin.

Now that we've sketched out some instances of the urban sustainability paradox, the problem is clear. Who can disagree with "neighborhood greening"; making cities more livable; or creating Complete Streets, "streets for people," or "livable streets"? The health, environmental, and economic arguments for green neighborhoods are compelling. However, the downside being experienced in cities around the world, as neighborhoods become greener, is a profound shift in neighborhood demographics. While not new, more recent concerns with the impact of street-level sustainability initiatives on

---

Photo 17.1
Photo by Kelly Grindstaff.

gentrification, and the displacement of lower income residents, who are often disproportionately residents of color, are a significant departure from earlier days when the focus of environmental justice scholarship and activism in cities was related primarily to disproportionate burdens: toxic threats and locally unwanted land uses. What are the new "neighborhood greening" challenges, and how can communities begin to understand and fight them?

FRONTIERS OF ENVIRONMENTAL INJUSTICE: GREEN GENTRIFICATION AND GREENLINING

Changing threats to neighborhoods under the logic of neoliberal cities have shifted the intellectual and activist focus of environmental justice from *dumping* of unwanted materials to the *displacement* of lower income people. This shift has necessitated different forms of scholarship, community awareness, and activism.[1] This has meant that the constituent parts of the move toward more *just sustainabilities* (Agyeman, 2013), such as environmental justice, economic justice, food justice, transportation justice, spatial justice, Right to the City, and affordable housing are coalescing in resistance to trends like "green," "environmental," and "super gentrification."[2] Animating these trends is a shift from the paradigm of *urban planning in the public interest* into one of *urban development in the private interest* (Brenner & Theodore, 2002; Hackworth, 2007; Peck et al., 2009; Raco, 2012; Weber, 2002). Put differently, the logic of neoliberal cities treats cities as economic machines and identifies the spaces within them most suitable for the extraction of profit. This shift has led geographer David Harvey (2016) to lament that "we have reached the insane side of urbanisation where we are not actually...building cities for people to live in. We are building cities for people to invest in." The investors, according to Harvey, are the ultra rich who "can't think of any other place to store their money [than] in private property."

This private-interest paradigm has become a primary driver of gentrification and displacement. Under the guise of the uber-narratives of sustainability, placemaking, Complete Streets, and, more recently, smart cities, urban planning has, in many respects, embraced New Urbanism. Beginning in the early 1980s, New Urbanist thinking heralded an expert-focused, design-led set of physical and structural changes in order to "green" neighborhoods by making them more walkable, cycleable, sociable, quiet, and "sticky" (i.e., places where people want to "hang around"). Indeed, New Urbanism is a major part of the "complicity of planning and design in furthering neoliberal urban redevelopment" (Hanlon, 2010, p. 95).

These sustainability characteristics are, after all, those that attract high-wealth urban dwellers with the potential to infuse money into a city's dwindling tax base and wider economy. Under this same paradigm, more deep-seated *sociocultural* questions such as who *belongs* in the changing neighborhood, who gets to decide what a street is (and what makes it "complete"), what a neighborhood or ultimately a city is, or, more important, what it can *become*, are conveniently omitted in favor of expert and physical, design-led changes. These questions of *belonging* and *becoming* are at the heart of the quest for more just sustainabilities.

In short, the ways in which Complete Streets narratives, policies, plans, and efforts are being envisioned and implemented are systematically reproducing and even amplifying many of the urban spatial and social inequalities and injustices that have characterized U.S. cities for the past century or more. Where we once had *redlining* (banks delineating areas where investment would be avoided, based on race), we now have *greenlining* (neighborhoods where sustainability "amenities" like bike lanes and farmers markets have an effect on housing prices and, in turn, neighborhood demographics). A key shift leading to the production of these new urban landscapes of contention and exclusion has been the flow of capital seeking safe investment havens, and the single-minded pursuit of economic development in planning processes. Whereas "planning" used to fall under the purview of cities, city officials, and planning professionals, it is increasingly developers, capital, and the potential for capital investments that has convinced many city officials that urban planning is essentially a strategy for upscale residential and economic development.

Shifting Activisms

The implications of this paradigm shift for all progressive activists, and in particular for those in low-income and communities of color, is that campaigns for equity and justice must now understand and target the sources and flows of capital. Yet the forces of globalization mean that even mid-sized cities in "middle America" are receiving capital investments from individuals and corporate entities the identities of which are increasingly difficult for activists to confirm. As one Boston-based activist explains, "[I]n the 80s we knew where the money was coming from, now it's more insidious with the flows of foreign, faceless capital."

Such trends compel us to ask what role movements for just sustainabilities have in the neoliberal city and how, or whether, these trends forge new partnerships between environmental justice activists and other movements such as those focused on housing, labor, immigration, education, or health. How can these movements, whether individually or collectively, intervene in the logics and the forces shaping the future of cities? In the next section we take a first step toward answering these questions by exploring through three vignettes how private interests are shaping urban planning.

URBAN DEVELOPMENT IN THE PRIVATE INTEREST: THREE VIGNETTES

Perhaps the best current example of the paradigm shift from *urban planning in the public interest* to *urban development in the private interest*, and the most controversial, is Sidewalk Toronto:

> [A] joint effort by Waterfront Toronto and Alphabet's Sidewalk Labs to create a new kind of mixed-use, complete community on Toronto's Eastern Waterfront, beginning with the creation of Quayside. Sidewalk Toronto will combine forward-thinking urban design and new digital technology to create people-centred neighbourhoods that achieve precedent-setting levels of sustainability, affordability, mobility, and economic opportunity. (Sidewalk Toronto, 2018)

Alphabet, a Google spinoff, is "a holding company that gives ambitious projects the resources, freedom, and focus to make their ideas happen." In effect, they are proposing to build a city from the ground up—not only physical structures and user-facing apps, but municipal protocols themselves—the "operating system" of cities, like zoning and procurement. Included in their vision are "subterranean utility channels filled with robots whisking away garbage" (Sauter, 2018). This is a fundamental change in the paradigm of urban planning, as we suggested earlier. As Agyeman and Claudel (2018) note:

> The proposal includes remarkable visions of an urban future, from autonomous cars to micro housing to adaptive weather technology. Many of these are desirable, efficient and exciting. And yet there are legitimate concerns about Alphabet's financial gains, about users' privacy, and about setting a global precedent. Will cities compete to host Alphabet's next city? How can the work of urban theorists inform our collective decision-making? How can a private company plan in the public interest?

The question we posed earlier—*urban planning in the public interest* versus *urban development in the private interest*—is central to academic and activist debates happening in Toronto and around the world.

Another example is the "Great Amazon Scramble." The online retailer and tech company Amazon announced in September 2017 that it was planning to build a second headquarters or HQ2. It was to be in North America, supplementing the current Seattle headquarters, but where? Amazon's announcement, together with a request for proposals from governments and economic development organizations, attracted a frenzied scramble that garnered serious attention from more than 200 cities in Canada, the United States, and Mexico. Slated to be home to 50,000 workers, Amazon's HQ2 was estimated to bring $5 billion in new construction to whichever city the tech giant eventually selected. Some states and cities offered huge swaths of land and the lure of big tax incentives totaling billions of dollars.

But what might be the hidden cost of HQ2 to the "lucky winner" of Amazon's largesse? Amazon employs around 45,000 people in Seattle, housed in more than 30 buildings close to downtown. This has led to a construction frenzy, but new housing hasn't kept up with the demand that has been created, leading to sharply rising housing costs, with rents now close to those in Boston and New York and home prices growing faster than in any other large U.S. city. In May 2018, with cities vying to host HQ2 watching closely, Seattle's city council voted for an "Amazon Tax," a head tax on the city's largest employers, expected to raise $75 million annually, to help build critically needed affordable housing. The rest was to go to homelessness support services. But after intense pressure from Amazon and Starbucks, the city council repealed the tax a month later, sending a clear message to cities vying for HQ2 regarding their prospects for negotiating with Amazon to support city services and infrastructure. *Philadelphia Inquirer* columnist Will Bunch, referring to Philadelphia's attempt to win the Great Amazon Scramble, expressed a fear of "soaring rents and a crisis of affordable housing, pricing out young artists and dreamers, and crushing any and all cultural diversity and vibrancy" (Bunch, 2017). This erasure is the very essence of gentrification.

Our third vignette focuses on San Jose and its relationship with Google, a relationship that the following headlines suggest is seeing increasing press attention and local community resistance: "Google Transformed Mountain View, Is San Jose next?" (Li, 2019), "San Jose Approves Land Sale for Google Expansion Despite Protest" (NBC Bay Area staff, 2018), and "Google Says It's Close to Owning Enough Downtown San Jose Properties for 'Viable' Development" (Avalos, 2018). By "viable" the company means "viable" transit-oriented development, a transit-oriented community near the city's Diridon Station, a transit hub for Santa Clara County and Silicon Valley. Transit-oriented development or TOD has long been a mantra for New Urbanists and is generally well regarded by urban planners. Typically, TOD is high density with a mix of uses and is designed to be more walkable, with less of a focus on automobile transportation. Despite Google's pledge that 15 to 20 percent of the homes would be "affordable," the potential Google project has created a backlash among locals who fear the incoming workers will push housing further out of their reach. At present, the median value of a single-family home in the San Jose metropolitan area, which includes parts of Silicon Valley, is $1.27 million, which makes it the most expensive metropolitan area in the country (National Association of Realtors, 2018). Local action group, Silicon Valley Rising, a coalition of labor, faith leaders, community-based organizations, and workers trying to persuade the tech industry to build an inclusive middle class in Silicon Valley, notes:

> The tech boom has driven San Jose rents sky-high. Between 2009 and 2015, the inflation-adjusted average rent for an apartment jumped by 32.2%. Yet over that same time, adjusted median incomes for renters have actually declined 2.8%. This has forced families to cram together in small apartments, move away to cheaper places, make unhealthy trade-offs between rent and other essentials like food and prescriptions, or sleep on the streets. (Silicon Valley Rising, n.d.)

What these three vignettes show is that a new generation of global companies sees the city, or at least the neighborhood, as a new frontier in need of "fixing," as being amenable to private-sector "solutions" in the face of public-sector "failures." Their aim is to sidestep the "messiness" of politics and to frame opposition to their grand schemes as Luddite or backward looking. But these companies are not merely out to remake cities into desirable places for their employees. Their forays into urban planning are also about the possibility of new ways of extracting value. By planning cities the way they want, they can ensure "smart city" efficiencies such as one-hour deliveries, while also gathering enormous amounts of data that can be sold and used for advertising and marketing purposes. Will they succeed? To emphasize the scale of the challenge, and the need to intervene at the level of market dynamics (rather than street-level contestations over space and the shape of urban sustainability), next we turn to bicycles and gardens as specific forms of urban transformation and resistance.

DIFFERING FORMS OF URBAN TRANSFORMATION AND DIFFERING FORMS OF RESISTANCE

As capital and corporate interests increasingly shape urban planning, new forms of resistance are required to demand environmental, spatial, and economic justice. The following sections illustrate this by drawing on recent research into urban transformations and the resistance they engender. The analysis aims to support our contention that struggles for just sustainabilities play out on a new frontier that requires an understanding of the multiple scales through which capital flows. As neoliberal urban logics co-opt the language of sustainability to transform streets, neighborhoods, and entire cities and regions in the direction of further capital accumulation, forms of resistance must similarly learn to scale and flow in dynamic ways.

Bicycles as Transportation and Transit: Can Retrofitting Streets Retrofit Justice?

From bisecting once-thriving neighborhoods with superhighways to facilitate the flow of cross-city traffic to investing in sleek suburb-serving light rail at the expense of bus service for car-free minimum wage earners, the legacy of nearly a century of automobile-centric urban development has not been kind to communities of color. Not only have automobile-centric policies disconnected communities of color, they have also exposed them to air pollution levels linked to heightened incidence of asthma (Cook et al., 2011; Li et al., 2011) and an average annual mortality rate due to asthma that is six times higher in black as compared to Hispanic and white children (Arroyo et al., 2017).

Reversing the car-normative paradigm in urban planning and development, one might assume, would begin to undo the injustices and inequalities produced by the automobile-centric century. Just such a reversal appears to be underway in what some describe as the Complete Streets revolution (*Transportation Alternatives*, 2007). Complete Streets are designed "to enable safe access for all users [so that] pedestrians, bicyclists, motorists and transit riders of all ages and abilities...[are]...able to safely move along and across a complete street" (McCann & Rynne, 2010, p. 3). According to Zavestoski and Agyeman (2014), the enthusiasm for Complete Streets has led mayors, community organizers, and planners to proclaim the potential of Complete Streets to do everything from address childhood obesity to nurture diversity.

Ultimately, Zavestoski and Agyeman (2014) are interested in problematizing Complete Streets by asking, "What mistakes might we be making in assuming that redesigning streets (with the goal of providing safe access to all users) can sufficiently address the broader historical, political, social, and economic forces shaping the socioeconomic and racial inequalities embedded in and reproduced by the spaces we call streets?" (p. 4). The resistance by the environmental justice movement to disproportionate burdens of toxic waste rests on the solid footing of normative claims that toxic waste is bad. But Complete Streets rhetoric tends to be woven into narratives of "smart" and "sustainable" cities employed not just by urban planners but also by Google, Amazon, and other companies entering the urban planning business. Consequently, movements for just sustainabilities must develop nuanced positions when questioning the implicit normative claims that Complete Streets, and the livability and sustainability they promise to deliver, are universally good.

Questioning might begin with an inquiry into whether neighborhoods in poorer parts of a city are as likely to see implementation of Complete Streets projects as neighborhoods in wealthier parts of a city. Or, before Complete Streets projects are implemented, it would be useful to know what level of involvement the surrounding community has in determining what its needs are and how they might be met through street-level redesign.[3] At yet another layer of complexity, movements for just sustainabilities must ask what the hidden agendas of property owners, corporate interests, real estate developers, and builders might be in seeing Complete Streets projects implemented.

Let's explore this terrain, metaphorically speaking, by bicycle. Complete Streets, by definition, accommodate bicycles as a form of transportation. This usually means, at a minimum, the inclusion of bicycle lanes or separate cycle paths. As such, bicycle advocacy organizations are generally supporters of Complete Streets policies and projects, which can be leveraged to extend bike routes and other infrastructure to increase the safety of cyclists.

But if bicycle infrastructure also happens to be seen by developers as the kind of amenity that attracts young, upwardly mobile Millennials and Gen Xers, then developers will begin investing in adjacent properties that can be redeveloped into live-work loft spaces, galleries, cafes, and craft breweries. If this is a likely outcome of building bike lanes,[4] then who are bike lanes really for? As a minister in Atlanta put it after a protected bike lane disrupted the flow of traffic and limited parking around his church, "Most people feel like these bike lanes are not for the people here.... It's for the people to come" (Torpy, 2017). In extreme cases, elected officials pursue economic growth for their cities by building bike lanes to attract the so-called Creative Class. "Our very public bike culture," the mayor of Minneapolis has been quoted as saying, "has been an enormous asset in attracting talented people...in advertising, financial services, the arts, [and] politics" (Hoffmann, 2016, p. 140). From an insider perspective in her previous position as equity initiative manager at the League of American Bicyclists, Lugo (2018) came to the conclusion that "the bike movement, or at least its policy arm, has decided that their goal of getting more people on bikes is not in conflict with the goal of making urban neighborhoods more expensive" (p. 146). She quotes Oregon Rep. Earl Blumenauer as describing "bike projects as a strategy to 'attract talent,' bringing 'the best and the brightest' to places like Portland" (p. 146).[5]

Based on this discussion, retrofitting cities whose streets were built for cars with bicycle infrastructure hardly appears to be retrofitting cities for justice. In fact, inasmuch as it is bound up with gentrification, the bicycle renaissance in American cities is most likely exacerbating preexisting inequalities. Next, we take a look at the challenge bike-share systems have had in serving diverse communities of users and strategies to retrofit bikeshare systems to achieve more equitable access. Bike sharing can take many forms, but the dominant form at present entails a public-private partnership to fund the installation of bike-share "stations" around a city where individuals can check out a bike for a short ride before depositing the bike at a station.

For reasons ranging from financial barriers to lack of stations in underserved neighborhoods, bike-share programs tend to be less utilized by low-income and minority populations. In Boston, now a majority-minority city, 85.7 percent of cyclists were white as of 2014, as were 87 percent of Hubway bike-share users. To their credit, many of these programs have acknowledged these challenges and taken steps like subsidizing bike-share program memberships, partnering with community organizations for outreach, and expanding into underserved neighborhoods.

The irony is that bike-share programs represent a retrofit to city transportation networks that prioritize automobiles with the potential to offer improved mobility especially to those who do not benefit from car-centric transportation infrastructure (i.e., those unable to afford a private automobile). Yet these very bike-share programs then had to be retrofitted themselves—from pricing, to marketing, to siting of stations—in order to deliver the promised benefits. Unfortunately, according to Hannig (2016), these efforts have had little effect in increasing ridership among low-income individuals and people of color. Next-generation bike-share programs with "dockless" bikes that can be left anywhere seem to have the "potential to bring bike-share to new areas, especially lower-income neighborhoods or neighborhoods of color that have been left out" (Schmitt, 2018). There is also anecdotal evidence that people of color are more comfortable using bikes in dockless bike-share systems than in the sometimes hard-to-figure-out systems requiring a bike to be removed from and returned to a dock (Sturdivant-Sani, 2018). The potential for dockless bikes to reach low-income communities seems to appeal to the Better Bike Share Partnership, a collaborative effort to fund programs that improve bike-share equity, as evidenced in its current round of grants, which saw two out of eight grantees focusing on dockless bike-share systems (*Bicycle Retailer and Industry News*, 2018). Dockless programs are not currently the norm, but many start-up micro mobility companies are piloting dockless bikes and scooters in U.S. cities, increasingly with pushback from concerned city officials.

This paradox is actually not a paradox at all, but rather evidence of a deeper problem: Most bike-share schemes, and many other urban sustainability initiatives, *were never designed with equity or social justice in mind.* They were designed around environmental and economic goals intended to reduce carbon dioxide emissions and stimulate urban renewal. When equity concerns are raised after these initiatives are pushed through, social justice is reduced to a retrofit strategy. Additionally, contesting sustainable transportation initiatives at this stage does nothing to disrupt the flows of capital that rely on the supposed value neutrality of markets to transform urban spaces in order to extract further profit.

Ideally, the question we should be asking is: How do we move equity and justice to the center in designing just, equitable, and sustainable cities, including questioning the origins of investments and the intentions of investors, rather than retrofitting them after our "real" goals have been met?[6]

Gardens for Food in Food Deserts: Are Equity and Justice Intentions Enough?

The labyrinthine terrain over which movements for just sustainabilities pursue equitable, just, and sustainable cities is illustrated in the challenges of avoiding unintended outcomes even when equity and justice *are* front and center.

Community gardens and urban agriculture more broadly exemplify this conundrum. Cadji and Alkon (2014), for example, analyze how Phat Beets Produce, a food justice organization committed to creating a healthier and more equitable food system in North Oakland, California, ultimately got priced out of its own diverse neighborhood when realtors began to promote the farmers market Phat Beets launched and operated to prospective homebuyers looking to "discover" an "up and coming" neighborhood. This might be considered what McClintock (2014) calls an internal contradiction of urban agriculture, which at once brands itself as against the industrial agri-food system but is also deeply embedded within the multiple processes of neoliberalization.

Finding a path that does not reproduce or feed the very mechanisms of capital accumulation that urban agriculture/food justice attempt to counter is a significant challenge. As Walker (2016) argues, "[U]rban agriculture has radical potential as a grassroots response to economic and environmental injustice, but has also been enrolled as a device by the local state in which the primary goal of sustainability planning becomes enhanced economic competitiveness" (p. 163).

Horst, McClintock, and Hoey (2017) point out an additional problem wherein "the precariousness of land access for urban agriculture is another limitation, particularly for disadvantaged communities" (p. 277). Reynolds's (2015) analysis of social injustices in New York City's urban agriculture system lead her to conclude that "failure to critically examine urban agriculture's role in either supporting or dismantling much broader social and political oppression may perpetuate an inequitable system that is legitimated through progressive narratives about the positive impact that urban farming and gardening can have on issues such as food access, education, job creation, and public health" (p. 241).

Against Green Gentrification: Strategies for Just Sustainabilities

Urban gardening and bicycle activism, Stehlin and Tarr (2017) argue, overvalue local strategies, which acts to "obfuscate the historical-geographical conditions in specific regions that created the conditions that activists hope to address through gardening and bicycling" (p. 1331). The problem with the bias toward the local, they maintain, is that "local debates over investment often succumb to the neoliberal false dichotomy that neighborhoods must either be left disinvested to remain affordable and livable places or accept the inevitable, naturalized forces of gentrification" (p. 1344). Stehlin and Tarr say the risk is that bicycle and garden advocacy too easily fall prey to both sides of the dichotomy by "reifying low-cost food and transit as resilient forms of poverty, and/or promising that they can attract capital back to the neighborhood" (p. 1344).

If bicycle and urban agriculture advocacy are to avoid such a trap, what tools do they have at their disposal for resisting the neoliberal vortex wherein capital accumulation is the only logic and metric of a city's livability? What forms of resistance can green neighborhoods cultivate without gentrifying them? Among scholars studying the relatively new phenomenon of green gentrification, a few strategies have been identified. Anguelovski (2015) calls for community-based coalitions and bottom-to-bottom networks in street, technical, and funder activism. Gould and Lewis (2017) call for communities to organize for anti-market and anti-racist public policy interventions. Community land trusts and inclusive zoning, among other interventions, must be part of the toolkit. Hutson (2015) focuses on the nurturing of community development tools, the negotiation of community benefit agreements with developers, the use of local ballot initiatives, and legal challenges as tools that "can lead to the design and implementation of healthy, sustainable urban communities" resistant to gentrification (p. 5). But local organizing is not enough, according to Hutson. Also required is a broad-based national community coalition with the support of progressive urban policies at the state and national levels.

No one wants to forego efforts to address existing injustices through improvements to transportation or food systems just because such efforts might produce unintended outcomes like displacement. Curren and Hamilton's (2012) concept of "just green enough" offers a workaround by proposing pursuing urban sustainability initiatives but in ways that deliver the benefits to neighborhoods without necessarily displacing them. Their research finds that partnerships between newcomers and longtime residents can lead, for example, to successful remediation of contaminated sites without triggering displacement, especially when government involvement creates opportunities for meaningful community participation. Much needs to be fleshed out with respect to how just-green-enough strategies can best be advanced. Curren and Hamilton's (2018) edited volume on the topic moves the dialogue forward by unpacking topics such as how a just-green-enough approach can be taken in the context of climate justice and community resilience (Sze & Yeampierre, 2018), how state policies can undermine local just-green-enough strategies (Bowen, 2018), how disaster relief strategies trumped just-green-enough efforts following devastating monsoon rains in Chennai in 2015 (Narayan, 2018), and how community identity was mobilized in Chicago to imagine a just-green-enough future following the closure of two heavily polluting coal-fired power plants (Kern, 2018).

The findings in the just-green-enough research coalesce around the challenges of implementing just-green-enough strategies at the local level when state and national governments fail to intervene in ways that can disrupt the flows of capital that want to find the spaces in which profit can be maximized regardless of the inequitable outcomes that might result. Trudeau (2018) analyzes this phenomenon by studying the institutional contexts that shaped New Urbanist projects in several U.S. cities. His particular focus is the ways that social equity is either marginalized or privileged in New Urbanist projects, which he measures by examining whether affordable housing is built into New Urbanist projects with stated environmental sustainability agendas. Trudeau finds that projects can be just green enough if they have institutional champions who bring "patient capital," by which he means investors who are committed to social equity from the outset and willing to forego the potential windfall of a speculative investment.

Immergluck (2017) arrives at a similar conclusion based on his analysis of development around Atlanta's BeltLine, a large-scale project converting 22 miles of abandoned railroad track into interconnected trails and parks ringing Atlanta. Larger in scale but with the same ambition of Manhattan's High Line, Atlanta's BeltLine is being planned and implemented over a 25-year period by a quasi-governmental organization and funded through a mix of public-sector financing and corporate and foundation philanthropy. Immergluck concludes that such massive sustainability initiatives are likely to drive environmental gentrification unless they adopt what he calls an "Affordability First" approach. For Immergluck (2017), this means "recogniz[ing] that when a project is of such a scale and impact that it has the potential to spur rapid increases in land and housing costs, provisions for preserving significant housing affordability must be put in place before other aspects of the project are considered."

Trudeau's "patient capital" and Immergluck's "Affordability First" concepts share a concern with prioritizing social equity at the earliest possible stages. Bike lanes and community gardens may symbolize changes in a neighborhood, but resisting these changes does little to impact the priorities set by city officials and investors looking to capitalize on such changes. Instead, Trudeau and Immergluck, along with the growing cadre of researchers in the just-green-enough approach, point to strategies such as reduced financing costs or property tax savings for developers who commit to affordable housing. Immergluck also proposes mandatory inclusionary housing policies. Each of these strategies requires a strong local community voice and presence in decision-making processes. In this sense, the challenges of 21st-century environmental justice activists addressing the urban sustainability paradox are rather similar to the toxics activism defining the early environmental justice movement. Both require a careful balancing of intense focus on resisting immediate local threats—whether those are the presence of toxic chemicals or toxic capital investments—with attention to laws at multiple levels of government that might be implemented to prevent further toxic exposure.

It is clear from the foregoing that neoliberalism has turned the paradigm of *urban planning in the public interest* into one of *urban development in the private interest* (Brenner & Theodore, 2002; Hackworth, 2007; Peck et al., 2009; Raco, 2012; Weber, 2002). New Urbanism, as an early manifestation of the neoliberal turn toward cities as economic engines, and now urban sustainability more broadly, increasingly appear to be narratives that justify the handing over of the urban planning reins to corporations, whether Amazon or Google as in the vignettes we shared, or real estate companies, banks, or development enterprises. As private interests use the rhetoric of sustainable, livable, and smart cities to gain more and more influence in urban planning and development, the environmental justice movement must adapt through a combination of drawing on lessons learned from the movement's early history and innovating new ways of identifying the earliest possible signs of speculative investment and then building coalitions capable of placing social equity—which is, after all, one of the three pillars of sustainability—at the center of such projects.

DEEPENING OUR UNDERSTANDING

1. *Gentrification* is a loaded term that is often left undefined. To explore local gentrification, you will use Google Street View's "time machine" feature to capture a picture of gentrification.

   Below is an example of two street views of a block of Broadway in Oakland, California. By going into Street View and then clicking the clock icon, you can reveal a slider that allows you to change your view to any point in time at which Google captured an image. The first image is from 2011, and the second image is from 2018. The used car lot is replaced by new commercial space that includes a market, Starbucks, and Chipotle. Also notice the bike-share station in front of the market.

© 2019 Google

© 2019 Google

a. Choose a street where you live, where you're from, or in a city with which you are familiar. You might also consider choosing a specific neighborhood where, based on your understanding, you believe green gentrification is occurring (e.g., a former industrial area turned into a park, a section of a street with separated cycleways, or a prominent community garden or farmers market).

b. Use the Google Street View time machine to explore how the streetscape has changed over time. Trying not to make any judgments about whether the changes are "good" or "bad," describe the following:

- What did the old street look like?
- What type of people, if any, might have been likely to frequent the businesses or live in the homes on the old street?
- What is different or new about the street today?
- What has been displaced or is missing?

- Given the style of the new additions, who do you think would feel most comfortable spending time on the street or living in the homes on the street as it exists today?

c. Try to imagine the forces that shaped the changes you observed. Gentrification can result from both push and pull factors. Disinvestment in a neighborhood might make it unsafe (push), thus driving down property values and opening up the potential for speculative investments. Or changes in the surrounding neighborhoods might drive up property values, making those who own more likely to consider selling their homes (pull).

d. As a result of these push and pull factors, which exist unevenly in space and time, gentrification often happens in subtle and isolated ways before the explosive changes occur that most people identify with a gentrified neighborhood. Go back to your chosen street and scan up and down the street and the surrounding blocks. Do you see any patterns? For example, in the images of Broadway in Oakland, the opposite side of the street did not change at all over the same time period.

e. Try to describe the unevenness of the gentrification you see. Given that gentrification happens unevenly and slowly at first, try to imagine the strategies that communities might employ to resist gentrification. In the example you chose, what would a just-green-enough strategy look like?

2. Extension: Explore resources for social equity in urban renewal.

   a. Go to your city government's website and try to find the existing codes and ordinances, if any, that communities could leverage to ensure that social equity is part of the decision-making process in any development projects or sustainability initiatives.

   b. Research existing community organizations to identify existing assets and assess the potential for a just-green-enough community coalition.

References

Agyeman, J. (2013). *Introducing just sustainabilities: Policy, planning and practice.* London: Zed Books.

Agyeman, J. (2005). *Sustainable communities and the challenge of environmental justice.* New York: New York University Press.

Agyeman, J., & Claudel, M. (2018, February 16). Streets and sidewalks in Alphabet's city. Medium. https://medium.com/cities-for-people/streets-and-sidewalks-in-alphabets-city-7200fb8db05b

Anguelovski, I. (2015). Tactical developments for achieving just and sustainable neighborhoods: The role of community-based coalitions and bottom-to-bottom networks in street, technical, and funder activism. *Environment and Planning C: Government and Policy, 33*(4), 703–725.

Anguelovski, I. (2016). From toxic sites to parks as (green) LULUs? New challenges of inequity, privilege, gentrification, and exclusion for urban environmental justice. *Journal of Planning Literature, 31*(1), 23–36.

Arroyo, A. J. C., Chee, C. P., & Wang, N. E. (2017). Where do children with asthma die? A national perspective from 2003 to 2014. *Journal of Allergy and Clinical Immunology, 139*(2), AB89.

Avalos, G. (2018, April 10). Google says it's close to owning enough downtown San Jose properties for "viable" development. *San Jose Mercury News.* https://www.mercurynews.com/2018/04/10/google-says-its-getting-closer-to-owning-enough-downtown-san-jose-properties-for-viable-development/

Bicycle Retailer and Industry News. (2018, March 9). Eight programs receive grants for equitable bike share and related research. https://www.bicycleretailer.com/industry-news/2018/03/09/eight-programs-receive-grants-equitable-bike-share-and-related-research#.WvYkpi_MwWo

Bowen, J. E. (2018). Environmental gentrification in metropolitan Seoul: The case of greenbelt deregulation and development at Misa Riverside City. In W. Curran & T. Hamilton (Eds.), *Just green enough: Urban development and environmental gentrification* (pp. 123–138). New York: Routledge.

Brenner, N., & Theodore, N. (2002). Cities and the geographies of "actually existing neoliberalism." *Antipode, 34*(3), 349–379.

Bunch, W. (2017, September 10). Should Philly screw over its schoolkids to make world's 2nd richest man even richer? *Philadelphia Inquirer.* http://www.philly.com/philly/columnists/will_bunch/should-philly-screw-over-its-schoolkids-to-make-worlds-2nd-richest-man-even-richer-20170910.html

Cadji, J., & Alkon, A. (2014). One day, the white people are going to want those houses again. In S. Zavestoski & J. Agyeman (Eds.), *Incomplete streets: Processes, practices and possibilities* (pp. 154–175). London, UK: Routledge.

Cook, A. G., Annemarie, J. B. M., Pereira, G., Jardine, A., & Weinstein, P. (2011). Use of a total traffic count metric to investigate the impact of roadways on asthma severity: A case-control study. *Environmental Health, 10*(1), 52.

Curran, W., & Hamilton, T. (2012). Just green enough: Contesting environmental gentrification in Greenpoint, Brooklyn. *Local Environment, 17*(9), 1027–1042.

Curran, W., & Hamilton, T. (Eds.). (2018). *Just green enough: Urban development and environmental gentrification.* New York: Routledge.

Flanagan, E., Lachapelle, U., & El-Geneidy, A. (2016). Riding tandem: Does cycling infrastructure investment mirror gentrification and privilege in Portland, OR and Chicago, IL? *Research in Transportation Economics, 60*, 14–24.

Gould, K. A., & Lewis, T. L. (2017). *Green gentrification: Urban sustainability and the struggle for environmental justice.* New York: Routledge.

Hackworth, J. (2007). *The neoliberal city: Governance, ideology, and development in American urbanism.* Ithaca, NY: Cornell University Press.

Hanlon, J. (2010). Success by design: HOPE VI, New Urbanism, and the neoliberal transformation of public housing in the United States. *Environment and Planning A, 42*, 80–98.

Hannig, J. (2016). Community disengagement: The greatest barrier to equitable bike share. In A. Golub, M. L. Hoffman, A. E. Lugo, & G. F. Sandoval (Eds.), *Bicycle justice and urban transformation: Biking for all?* (pp. 203–216). New York: Routledge.

Harvey, D. (2016, December 13). David Harvey: "There is no way you can change the world without changing your ideas!" LeftEast. http://www.criticatac.ro/lefteast/david-harvey-interview-2016/

Hoffmann, M. L. (2016). *Bike lanes are white lanes: Bicycle advocacy and urban planning.* Lincoln, NE: University of Nebraska Press.

Hoffmann, M. L., & Lugo, A. (2014). Who is "world class"? Transportation justice and bicycle policy. *Urbanities, 4*(1), 45–61.

Horst, M., McClintock, N., & Hoey, L. (2017). The intersection of planning, urban agriculture, and food justice: a review of the literature. *Journal of the American Planning Association, 83*(3), 277–295.

Hutson, M. A. (2015). *The urban struggle for economic, environmental and social justice: Deepening their roots.* New York: Routledge.

Immergluck, D. (2017, September 1). Sustainable for whom? Large-scale sustainable urban development projects and "environmental gentrification." Shelterforce: The Voice of Community Development. https://shelterforce.org/2017/09/01/sustainable-large-scale-sustainable-urban-development-projects-environmental-gentrification/

Kern, L. (2018). Mobilizing community identity to imagine just green enough futures: A Chicago case study. In W. Curran & T. Hamilton (Eds.), *Just green enough: Urban development and environmental gentrification* (pp. 167–180). New York: Routledge.

Lees, L. (2003). Super-gentrification: The case of Brooklyn Heights, New York City. *Urban Studies, 40*(12), 2487–2509.

Li, R. (2019, January 8). Google transformed Mountain View, is San Jose next? *San Francisco Chronicle.* https://www.sfchronicle.com/business/article/Google-transformed-Mountain-View-is-San-Jose-13515691.php

Li, S., Batterman, S., Wasilevich, E., Elasaad, H., Wahl, R., & Mukherjee, B. (2011). Asthma exacerbation and proximity of residence to major roads: A population-based matched case-control study among the pediatric Medicaid population in Detroit, Michigan. *Environmental Health, 10*(1), 34.

Lugo, A. (2018). *Bicycle/race: Transportation, culture, & resistance.* Portland, OR: Micocosm Publishing.

McCann, B., & Rynne, S. (2010). *Complete Streets: Best policy and implementation practices.* American Planning Association, Planning Advisory Service Report Number 559.

McClintock, N. (2014). Radical, reformist, and garden-variety neoliberal: Coming to terms with urban agriculture's contradictions. *Local Environment, 19*(2), 147–171.

Narayan, P. (2018). Displacement as disaster relief: Environmental gentrification and state informality in developing Chennai. In W. Curran & T. Hamilton (Eds.), *Just green enough: Urban development and environmental gentrification* (pp. 139–150). New York: Routledge.

National Association of Realtors. (2018, February 13). Fourth quarter home prices up 5.3 percent; nearly two-thirds of markets at all-time high. https://www.nar.realtor/newsroom/fourth-quarter-home-prices-up-53-percent-nearly-two-thirds-of-markets-at-all-time-high

NBC Bay Area staff. (2018, December 4). San Jose approves land sale for Google expansion despite protest. https://www.nbcbayarea.com/news/local/Vote-Future-Google-San-Jose-Expansion-501877221.html

Peck, J., Theodore, N., & Brenner, N. (2009). Neoliberal urbanism: Models, moments, mutations. *SAIS Review of International Affairs, 29*(1), 49–66.

Raco, M. (2012). Neoliberal urban policy, aspirational citizenship and the uses of cultural distinction. In T. Tasan-Kok & G. Baeten (Eds.), *Contradictions of Neoliberal Planning.* Dordrecht, Netherlands: Springer.

Reynold, K. (2015). Disparity despite diversity: Social injustice in New York City's Urban Agriculture System. *Antipode, 47*(1), 240–259.

Sauter, M. (2018, February 13). Google's guinea-pig city. *The Atlantic.* https://www.theatlantic.com/technology/archive/2018/02/googles-guinea-pig-city/552932/

Schmitt, A. (2018). Is the dockless bike-share revolution a mirage? Streetsblog USA. https://usa.streetsblog.org/2018/02/07/is-the-dockless-bike-share-revolution-a-mirage/

Sidewalk Toronto. (2018). https://www.sidewalktoronto.ca

Silicon Valley Rising. (n.d.). Soaring rents, falling wages. Working Partnerships USA. https://siliconvalleyrising.org/files/SoaringRentsFallingWages.pdf

Stehlin, J. (2015). Cycles of investment: Bicycle infrastructure, gentrification, and the restructuring of the San Francisco Bay Area. *Environment and Planning A, 47*(1), 121–137.

Stehlin, J. G., & Tarr, A. R. (2017). Think regionally, act locally? Gardening, cycling, and the horizon of urban spatial politics. *Urban Geography, 38*(9), 1329–1351. doi:10.1080/02723638.2016.1232464

Sturdivant-Sani, C. (2018, January 9). Can dockless bikeshare pump up cycling's diversity? CityLab. https://www.citylab.com/transportation/2018/01/can-dockless-bikeshare-pump-up-cyclings-diversity/549629/

Sze, J., & Yeampierre, E. (2018). Just transition and just green enough: Climate justice, economic development and community resilience. In W. Curran & T. Hamilton (Eds.), *Just green enough: Urban development and environmental gentrification* (pp. 61–73). New York: Routledge.

Torpy, B. (2017, August 14). Torpy at large: Bike lanes a gentrifying tool? *Atlanta Journal-Constitution.* https://www.myajc.com/news/local/torpy-large-bike-lanes-gentrifying-tool/HP3biL52H9nCuyNhSWo0DO/

Transportation Alternatives. (Fall 2007). A lot can happen between the lines: Completing NYC streets. http://transalt.org/files/news/magazine/2007/fall/10-13.pdf

Trudeau, D. (2018). Patient capital and reframing value: Making new urbanism just green enough. In W. Curran & T. Hamilton (Eds.), *Just green enough: Urban development and environmental gentrification* (pp. 227–238). New York: Routledge.

Walker, S. (2016). Urban agriculture and the sustainability fix in Vancouver and Detroit. *Urban Geography, 37*(2), 163–182.

Weber, R. (2002). Extracting value from the city: Neoliberalism and urban redevelopment. *Antipode, 34*(3), 519–540.

Zavestoski, S., & Agyeman, J. (Eds.). (2014). *Incomplete streets: Processes, practices, and possibilities.* New York: Routledge.

Notes

1. In some ways, this shift addresses an approach to environmental justice that Agyeman (2005) advocated more than a decade ago when he urged that "environmental justice should not only be reactive to environmental 'bads'...but that it should also be proactive in the distribution and achievement of environmental 'goods'—for instance....creating a sustainable community with a higher quality of life" (p. 26). Anguelovski (2016), in her analysis of the evolution of the environmental justice movement from toxics resistance to mobilizing for environmental goods and then to resisting environmental gentrification, builds on this, concluding "that locally unwanted land uses can be reconceptualized from contamination sources to new green amenities because of the displacement they seem to trigger or accelerate" (p. 23).
2. Lees (2003) describes this as "intense investment and conspicuous consumption by a new generation of super-rich 'financifiers' fed by fortunes from the global finance and corporate service industries" (p. 2487).
3. For example, although Sidewalk Labs is organizing a year-long public engagement process filled with public roundtables, Sidewalk Toronto "pop-up stations," and even a "design jam" with architects and planners, Sauter (2018) points out that "it's unclear how much of an impact ordinary Torontonians will be able to have on the project through these outreach mechanisms, many of which sound more like marketing efforts than public meetings."
4. The causal relationship between bike lanes and gentrification is a complicated one. Stehlin (2015) provides a comprehensive

conceptual explanation of the relationship. Empirical linkages have been established by Flanagan, Lachapelle, and El-Geneidy (2016), who found evidence in Portland, Oregon, and Chicago, Illinois, that cycling infrastructure investments were correlated with areas of increasing or existing privilege, and by Herrington and Dann (2016), who found statistically significant relationships in Portland between high levels of gentrification (e.g., increasing whiteness and educational attainment) and increased bicycle ridership.

5. Hoffmann and Lugo (2014) document extensively the praising of the potential for bicycle infrastructure projects for contributing to rising property values. Hoffmann (2016) and Lugo (2018) also offer two of the most thorough critiques of how the mainstream bicycle advocacy movement uses cultural privilege and power in ways that make people of color and low-income cyclists feel unwelcome in the movement and to frame new urban bicycling infrastructure as a vehicle of urban renewal and neighborhood improvement. "In an era of rising housing costs, gentrification, and displacement," writes Lugo (2018), "the bike movement chose an advocacy strategy blatantly promoting neighborhood change" (p. 141).

6. Early findings suggest the perceptions that underserved communities have of bike share are significant and can be overcome by building relationships early on that lead to meaningful engagement in decision-making processes (Hannig, 2016).

18

BLACK LIVES MATTER AS AN ENVIRONMENTAL JUSTICE CHALLENGE

David N. Pellow

I begin with the cases of Trayvon Martin, Rekia Boyd, and Stephon Clark—all African Americans who died at the hands of police or other persons, and whose deaths remain potent rallying points for Black Lives Matter supporters. The research questions I pose here are (a) how might the scourge of police violence directed at African Americans be more effectively understood and challenged through an environmental justice framework, and (b) how might an environmental justice framework be enhanced by addressing the scourge of police violence directed at African American communities? I explore these questions through examining the logic of institutional racist violence (which includes both state-based and corporate-driven practices and policies) and how this social phenomenon undergirds the Black Lives Matter cause and the seemingly unrelated problem of environmental racism.

MY PERSONAL PERSPECTIVE ON RACE AND ENVIRONMENTAL CONCERNS

I became interested in environmental justice and racial justice in my early years growing up in the U.S. South. Two of my parents were deeply involved in the civil rights movement in Nashville, Tennessee, and worked diligently to confront institutional and interpersonal racism and racist violence through organizing, protests, and other means when they were college students. All of my parents have been trailblazers in their fields, as educators and community leaders in Tennessee and Georgia.

I became interested in environmental issues as a child, since my father would take my brothers and me on camping trips and nature hikes and encouraged us to learn the names of trees and a range of flora and fauna. We also went swimming regularly in lakes and rivers in Tennessee and North Carolina, particularly in the Great Smoky Mountains National Park. I often noticed the lack of racial and cultural diversity among those populations sharing these spaces with us, realizing only later that this was a long-standing national and historical trend rooted in exclusionary policies and even racist violence (Finney, 2014).

When I was an undergraduate student I was introduced to Robert Bullard's groundbreaking research on environmental racism and environmental justice movements in the American South, and everything clicked for me. I met him at a conference two years later and have considered him a mentor ever since.

ENVIRONMENTAL JUSTICE STUDIES AND BLACK LIVES MATTER

From that moment until now, I have studied the relationship between social inequality and environmental quality, zeroing in on the myriad ways in which

---

Photo 18.1
S. Amos/Shutterstock

communities that are already marginalized because of their social status are also much more likely to face additional threats from environmental harm associated with industrial and government activities (Bullard, 2000; Bullard, Mohai, Saha, & Wright, 2007; Bullard & Wright, 2012). These threats include health risks associated with living in proximity to landfills, waste incinerators, oil refineries, mining operations, flood plains, and highways; contaminated water, air, and land; lack of access to healthy and affordable food; and greater vulnerability to the drivers and impacts of anthropogenic climate change (Taylor, 2000, 2009).

The scholarship on these uneven spatial, political, and cultural geographies is known as environmental justice studies, an interdisciplinary body of work that spans the humanities, social sciences, and sciences. More broadly, environmental injustices are embodied in the practices, discourses, policies, and histories of colonialism, conquest, enslavement, and the domination of land, territories, forests, waterways, and ecosystems by nation-states, empires, corporations, and other institutions that have been the hallmark of modern human history for the past 500 years (Pellow & Park, 2002; Voyles, 2015; Whyte, 2017) but that go back much further in time (Nibert, 2011; Nibert & Fox, 2002). The vast majority of research in environmental justice studies examines the spatial relationship, or proximity, between race, class, and some type of environmental hazard.

The Black Lives Matter (BLM) movement (now variously referred to as the Movement for Black Lives and the Black Lives Matter Global Network) burst onto the U.S. and global cultural scene in 2013 after the killing of 17-year-old Trayvon Martin on the evening of February 26 of the previous year. Martin was an unarmed African American boy in Sanford, Florida, who was killed by a self-appointed neighborhood watch volunteer, George Zimmerman. Zimmerman was eventually acquitted by a grand jury, reflecting the ways in which, even when a lone individual kills an unarmed African American person, the state and the criminal legal system all too frequently reinforce the message that such acts of violence are acceptable.

Less than a month after Martin's killing, on March 21, 2012, Rekia Boyd, a 22-year-old African American woman, was killed by Dante Servin, an off-duty police officer in Chicago (*Chicago Tribune* Editorial Board, 2015). Servin got into a verbal altercation with a group of young persons (of which Boyd was a part) during a party at Douglas Park on the city's West Side (this is also significant because African Americans and other marginalized populations have had to struggle for centuries for access to green spaces like parks; see Taylor, 2009, and Finney, 2014). Servin claimed that he drew his gun after one of the group—Antonio Cross—pulled a gun on him. He then shot Boyd in the head, killing her. The only thing found in Cross's possession was a cell phone. Although Servin was initially charged with involuntary manslaughter, all charges were later dropped. Servin eventually resigned from the police force before a hearing was to be held to decide whether he should be fired.

In March 2018, a 22-year-old unarmed African American man, Stephon Clark, was shot by Sacramento police officers in a hail of 20 bullets (Del Real, 2018). Police had been called to the scene after receiving a report of someone breaking car windows. As has happened in many similar cases, the officers believed Clark was armed but found only a cell phone on his person. As with the Trayvon Martin and Rekia Boyd cases and many others, BLM activists and supporters took to the streets, the airwaves, social media, and sites of government power to register outrage and demand justice in the face of state-sanctioned and vigilante violence.

BLM co-founder Alicia Garza explained what the movement stands for: "Black Lives Matter is an ideological and political intervention in a world where Black lives are systematically and intentionally targeted for demise. It is an affirmation of Black folks' contributions to this society, our humanity, and our resilience in the face of deadly oppression" (Garza, 2014). In BLM's 2017 report on the first four years of organizing, the authors state: "Black lives Matter began as a call to action in response to state-sanctioned violence and anti-Black racism. Our intention from the very beginning was to connect Black people from all over the world who have a shared desire for justice to act together in their communities. The impetus for that commitment was, and still is, the rampant and deliberate violence inflicted on us by the state" (BLM, 2017).

Black Lives Matter has become one of the most important sources and sites of racial justice struggle in the early 21st century, growing from a series of protest actions focused on particular incidents of state-based and vigilante antiblack violence to a network and organization with chapters around the United States and beyond. The movement has led street protests in cities across the nation, shutting down traffic on interstate highways, occupying police station offices, disrupting politicians' campaign speeches, and disrupting professional basketball games.

In addition to protesting police and other forms of institutional violence against black people, BLM has promoted the production of art, a healing and mental health promotion project, free subway rides for people living in New York City, fundraisers to bail out black mothers on Mother's Day, and a partnership with the Sacramento Kings basketball team and management called "Build Black" to finance educational opportunities for youth in underserved communities (Cacciola, 2018). BLM activists also succeeded in getting Democratic presidential candidates Bernie Sanders and Hillary Clinton to address and endorse key demands during their 2016 presidential campaigns.

An ominous sign of the movement's success is the view held by police departments, the FBI, and the White House that it represents a threat to the status quo of unquestioned dominance of law enforcement and white supremacy. This is evidenced by Donald Trump's public denunciation of racial justice protesters, including National Football League players who bravely decided to kneel during the singing of the national anthem at the start of games; his celebration of the "Blue Lives Matter" slogan to indicate his uncritical support of police; and his reversal of an Obama-era limitation on the level of military surplus equipment that municipal police departments could access (Goldman, 2017; Keller, 2018; Wagner & Maske, 2018). Similarly, the FBI's Counterterrorism Division report from August 2017 lumped together six instances of attempted and actual violence against police officers and/or prosecutors with the BLM movement, labeling them all "Black Identity Extremists" (Beydoun & Hansford, 2017; FBI, 2017).

Scholars and activists have long recognized that African Americans, like numerous other groups, are subjects of colonialism—systems of power that seek to control populations and their lands and territories via externally imposed domination for the benefit of those controlling that system (Blauner, 1972; Von Eschen, 1997). For that reason, the BLM movement should be seen as building on the long history of anticolonial/decolonial resistance movements among Indigenous people, people of color, and their allies for the past half millennium (Dillon & Sze, 2018).

Racist violence by police and other persons against unarmed African Americans is a regular occurrence in the United States. Environmental racism—the disproportionate targeting of communities of color through socioenvironmental violence—is also a fundamental part of the American landscape. Accordingly, there are two guiding questions in this chapter concerning these matters. First, how might the phenomenon of police, intrapersonal, and institutional brutality against African Americans be linked to and expand our understanding of environmental justice concerns? Second, how can struggles for environmental justice be made more effective and transformative through attention to antiblack police violence? I explore these questions through the previously described cases and through the lens of critical environmental justice—a framework that pushes traditional environmental justice studies analysis in new directions that allow us to broaden and deepen our understanding of socioenvironmental injustices and to develop more transformative responses. Few scholars have connected the BLM movement and police violence to environmental justice politics (for groundbreaking exceptions, see Dillon & Sze, 2016, and Dillon & Sze, 2018), but there is much to be gained from doing so.

CRITICAL ENVIRONMENTAL JUSTICE AND THE BLACK LIVES MATTER MOVEMENT

I developed the critical environmental justice studies framework as a way of contextualizing environmental justice studies and the problem of environmental racism, and as a tool for rethinking the BLM movement. The preceding section presented an overview of traditional (or first-generation) environmental justice research approaches. In this section I expand on those insights and reveal the value of a critical environmental justice studies framework for deepening and extending the reach and scope of environmental justice research and methods built upon the following pillars:

- Expanding the range of social categories under consideration. Most environmental justice studies research focuses primarily on the relationship between race or class and environmental threats. But gender; sexuality; nationality, citizenship, or immigration status; Indigeneity; disability; and age also shape people's risk of environmental harm and their access to safe, healthy, and nurturing environments. Critical environmental justice urges scholars to expand the scope and range of social categories so as to develop more comprehensive analyses of the problem of environmental injustice and possible solutions to this challenge. Furthermore, critical environmental justice embraces the task of expanding this range of categories beyond the human, to explore the ways in which different species play roles in environmental justice struggles.
- Grappling with the deeply entrenched power and inequalities that emanate from states, corporations, and other institutions in

constraining both the lives of marginalized communities and the scholarship in environmental justice studies. My review of the environmental justice studies literature concludes that, although scholars are generally critical of the abuse of power by dominant institutions with respect to the perpetration of environmental injustices, they generally do not question the existence of those institutions. In other words, the approach is reformist rather than abolitionist, and this is concerning for me and many other scholars who view the existence of certain institutions as an existential threat to social and environmental justice, and ecological sustainability.

- Attending to the value and power of multiscalar analyses and methods for environmental justice research. Most environmental justice studies scholarship focuses on one scale of analysis (for example, a neighborhood, city, region, or nation), which is useful but limited, since environmental injustices simultaneously function at multiple spatial and temporal scales. Attending to this complexity is critical for developing more rigorous analyses of these challenges and their potential resolutions.
- Exploring the implications of challenging "racial expendability" with an embrace of "indispensability" in multi-species environmental justice scholarship and politics. Although environmental justice scholarship and politics decry the impacts and harms of environmental injustice, there is rarely a space for leaders and authors to explicitly proclaim that we are all key actors in interdependent webs of social and ecological networks and that each person, being, and thing in those multi-species communities plays a key role in our everyday existence. The phrase "Black Lives Matter" is rather timid in this regard because it implies that black people should have the right to exist, but it falls far short of stating that black people are indispensable. Moreover, the "Black Lives Matter" versus the "All Lives Matter" debate disallows a more important discussion of the specific, unique, and particular ways in which various populations—both human and nonhuman—play key roles in the daily functioning of our multi-species societies. Critical environmental justice studies boldly declares that we are all indispensable to creating environmentally and socially just and resilient futures, with an eye toward the ways in which all groups have distinct experiences with privilege and inequalities and can participate in bringing those futures into being.

With respect to my sources of evidence for this chapter, I draw on key statements from the Black Lives Matter website, from media and online interviews and statements from BLM leaders and supporters, and the 2017 BLM report titled *Celebrating Four Years of Organizing to Protect Black Lives.* These sources are viewpoints and platforms from the movement's leadership, representing part of the organization's primary public record.

First Pillar: Expanding the Range of Social Categories

The BLM movement has done an admirable job of taking intersectionality seriously, advancing an inclusive vision and practice of social change. This could be instructive for the field of environmental justice studies, which tends to focus primarily on racial or class inequalities. As BLM co-founder Alicia Garza (2014) writes, "Black Lives Matter affirms the lives of Black queer and trans folks, disabled folks, Black-undocumented folks, folks with records, women and all Black lives along the gender spectrum. It centers those that have been marginalized within Black liberation movements. It is a tactic to (re)build the Black liberation movement." Even so, I find that both BLM and environmental justice studies could be more expansive and embrace an analysis of species and the roles they play in shaping the lives of both humans and nonhumans facing threats of institutional violence. Consider the following example: A 2015 *Baltimore Sun* investigation of the Baltimore (Maryland) Police Department (serving the same town where Freddie Gray was killed by police) detailed numerous incidents in which vulnerable people and nonhumans were the subjects of brutal and sometimes lethal force at the hands of police in that city. The report noted, for example, that in a series of cases, not only were young African American males the targets of such violence, but so were elderly people, women, children, and nonhuman animals:

> A grandmother's bones were broken. A pregnant woman was violently thrown to the ground.... Even animals couldn't escape the brutality of the Baltimore police.... Officer Thomas Schmidt, a 24-year veteran assigned to the Emergency Services unit, was placed on paid administrative leave after police say he held down a Shar-Pei while a fellow officer, Jeffrey Bolger, slit the dog's throat. A month later, a Baltimore police officer [pleaded] guilty to a felony animal cruelty charge after he *fatally beat and choked his girlfriend's Jack Russell terrier*. That very same year,...[f]our investigators from agencies outside Baltimore were working to determine who *left a dead rat on the car windshield of an officer who was cooperating with prosecutors on a police brutality case*....[The prior year] [t]here was a murder-suicide, with a policeman killing a firefighter, his girlfriend, and himself. There was a different officer who killed himself in jail after being charged with killing his fiancée. (Friedersdorf, 2015, emphasis added)

The *Baltimore Sun* investigation appears to reflect and support what scholarly studies have long revealed: that there is a well-documented link between the use of violence against nonhuman animals and efforts to exert control over other humans, whether in the destruction of livestock and other food sources during wartime and conquests, or through domestic violence directed primarily at women, children, and nonhuman companion animals or "pets." Consider, for example, the intersecting violence and oppression from the preceding excerpt, which includes the murder of a rat to intimidate a police officer working on a police brutality case and the murder of a dog to intimidate a girlfriend. The use of violence against nonhumans as a means of control over other humans is a time-honored tradition that includes, for example, the U.S. Army's efforts to exterminate the buffalo as a means of starving and conquering the Plains Indians in the 19th century (Smits, 1994).

The *Baltimore Sun* report also mirrors the scholarly research on intersecting oppressions associated with domestic violence (Lockwood & Ascione, 1999). For example, in 2011, there were 1.3 million reported incidents of intimate partner violence in the United States and more than 70 percent of those cases are believed to have included violence directed at a companion animal (Komorosky, Woods, & Empie, 2015). Furthermore, a number of studies conclude that people who desire to hurt or control their domestic partners or children frequently achieve that goal through violence directed at a companion animal (Upadhya, 2014) and that, compounding the problem, children in those environments grow up with a greater than average probability of hurting a companion animal as well (Currie, 2012).

Second Pillar: Attending to the Value and Power of Scale

The second pillar of critical environmental justice studies applies multiscalar methodological and theoretical approaches to studying environmental justice issues in order to better comprehend the complex spatial and temporal causes, consequences, and possible resolutions of environmental justice struggles. Environmental justice studies scholarship is characterized largely by a focus on just one scale (for example, a local case study, or a citywide or national study), so a critical environmental justice studies framework would explore how environmental justice struggles function at multiple spatial/geographic and temporal scales. For example, we are interested in how environmental justice struggles manifest at the microscale of the body and in interpersonal interactions in a single family or community, on up to the global scale. In terms of temporal dynamics, the roles of time and history should be examined as well. Multiscalar analyses are important for studying and responding to environmental and social injustices because they can make visible relationships, impacts, and possibilities that might have previously escaped our attention. For example, much of the focus on climate change has been at the *global* scale and on the *future* consequences of a fossil fuel economy, which are of great importance, but that spatial and temporal fixation can obscure the everyday, ongoing effects of climate change on *local* communities (generally Indigenous, low-income, and people of color) who are facing extreme weather events, crop failures, heat waves, and coal-fired power plant pollution *right now*.

Black Lives Matter provides an excellent case through which to consider multiscalar analyses. Let's start with the microscale of race and racism. Research on social cognition and interpersonal interactions has found that "implicit bias" exists with such great intensity in the United States across racial groups that white research subjects tend to perceive threats to their well-being when they see black and brown people when no such threat exists (Kang, 2005). This research on implicit bias is highly consequential for the kinds of everyday microsociological interactions between and among individuals across the racial spectrum, especially in the case of gun violence in the name of white "self defense" and police brutality because when people of color are viewed as threats to the security and well-being of others, they can end up on the receiving end of lethal violence.

Race and scale intertwine to reveal also that when black people respond to racism (whether by police or via environmental racism), their actions are frequently viewed as a menace and a threat that is disproportionate and outsized. We can see this, for example, in the militarized response by police departments when interacting with the BLM movement, with perhaps the most egregious instance occurring in Ferguson, Missouri, in 2014. Under the Pentagon's Excess Property (or 1033) Program, the U.S. Department of Defense provided billions of dollars worth of military equipment to local police departments, a legacy of the surplus from foreign wars (Redden, 2015). The 1033 Program came under fire by civil rights activists, supporters of the BLM movement, and even military veterans when disturbing images of police and protesters clashing in Ferguson, Missouri, in response to the police killing of Michael Brown seemed to be indistinguishable from media images of civilians being repressed by an occupying military force in some faraway land. Such conflicts suggest that we should also pay attention to global-scale dynamics that reveal linkages between the militarized oppression of African American communities and the fact that the U.S. military exercises control over people of color in many other parts of the world, such as Niger, Afghanistan, Iraq, Syria, Pakistan, Yemen, Palestine, Japan, and South Korea. This is also a clear environmental justice concern because the U.S. military is one of the largest sources of pollution on earth (Nazaryan, 2014) *and* because militarism and *masculinist* politics tend to go hand in hand and both tend to result in socially and ecologically harmful practices.

Finally, BLM's work speaks to the myriad ways that scale can be thought of and articulated temporally. In fact, the entire point of the BLM movement is, in some ways, an intervention to remind and challenge the United States that blatant acts of antiblack violence are not a thing of the past and are still rampant today. A number of scholars contend that it is helpful to understand the U.S. criminal legal system not as a phenomenon that regrettably produces discriminatory outcomes for communities of color, but that the system was actually designed to do so, as an infrastructure of racial punishment that is rooted in the history and logic of chattel slavery (Escobar, 2016; Gilmore, 2007; Rodriguez, 2006). And although one might offer a counterargument that slavery ended with the Emancipation Proclamation, the 13th Amendment to the U.S. Constitution contains an exception clause that allows enslavement "as a punishment for crime." And given that African Americans make up a disproportionate percentage of the U.S. prison population, there are historical and contemporary reasons for viewing the U.S. legal system as a mechanism for the devaluation of black life in the United States.

Third Pillar: Addressing Entrenched Inequalities and Institutional Power

The third pillar of critical environmental justice studies is the view that social inequalities are not aberrations or abstractions but rather are deeply embedded in society and reinforced by entrenched institutional power. Therefore, the current social order stands as a fundamental obstacle to social and environmental justice. A logical conclusion of this observation is that social change movements may be better off being cautious about engaging the state and other dominant institutions as reliable partners in their efforts to create social change.

Since Black Lives Matter's aim is to confront state-sanctioned racist violence, I believe it would be productive to link this movement to the challenge of environmental racism. Like police brutality, environmental racism is often a form of state-sanctioned violence that occurs when government agencies and state-regulated firms target communities of color with pollution and other environmental impacts. For that reason, BLM could broaden its appeal and reach by extending its focus to this topic. Similarly, I believe that environmental justice scholarship could benefit immensely from paying closer attention to the ways in which environmental racism is a form of state-sanctioned violence (rather than primarily a matter of civil rights discrimination or a lack of sufficient regulation). BLM activists and environmental justice scholars might then be encouraged to envision practices that could result in environmental justice for communities *beyond the state*, or at least by minimizing our reliance on state institutions. In fact, some BLM activists urge us to think about how to make our communities safe "beyond policing" (Tometi, 2015), which is a good start. The BLM movement, the environmental justice movement, and environmental justice scholarship generally share a consensus that we should look to the state and the regulatory-legal apparatus for relief, recognition, and justice. I find that perspective limiting and at odds with what the empirical evidence suggests.

Police brutality and the prison system are forms of state control to which African American communities are routinely subjected. Black Lives Matter is explicitly focused on this problem, but we have yet to see many advocates forge a connection between racist law enforcement policies and environmental racism. I think we can make those linkages because state-sanctioned violence is a system of control over a community's movements and its members' access to a range of resources and spaces. State-sanctioned violence is therefore also a practice that reflects and results in the differential valuation of people, their knowledge systems, and cultural practices, with certain populations rendered subordinate to more powerful groups. Therefore, if we think of environmental racism as an example of these processes, then we can more effectively theorize it as a form of state violence, a perspective that is generally absent from most environmental justice scholarship (for exceptions, see Downey, 2015; Pulido, 2016; and Smith, 2005). Moreover, a critical environmental justice studies approach draws from scholarship that views states as inherently authoritarian, exclusionary, and violent, so that environmental justice scholarship and advocacy that relies on state reforms will be dramatically limited in scope and possibility.

BLM activists have regularly sought to participate in and sponsor government-driven initiatives to reduce police brutality and improve the plight of black communities. BLM is therefore *not* envisioning how we might build safer communities *beyond* the state but rather how we might do so with *greater* state involvement. BLM co-founder Patrice Cullors and her colleague Darnell Moore have a vision of social change that includes divesting from certain punitive state functions and investing in other, supposedly progressive functions. They write:

> [W]e will advocate for a decrease in law-enforcement spending at the local, state and federal levels and a reinvestment of that budgeted money into the black communities most devastated by poverty in order to create jobs, housing and schools. This money should

> be redirected to those federal departments charged with providing employment, housing and educational services.... Our group is proof that dedicated and skilled black folks can work together to end state violence, homelessness, joblessness, imprisonment and more inside black communities. (Moore & Cullors, 2014)

This statement reflects generations of progressive thinking about social change, but it is also potentially counterproductive because while Moore and Cullors and other BLM leaders seek to shift government resources away from overtly punitive, criminal legal purposes to public services and basic needs, this perspective ignores the reality that reinforcing state power for these progressive ends will also reinforce state power more generally, including its repressive dimensions. That is, through this policy framework, BLM fails to offer a transformative vision of change by *channeling* rather than *challenging* the state. The following quote from one BLM supporter offers a related piece of evidence of the pitfalls of this orientation: "In order for world powers to maintain slavery and post-slavery racist legislation, it was crucial for the people to believe that the uncivilized nature of Africans required excessive force, policing, and restraint" (Lyiscott, 2014). This statement, like much of the BLM-related discourse, contends that it is *excessive* state force, not state force itself, that is the main obstacle to social justice.

However, having examined the historical record of racial violence and nation-state formation, many scholars conclude that repression and violence are not practices that states engage in on an occasional basis; they are at the core of what states do and why they exist. Modern nation-states claim a "monopoly on the legitimate use of physical force" (Weber et al., 2004). As philosopher Charles Mills (1999) writes, "To understand the long, bloody history of police brutality against blacks in the United States, for example, one has to recognize it not as excesses by individual racists but as an organic part of this political enterprise" (pp. 84–85). Thus, people, organizations, and movements aiming to confront state-sanctioned violence such as police brutality might consider the implications of the anarchist slogan "Police existence is brutality," which means that policing as an institution is inherently authoritarian and violent.

Fourth Pillar: Indispensibility Over Expendability

The fourth pillar of the critical environmental justice studies framework centers on the concept of indispensability, with a particular emphasis on *racial* and *socio-ecological indispensability*. Many ethnic studies scholars today argue that because many people of color are viewed as threats to white supremacy and cultural purity, they are deemed expendable by the state and the legal system. Concretely, that means that members of these populations are frequently targeted with institutional violence as a means of achieving racial dominance (Gilmore, 2007; Márquez, 2014). This "racial expendability" undergirds and supports antiblack police brutality and environmental racism. It is also closely related to "ecological expendability," which deems nonhuman species, spaces, and ecosystems either as threats to human dominance or as a means of enabling it, so that control over, or the consumption and destruction of, nonhuman bodies and habitats is necessary for the maintenance of human supremacy or "dominionism."

A critical environmental justice studies perspective responds to these dominant cultural and institutional orientations with an understanding that—in opposition to white and human supremacy—humans, nonhumans, and ecosystems (particularly those on the receiving end of exclusion, marginalization, and othering) should be seen not as expendable but rather as *indispensable* to our collective futures. This is what I term *racial indispensability* (when referring to people of color) and *socio-ecological indispensability* (when referring to broader communities within and across the human/more-than-human spectrum). Racial indispensability is intended to challenge the logic of racial expendability and is the idea that institutions, policies, and practices that support and perpetrate antiblack racism intended to destroy African American bodies suffer from the flawed assumption that the future of African Americans is somehow delinked from the future of white communities.

This perspective reflects what has been described as one of the first rules of ecology: that everything in the universe is linked to everything else, so that whatever impacts one being, thing, or element affects all of us. Violence against people of color harms white people and it impacts the more-than-human world, and vice versa, so critical environmental justice studies affirms that black lives, the lives of people of color, are *indispensable*. Going a step beyond that claim, the concept of *socio-ecological indispensability* reflects the critical environmental justice studies view that the well-being of all living things and of all species is *indispensible* as well. This is both a socio-ecological reality and an affirmation of a politics of solidarity and coalition building that loudly proclaims, "All of us or none!"

Black Lives Matter activists write and speak out on this issue regularly. In 2015, BLM issued a "State of the Black Union" in which they wrote, "None of us are free until all of us are free" (BLM, 2015). This is a variation on a quote that has been attributed to the poet Emma Lazarus, the Reverend Dr. Martin Luther King Jr., and many others. It is also articulated powerfully in Barry Mann's rhythm and blues song "None of Us Are Free," which includes the chorus "None of us are free, none of us are free, if one of us is chained, none of us are free." In its 2017 report, Black Lives Matter writes,

> We organize because it is a matter of life or death. Our ancestors and movement elders dedicated their lives to organizing because they envisioned a day when their grandchildren could thrive outside the confines of oppression. Organizing is building and leveraging people power in order to disrupt systems that threaten our lives and the lives of others, and to build our own life-affirming systems. This can only be done at the intersections of identities, as we cannot get free until the most marginalized of us do. (BLM, 2017)

BLM co-founder Alicia Garza echoes and articulates this idea:

> Black Lives Matter doesn't mean your life isn't important—it means that Black lives, which are seen as without value within White supremacy, are important to your liberation. Given the disproportionate impact state violence has on Black lives, we understand that when Black people in this country get free, *the benefits will be wide reaching and transformative for society as a whole....* We remain in

> active solidarity with all oppressed people who are fighting for their liberation and we know that our destinies are intertwined.... And, perhaps more importantly, when Black people cry out in defense of our lives, which are uniquely, systematically, and savagely targeted by the state, we are asking you, our family, to stand with us in affirming Black lives. Not just all lives. Black lives.... Our collective futures depend on it. (Garza, 2014, December 6, emphasis added)

Indispensability is a challenge to the logic of expendability because it responds to the brutality of such an idea by calling for multi-racial, intersectional solidarities and a vision that insists that our collective futures must include "all of us or none." This is a powerful way to apply our knowledge of social and environmental injustices toward a clarion call for a vision of society that is socially just and ecologically sustainable.

SUMMARY AND CONCLUSION

The field of environmental justice studies presents a novel and generative space for thinking about new ways to engage a range of social movements in the 21st century. Black Lives Matter offers one of those opportunities because it is a movement focused on the scourge of racist state-sanctioned violence, which is a concern that can readily be extended to the analysis and politics of environmental justice. In this chapter, I sought to link these two phenomena through the use of a critical environmental justice framework in order not only to find common ground between environmental justice and Black Lives Matter but also to push both perspectives toward more rigorous analytical and political possibilities.

DEEPENING OUR UNDERSTANDING

Organize into small groups to address the following tasks, and have one person in each group serve as a recorder and another as a reporter:

1. Compare and contrast how a traditional environmental justice studies framework versus a critical environmental justice studies framework might explain and address the social and ecological drivers and consequences associated with militarization and warfare.
2. Compare and contrast how a traditional environmental justice studies framework versus a critical environmental justice studies framework might explain and address the social and ecological drivers and consequences associated with the prison industrial complex.
3. Discuss what your group thinks might be some important intersections among militarization, warfare, and the prison industrial complex.

(You may have time to do an Internet search for scholarly sources and media reports on these topics to fill in any gaps in knowledge.)

Report back on your group's discussion and debate about which of the two frameworks (traditional environmental justice versus critical environmental justice) is more persuasive (and why) and on how warfare and prison systems might be related.

REFERENCES

Beydoun, K. A., & Hansford, J. (2017, November 15). The FBI's dangerous crackdown on "black identity extremists." *New York Times*. https://www.nytimes.com/2017/11/15/opinion/black-identity-extremism-fbi-trump.html

Blauner, R. (1972). *Racial oppression in America*. New York: HarperCollins.

Bullard, R. D., & Wright, B. (2012). *The wrong complexion for protection: How the government response to disaster endangers African American communities*. New York: New York University Press.

Bullard, R. D. (2000). *Dumping in Dixie: Race, class, and environmental quality*, 3rd ed. Boulder, CO: Westview Press.

Bullard, R. D., Mohai, P., Saha, R., & Wright, B. (2007). *Toxic wastes and race at twenty, 1987–2007*. Cleveland, OH: United Church of Christ Justice and Witness Ministries.

Cacciola, S. (2018, March 29). Kings partner with Black Lives Matter Sacramento and build black. *New York Times*. https://www.nytimes.com/2018/03/29/sports/sacramento-kings-black-lives-matter.html

Chicago Tribune Editorial Board. (2015, April 22). Rekia Boyd shooting was "beyond reckless," so cop got a pass. *Chicago Tribune*. https://www.chicagotribune.com/opinion/editorials/ct-cop-verdict-servin-edit-0423-20150422-story.html

Currie, C. L. (2012). Corrigendum to "Animal cruelty by children exposed to domestic violence." *Child Abuse & Neglect, 36*(11–12), 800.

Del Real, J. A. (2018, March 28). 20 Shots in Sacramento: Stephon Clark killing reignites a furor. *New York Times*. https://www.nytimes.com/2018/03/28/us/sacramento-stephon-clark.html

Dillon, L., & Sze, J. (2016, Fall/Winter). Police power and particulate matters: Environmental justice and the spatialities of in/securities in U.S. cities. *English Language Notes, 54*(2).

Dillon, L., & Sze, J. (2018). Equality in the air we breathe: Police violence, pollution, and the politics of sustainability. In J. Sze (Ed.), *Sustainability: Approaches to environmental justice and social power* (pp. 246–270). New York: New York University Press.

Downey, L. (2015). *Inequality, democracy and the environment*. New York: New York University Press.

Escobar, M. (2016). *Captivity beyond the state: Criminalization experiences of Latina (im)migrants*. Austin: University of Texas Press.

Federal Bureau of Investigation. (2017, August 3). *Black identity extremists likely motivated to target law enforcement officers*. Counterterrorism Division. Washington, DC: Department of Justice.

Finney, C. (2014). *Black faces, white spaces: Reimagining the relationship of African Americans to the great outdoors*. Chapel Hill: University of North Carolina Press.

Friedersdorf, C. (2015, April 22). The brutality of police culture in Baltimore. *The Atlantic*. https://www.theatlantic.com/politics/archive/2015/04/the-brutality-of-police-culture-in-baltimore/391158/

Garza, A. (2014). Blacklivesmatter.com.

Gilmore, R. W. (2007). *Golden gulag: Prisons, surplus, crisis, and opposition in globalizing California*. Berkeley: University of California Press.

Goldman, A. (2017, August 28). Trump reverses restrictions on military hardware for police. *New York Times.* https://www.nytimes.com/2017/08/28/us/politics/trump-police-military-surplus-equipment.html

Kang, J. (2005). Trojan horses of race. *Harvard Law Review, 118*, 1489–1593.

Keller, J. (2018, May 23). The triumph of Trump's law and order politics. *Pacific Standard.* https://psmag.com/news/the-triumph-of-trumps-law-and-order-politics

Komorosky, D., Woods, D. R., & Empie, K. (2015). Considering companion animals: An examination of companion animal policies in California domestic violenc shelters. *Society & Animals, 23*, 298–315.

Lockwood, R., & Ascione, F. (Eds.). (1999). *Cruelty to animals and interpersonal violence: Readings in research and in application.* West Lafayette, IN: Purdue University Press.

Lyiscott, J. (2014, September 17). Stop hitting yourself. https://www.theglamsavvylife.com/stop-hitting-yourself/

Márquez, J. (2014). *Black-brown solidarity: Racial politics in the new Gulf South.* Austin: University of Texas Press.

Mills, C. (1999). *The racial contract.* Ithaca, NY: Cornell University Press.

Moore, D., & Cullors, P. (2014, September 4). Five ways to never forget Ferguson—and deliver real justice for Michael Brown. *The Guardian.* theguardian.com/commentisfree/2014/sep/04/never-forget-ferguson-justice-for-michael-brown

Nazaryan, A. (2014, July 17). The U.S. Department of Defense is one of the world's biggest polluters. *Newsweek.* http://www.newsweek.com/2014/07/25/usdepartment-defence-one-worlds-biggest-polluters-259456.html

Nibert, D. (2011). *Animal oppression and human violence: Domesecration, capitalism, and global conflict.* New York: Columbia University Press.

Nibert, D., & Fox, M. W. (2002). *Animal rights/human rights: Entanglements of oppression and liberation.* New York: Rowman & Littlefield.

Pellow, D. N., & Park, L.S. (2002). *The Silicon Valley of dreams: Environmental injustice, immigrant workers, and the high-tech global economy.* New York: New York University Press.

Pulido, L. (2016). Geographies of race and ethnicity II: Environmental racism, racialcapitalism and state-sanctioned violence. *Progress in Human Geography*, 1–10. doi:10.1177/0309132516646495

Redden, Molly. (2015, August 12). The Pentagon just realized it gave too much military equipment to the Ferguson Police. *Mother Jones.* https://www.motherjones.com/politics/2015/08/pentagon-forces-ferguson-return-two-humvees-police-militarization-program/

Rodriguez, D. (2006). *Forced passages: Imprisoned radical intellectuals and the U.S. prison regime.* Minneapolis: University of Minnesota Press.

Smith, A. (2005). *Conquest: Sexual violence and American Indian genocide.* Cambridge, MA: South End Press.

Smits, D. (1994). The frontier army and the destruction of the buffalo: 1865–1883. *The Western Historical Quarterly, 25*(3), 312–338.

Taylor, D. (2000). The rise of the environmental justice paradigm: Injustice framing and the social construction of environmental discourses. *American Behavioral Scientist, 43*, 508–580.

Taylor, D. (2009). *The environment and the people in American cities, 1600s–1900s: Disorder, inequality, and social change.* Durham, NC: Duke University Press.

Tometi, O. (2015, July 24). Interview on Democracy Now!

Upadhya, V. (2014). The abuse of animals as a method of domestic violence. *Emory Law Journal, 63*, 1163–1209.

Von Eschen, P. (1997). *Race against empire: Black Americans and anticolonialism, 1937–1957.* Ithaca, NY: Cornell University Press.

Voyles, T. B. (2015). *Wastelanding: Legacies of uranium mining in Navajo country.* Minneapolis, MN: University of Minnesota Press.

Wagner, J., & Maske, M. (2018, May 24). Donald Trump on national anthem: Players who protest "maybe shouldn't be in the country." *Chicago Tribune*. https://www.chicagotribune.com/sports/ct-trump-national-anthem-20180524-story.html

Weber, M., Owen, D. S., Strong, T. B., & Livingstone, R. (2004). *The vocation lectures: Science as a vocation, politics as a vocation.* Cambridge, MA: Hackett.

Whyte, K. P. (2017, Spring). The Dakota Access Pipeline, environmental injustice, and U.S. colonialism. *Red Ink, 19*(1), 154–169.

CONCLUSION

Michael Mascarenhas

This book traces the lineage of environmental justice from the civil rights movement to Black Lives Matter, as I feel it is important that we understand both the historical roots and the current manifestations of environmental inequality and environmental racism in the United States and around the world. Much has changed since the civil rights movement. In fact, since then, the wealth gap in the United States has grown steadily. And in an era of *Citizen's United*, where campaign contributions are effectively limitless, it is difficult to know where the monied influence is coming from and who is behind the forces that produce environmental injustice and environmental racism. For example, dark monied interests like the Koch brothers and DeVos family have infiltrated every aspect of government policy that affects their corporate interests from energy to public infrastructure. Betsy DeVos is the current U.S. secretary of education. Today conservative think tanks write more policy than those we elect for public service. And they have gotten what they wanted: tax cuts, control of the judicial system from lower courts to the Supreme Court, excessive influence by foundations, increased militarization and criminalization, and support by the nation's premier institutions of higher education. These are formidable foes to grassroots efforts trying to participate in the political process, let alone understand the root causes of their uneven hardships.

In addition to providing a critical analysis of the abuse of power by dominant institutions with respect to the perpetration of environmental injustices, we also need a deeper exploration of the conditions under which contemporary race, class, and gendered politics are established, justified, and reproduced in the United States and around the world. Such an effort, the eminent scholar Paul Gilroy (1987) writes, would require scholars concerned with matters of inequality to root these particular findings of disproportionality within the historical processes and encompassing systems in which it is found. Such analysis would reveal not only the changing patterns in racist and sexist ideology and practice but also the manner in which these "fit into the transformation" of state institutions "and political culture at a time of extensive social and economic change." "The true responsibility for the existence of these deplorable conditions," Martin Luther King Jr. proclaimed, "lies ultimately with the larger society, and much of the immediate responsibility for removing the injustices can be laid directly at the door of the federal government."

These new relations—political, economic, and environmental—require us to think more critically about efforts to achieve environmental justice. We have suggested some strategies—theoretical and methodological—in this book, but we have only scratched the surface of possibilities available to us. Freedom and justice are indeed constant struggles, and we need to be vigilant of the emergent threats—new and old—to social and environmental justice, and ecological sustainability, as all of our lives depend on our doing so. I end with an important thought from the inspirational poet and writer Audre Lorde (2007), "[W]e must face with clarity and insight the lessons to be learned from the oversimplification of any struggle for self-awareness and liberation, or we will not rally the force we need to face the multidimensional threats to our survival" (p. 138). I hope this book provides a little of both clarity and insight.

References

Gilroy, P. (1987). *There ain't no black in the Union Jack*. New York: Routledge.

Lorde, A. (2007). *Sister outsider*. Berkeley, CA: Crossing Press.

INDEX

F